D0890096

Individualized Medicine

Advances in Predictive, Preventive and Personalised Medicine

Volume 7

Series Editor:

Olga Golubnitschaja

Managing Editor:

Kristina Yeghiazaryan

For further volumes:
http://www.springer.com/series/10051

Tobias Fischer • Martin Langanke • Paul Marschall
Susanne Michl
Editors

Individualized Medicine

Ethical, Economical and Historical Perspectives

 Springer

Editors
Tobias Fischer
University Medicine Greifswald
Department for Ethics
Theory and History in Life Sciences
Greifswald
Germany

Martin Langanke
University of Greifswald
Faculty of Theology
Greifswald
Germany

Paul Marschall
University of Greifswald
Department of Health Care Management
Greifswald
Germany

Susanne Michl
University Medical Center of the Johannes
Gutenberg University Mainz
Institute for History, Philosophy and Ethics
of Medicine
Mainz
Germany

ISSN 2211-3495   ISSN 2211-3509 (electronic)
ISBN 978-3-319-11718-8   ISBN 978-3-319-11719-5 (eBook)
DOI 10.1007/978-3-319-11719-5
Springer Cham Heidelberg New York Dordrecht London

Library of Congress Control Number: 2014955632

Printed on acid-free paper

Springer is part of Springer Science+Business Media (www.springer.com)

What this Book Series is About...

Current Healthcare: What Is Behind the Issue?

For many acute and chronic disorders, the current healthcare outcomes are considered as being inadequate: global figures cry for preventive measures and personalised treatments. In fact, severe chronic pathologies such as cardiovascular disorders, diabetes and cancer are treated after onset of the disease, frequently at near end-stages. Pessimistic prognosis considers pandemic scenario for type 2 diabetes mellitus, neurodegenerative disorders and some types of cancer over the next 10-20 years followed by the economic disaster of healthcare systems in a global scale.

Advanced Healthcare Tailored to the Person: What Is Beyond the Issue?

Advanced healthcare promotes the paradigm change from delayed interventional to predictive medicine tailored to the person, from reactive to preventive medicine and from disease to wellness. The innovative Predictive, Preventive and Personalised Medicine (PPPM) is emerging as the focal point of efforts in healthcare aimed at curbing the prevalence of both communicable and non-communicable diseases such as diabetes, cardiovascular diseases, chronic respiratory diseases, cancer and dental pathologies. The cost-effective management of diseases and the critical role of PPPM in modernisation of healthcare have been acknowledged as priorities by global and regional organisations and health-related institutions such as the Organisation of United Nations, the European Union and the National Institutes of Health.

Why Integrative Medical Approach by PPPM as the Medicine of the Future?

PPPM is the new integrative concept in healthcare sector that enables to predict individual predisposition before onset of the disease, to provide targeted preventive measures and create personalised treatment algorithms tailored to the person.

The expected outcomes are conducive to more effective population screening, prevention early in childhood, identification of persons at-risk, stratification of patients for the optimal therapy planning, prediction and reduction of adverse drug-drug or drug-disease interactions relying on emerging technologies, such as pharmacogenetics, pathology-specific molecular patters, sub/cellular imaging, disease modelling, individual patient profiles, etc. Integrative approach by PPPM is considered as the medicine of the future. Being at the forefront of the global efforts, the European Association for Predictive, Preventive and Personalised Medicine (EPMA, http://www.epmanet.eu/) promotes the integrative concept of PPPM among healthcare stakeholders, governmental institutions, educators, funding bodies, patient organisations and in the public domain.

Current Book Series, published by Springer in collaboration with EPMA, overview multidisciplinary aspects of advanced bio/medical approaches and innovative technologies. Integration of individual professional groups into the overall concept of PPPM is a particular advantage of this book series. Expert recommendations focus on the cost-effective management tailored to the person in health and disease. Innovative strategies are considered for standardisation of healthcare services. New guidelines are proposed for medical ethics, treatment of rare diseases, innovative approaches to early and predictive diagnostics, patient stratification and targeted prevention in healthy individuals, persons at-risk, individual patient groups, sub/populations, institutions, healthcare economy and marketing.

Prof. Dr. Olga Golubnitschaja

Book Series Editor

Dr. Golubnitschaja, Department of Radiology, Medical Faculty of the University in Bonn, Germany, has studied journalism, biotechnology and medicine and has been awarded fellowships for biomedical research in Paediatrics and Neurosciences (Medical Centres in Austria, Russia, UK, Germany, the Netherlands, and Switzerland). She is well-cited in the research fields of "gene hunting" and "subtractive hybridisation" applied to predictive prenatal and postnatal diagnostics published as

O. Labudova in years 1990-2000. Dr. Golubnitschaja is an expert in molecular diagnostics actively publishing in the fields of perinatal diagnostics, Down syndrome, diabetes mellitus, hyperhomocysteinemia, cardiovascular disease, neurodegenerative pathologies and cancer. She is the *cofounder* of the theory of multi-pathway organ-related blood fingerprinting with specific molecular patterns at epi/genomic, transcriptional and post/translational levels and author of fundamental works in *integrative medicine*. Dr. Golubnitschaja holds appointments, at the rank of Professor, at several European Universities and in International Programmes for Personalised Medicine and is author of more than 300 international publications in the field. Awards: National and International Fellowship of the Alexander von Humboldt-Foundation; Highest Prize in Medicine and Eiselsberg-Prize in Austria; She is *Secretary-General* of the "European Association for Predictive, Preventive and Personalised Medicine" (EPMA in Brussels, www.epmanet.eu), Editor-in-Chief of *The EPMA-Journal* (BioMed Central in London); Book Editor of *Predictive Diagnostics and Personalized Treatment: Dream or Reality,* Nova Science Publishers, New York 2009; Book Co-editor *Personalisierte Medizin,* Health Academy, Dresden 2010; Book Series Editor *Advances in Predictive, Preventive and Personalised Medicine,* Springer 2012; *European Representative* in the EDR-Network at the NIH/NCI, http://edrn.nci.nih.gov/; *Advisor and Evaluator* of projects dedicated to personalised medicine at the EU-Commission in Brussels, NIH/NCI, Washington, DC, USA, and Foundations and National Ministries of Health in several countries worldwide.

Preface

In 2012 the editors developed the idea of bringing together the research results from the disciplines of history, concept-based ethics, applied research ethics and health economics into one volume on "Individualized Medicine". These different disciplines constitute one project area as an integral part of the overall "Greifswald Approach to Individualized Medicine" (GANI_MED) research consortia. From their respective disciplinary affiliations, the editors and authors of the different sections hope to contribute to the ongoing debate on ethical, economical and societal implications of Individualized Medicine (IM) by focusing on the specific Greifswald approach. To address the challenges of this new approach in medicine, IM is first and foremost a research program capable of converging persons with different backgrounds and disciplines. This volume is also the result of intense interdisciplinary discussions and close cooperation with researchers from the clinic and information technology.

Our first thanks go therefore to all the Greifswalder researchers of any discipline and background for their readiness to cooperate and discuss these issues. Without this interdisciplinary exchange based on complimentary expertise, it would have been impossible to carry out some of the research projects, particularly to find together ethically adequate solutions to clinical and IT problems, to inform our conceptual understanding and to make economical analysis possible. Our special thanks go to Hans-Jörgen Grabe, the project leader and his assistants Claudia Richardt and Vivian Werner, who supported this publication from conception to completion.

We would like to express our special thanks to Olga Golubnitschaja, the competent and experienced series editor of the "Advances in Predictive, Preventive and Personalised Medicine" for admitting the contributions for publication. We thank Martijn Roelandse and Tanja Koppejan from Springer Biomedical, who have cleared up a number of editorial questions and provided us with useful help to go through the editing process.

An edited volume would be impossible to write without the help of many people who have proof-read, translated or commented upon manuscripts. First of all, James Wells has provided for editorial consistency by going through the whole volume. We are very thankful for all his stylistic improvements. Daniela Berner, Claudia

Gräfe, Stefan Kirschke and Sally Werner scrupulously read different parts of the volume and largely contributed to harmonize the manuscripts in form and style.

Last but not least, we are thankful to all the writers of this volume. The editors thank all who have contributed to the success of this project. Our thanks especially go out to them for their time, hard work and for their understanding for the gentle pressure which we sometimes had to put on them in order to keep to the tight time plan.

*

This work is part of the research project Greifswald Approach to Individualized Medicine (GANI_MED). The GANI_MED consortium is funded by the Federal Ministry of Education and Research and the Ministry of Cultural Affairs of the Federal State of Mecklenburg-West Pomerania (support codes: 03IS2061A & 03IS2061E). Further grants contribute to GANI_MED: IntegraMent (Federal Ministry of Education and Research); German Asthma and COPD Network (COSYCONET; BMBF 01GI0883).

July 2014   Greifswald

Contents

Contributors

Heinrich Assel Theologische Fakultät, Lehrstuhl für Systematische Theologie, Ernst-Moritz-Arndt-Universität Greifswald, Greifswald, Germany

Thomas Bahls Institut für Community Medicine Abt. VC, Universitätsmedizin Greifswald, Greifswald, Germany

Marcus Dörr Klinik und Poliklinik für Innere Medizin B, Universitätsmedizin Greifswald, DZHK (German Center for Cardiovascular Research), partner site Greifswald, Germany

Pia Erdmann Theologische Fakultät, Lehrstuhl für Systematische Theologie, Ernst-Moritz-Arndt-Universität Greifswald, Greifswald, Germany

Jakob Fasold Theologische Fakultät, Lehrstuhl für Systematische Theologie, Ernst-Moritz-Arndt-Universität Greifswald, Greifswald, Germany

Stephan B. Felix Klinik und Poliklinik für Innere Medizin B, Universitätsmedizin Greifswald, DZHK (German Center for Cardiovascular Research), partner site Greifswald, Germany

Tobias Fischer Department für Ethik, Theorie und Geschichte der Lebenswissenschaften, Universitätsmedizin Greifswald, Greifswald, Germany

Steffen Flessa Rechts- und Staatswissenschaftliche Fakultät, Lehrstuhl für Allgemeine Betriebswirtschaftslehre und Gesundheitsmanagement, Ernst-Moritz-Arndt-Universität Greifswald, Greifswald, Germany

Lars Geidel Institut für Community Medicine Abt. VC, Universitätsmedizin Greifswald, Greifswald, Germany

Hans-Jörgen Grabe Klinik und Poliklinik für Psychiatrie und Psychotherapie, Universitätsmedizin Greifswald und HELIOS Klinik Stralsund, Stralsund, Germany

Wolfgang Hoffmann Institut für Community Medicine Abt. VC, Universitätsmedizin Greifswald, Greifswald, Germany

Heyo K. Kroemer Universitätsmedizin Göttingen, Göttingen, Germany

Martin Langanke Theologische Fakultät, Lehrstuhl für Systematische Theologie, Ernst-Moritz-Arndt-Universität Greifswald, Greifswald, Germany

Timm Laslo Rechts- und Staatswissenschaftliche Fakultät, Lehrstuhl für Allgemeine Betriebswirtschaftslehre und Gesundheitsmanagement, Ernst-Moritz-Arndt-Universität Greifswald, Greifswald, Germany

Wolfgang Lieb Institut für Epidemiologie, Christian-Albrechts-Universität zu Kiel, Kiel, Germany

Wenke Liedtke Theologische Fakultät, Lehrstuhl für Systematische Theologie, Ernst-Moritz-Arndt-Universität Greifswald, Greifswald, Germany

Roberto Lorbeer Institute of Clinical Radiology, Klinikum der Ludwig-Maximilians-Universität München, Germany

Paul Marschall Rechts- und Staatswissenschaftliche Fakultät, Lehrstuhl für Allgemeine Betriebswirtschaftslehre und Gesundheitsmanagement, Ernst-Moritz-Arndt-Universität Greifswald, Greifswald, Germany

Henriette E. Meyer zu Schwabedissen Departement Pharmazeutische Wissenschaften, Biopharmazie, Universität Basel, Basel, Switzerland

Susanne Michl Institut für Geschichte, Theorie und Ethik der Medizin, Universitätsmedizin der Johannes Gutenberg-Universität Mainz, Mainz, Germany

Konrad Ott Philosophisches Seminar, Christian-Albrechts-Universität zu Kiel, Kiel, Germany

Johann-Christian Põder Theologische Fakultät, Lehrstuhl für Systematische Theologie, Ernst-Moritz-Arndt-Universität Greifswald, Greifswald, Germany

Uwe Völker Institut für Genetik und Funktionelle Genomforschung, Universitätsmedizin Greifswald, DZHK (German Center for Cardiovascular Research), partner site Greifswald, Germany

Henri Wallaschofski Facharzt für Innere Medizin Endokrinologie Diabetologie Andrologie, Erfurt, Germany

Kerstin Weitmann Institut für Community Medicine Abt. VC, Universitätsmedizin Greifswald, Greifswald, Germany

The Editors

Dr. Tobias Fischer M.A. is a philosopher and ethicist. He is lecturer for bioethics, scientific coordinator of the Department of Ethics, Theory and History of Life Sciences at the University Medicine Greifswald, Germany and member of the GANI_MED consortium. Moreover, Dr. Fischer is subdivision head of the BMBF-funded project MENON, which deals with ethical and economic implications of the systems medicine. His research interests include the theory and ethics of personalized medicine, reproductive medicine and the application of ethics in clinical practice.

Dr. Martin Langanke M.A. is a philosopher and ethicist. He serves as a lecturer for ethics at the Faculty of Theology, Ernst-Moritz-Arndt-University Greifswald, Germany. He is member of the GANI_MED consortium at Greifswald University and general coordinator of the BMBF-funded project MENON, which deals with ethical and economic implications of the systems medicine. Dr. Langanke's research interests include the foundations of ethics, philosophical anthropology, theory and ethics of personalized medicine, medical research ethics and animal ethics.

Dr. Paul Marschall is an economist and political scientist. He is lecturer for health economics and public finance at the Ernst-Moritz-Arndt-University Greifswald, at the Charité, Berlin (Germany) and in some international courses. Dr. Marschall was interim professor for economics at the University of Bayreuth, Germany. He is member of the GANI_MED consortium at Greifswald University. His research interest includes health economic evaluation, health services research and global health.

Dr. Susanne Michl M.A. is an historian. She is lecturer for medical history and ethics at the Institute for History, Philosophy and Ethics of Medicine of the Johannes-Gutenberg University Medical Center in Mainz and member of the local ethics committee. She was member of the GANI-MED consortium at Greifswald University and interim professor at the Department of Medical Ethics and History of Medicine at the University Medical Center Goettingen. Her research interests include history of pharmacogenetics, social and historical studies of personalized medicine, war and medicine, trauma narratives and clinical ethics.

Chapter 1
Introduction

Tobias Fischer, Martin Langanke, Paul Marschall and Susanne Michl

In October 2009 the University Medicine Greifswald launched the "Greifswald Approach to Individualized Medicine" (GANI_MED) to implement individualized diagnostic and therapeutic strategies in clinical settings. It was the first attempt to establish an Individualized Medicine (IM) program at a German university hospital. IM is a new approach in the context of prevention, tailored diagnostic and treatment of patients with regard to their individual characteristics. Since the completion of the human genome project in 2003, the approach of Individualized Medicine, along with other similar designations such as Personalized Medicine (PM) or Stratified Medicine (SM), has led to controversies about its clinical and economical potentials, as well as its societal implications and ethical requirements. History, applied research ethics, concept-based ethics and health-economics are integral parts of the GANI_MED consortium. In contrast to other comprehensive presentations of IM published so far, this anthology focuses on this research area. However, this volume includes some further contributions out of the GANI_MED context, contributing to an understanding of these works. In this book the research from these disciplines is

T. Fischer (✉)
Department für Ethik, Theorie und Geschichte der Lebenswissenschaften, Universitätsmedizin Greifswald, Walther-Rathenau-Str. 48, 17475 Greifswald, Germany
e-mail: tobias.fischer@uni-greifswald.de

M. Langanke
Lehrstuhl für Systematische Theologie, Theologische Fakultät, Ernst-Moritz-Arndt-Universität Greifswald, Am Rubenowplatz 2–3, 17487 Greifswald, Germany
e-mail: langanke@uni-greifswald.de

P. Marschall
Rechts- und Staatswissenschaftliche Fakultät, Lehrstuhl für Allgemeine Betriebswirtschaftslehre und Gesundheitsmanagement, Ernst-Moritz-Arndt-Universität Greifswald, Friedrich-Loeffler-Str. 70, 17487 Greifswald, Germany
e-mail: paul.marschall@uni-greifswald.de

S. Michl
Institut für Geschichte, Theorie und Ethik der Medizin, Universitätsmedizin der Johannes Gutenberg-Universität Mainz, Am Pulverturm 13, 55131 Mainz, Germany
e-mail: susmichl@uni-mainz.de

© Springer International Publishing Switzerland 2015
T. Fischer et al. (eds.), *Individualized Medicine,* Advances in Predictive, Preventive and Personalised Medicine 7, DOI 10.1007/978-3-319-11719-5_1

presented. It draws attention to the fact that the implementation of individualized approaches into medical research and clinical practice is linked to changes in health care systems and societal values.

One of the core research fields of IM is pharmacogenetics. Its approach serves as an example of how to tailor individualized therapies for different patients or patient groups. Each person differs according to his or her individual heredity, health related behavior and metabolism. Furthermore, especially the elderly suffer from many diseases. To this day, drugs are prescribed according to the clinical picture of the disease without considering the individual make-up of different patients. As a result, medications with proven efficacy often induce (adverse) side-effects or fail to exert any effect in a subgroup of patients. Besides the impact on personal well-being, this approach leads to economic costs that could be saved by more targeted therapies. The advances of pharmacogenetic research since the late 1950s has provided deeper insights into the future potentials of individually tailored drug therapies.

This need for more tailored therapies must be understood in a broader sense, including a wider range of biomedical initiatives and application fields. It is the commitment of GANI_MED to test promising concepts of individualization for their applicability in this field. The GANI_MED consortium includes experts from various national and international academic institutions as well as industrial partners that help to lay the foundations for the implementation of individualized therapies into different clinical fields.

GANI_MED was designed as a clinical-epidemiological project. In contrast to other research projects, that invoke similar concepts, GANI_MED focuses on disease phenotypes outside of oncology, especially common diseases. GANI_MED is grounded in the understanding of the importance of biomarkers. These are molecular and non-molecular parameters that allow, individually or in combination, to predict courses of diseases and/or the response of therapies by using stratification techniques. GANI_MED aims at identifying adequate biomarker candidates with the help of association studies. Additionally, the translation of IM based approaches into clinical routine is an important objective of this research program.

Between the introduction (chap. 1) and the conclusion (chap. 16), there are six parts that present a collection of findings from research projects from different disciplines.

Critics have often called "Individualized Medicine" (or "Personalized Medicine"—the term more commonly used in English-speaking countries) a "misnomer" or a false and fraudulent labeling. Because the term includes a wide variety of meanings in medical and popular writings, there is substantial doubt, not only about the usefulness of the term in particular, but moreover about the potential of an individualized approach in general. Given the vagueness of the concept, a research consortium labeled as IM or PM needs to strive for terminological precision not only of IM or PM, but also related terms such as "research approach", "health care practice", "biomarker", "prediction" and "stratification". Within the GANI_MED joint project an interdisciplinary working group was established to develop a precise definition of IM and to demarcate it from other uses and designations. Chapter 2 is the result of the discussion led by experts from the field of ethics and theory of science (**Martin Langanke, Pia Erdmann, Tobias Fischer, Hein-**

rich Assel), medicine (**Wolfgang Lieb**, **Marcus Dörr**, **Heyo Kroemer**) and health economics (**Steffen Flessa**). In addition to this preliminary, conceptual and terminological part, in chapter 3, **Hans-J. Grabe** and **Henri Wallaschofski** outline what this understanding of IM means when applied to the almost 4000 patients recruited in nine cohorts at Greifswald University Hospital (heart failure cohort, cerebrovascular disease cohort, periodontal disease cohort, renal and renovascular disease cohort, metabolic syndrome risk cohort, fatty liver disease cohort, cohort of adverse medication effects, cohort of pulmonary diseases, cohort of sepsis). To integrate IM into clinical practice, a central data management structure has been implemented guaranteeing the standardization of protocols for the assessment of medical history, laboratory biomarkers, and the collection of various biosamples for biobanking.

The need for a precise definition of IM becomes apparent against the background of the fluid and contradictory ways in which the concept of "Personalized Medicine" has been used over the last 15 years. Drawing on the definition of IM in the context of the GANI_MED research consortia, Part II considers "Personalized Medicine"—the term used in the sample of writings analyzed in this part—as a societal phenomenon that involves different biomedical initiatives in and outside universities and research units. Since the late 1990s, the term "Personalized Medicine" has been coined to describe and enable collaborations between different stakeholders. As a rhetorical frame, it constitutes an imaginary framework of visions, expectations and claims for a better, more patient-centered and more efficient health care system. Instead of deciding whether "Personalized Medicine" is more "hype" or "hope", scholars from the social studies of technology and science emphasize that the expectations revolving around new technology are not only accessory parts of scientific inventions or innovation networks. On the contrary, they regard them as essential in shaping these technologies. In Part II on "Perspectives of Socio-Cultural and Historical Studies", **Susanne Michl** draws on this "sociology of expectation" by analyzing specific ways in which the individualized approach to medicine has been framed in current medical and popular writings since the invention of the term "Personalized Medicine" (chap. 4) as well as in past developments of the narrower field of pharmacogenetic research (chap. 5). The focus is on narratives connecting past performances, present states and the future promises. In a historical perspective, the author draws attention to alternative, partly forgotten, partly surviving, conceptual frameworks of research projects centered upon the concepts of "individuality" or "variability." Such observations are also relevant for our understanding of pharmacogenetics research since the late 1950s and the rediscovery of the work of Archibald Garrod at the beginning of the last century.

Part III outlines clinical examples to illustrate how integrated analysis of biomarkers leads to significant improvement of therapeutic outcomes for a subgroup of patients.

Marcus Dörr, **Uwe Völker** and **Stephan B. Felix** (chap. 6) deal with the hemodynamic effects of a novel treatment option (immunoadsorption with subsequent IgG substitution) for Dilated Cardiomyopathy (DCM), which is one of the most common causes of heart failure. The study demonstrates the potentials of a biomarker-based research for DCM patients whose response to the therapy of immunoadsorption is predicted by the combination of two biomarkers (negative inotropic activity of antibodies in the blood and gene expression patterns derives from myocardial biopsies). Within the multi-disciplinary GANI_MED this clinical example led to an in-depth, health-economical analysis.

In chapter 7 of Part III **Henriette E. Meyer zu Schwabedissen** provides examples from pharmacogenetics, one of the major and most promising fields of IM. The part focuses on the underlying mechanisms of gene-drug associations, their clinical significance as well as the current status of clinical implementation. The first example, the association between CYP2D6 and tamoxifen in the treatment of ER-positive breast cancer, concerns a member of the protein family of cytochrome P450 enzymes that is historically one of the first findings paving the way to advances in pharmacogenetics in general. The contribution sheds light on the strengths, but also on the difficulties of pharmacogenetic findings and their successful clinical implementation. In addition to the optimization of treatment outcomes, genetic findings have contributed to the development of new drug therapies and their clinical approval, such as the development of CCR5 antagonists and inhibitors of the bcr-abl tyrosine kinase and the novel drug ivacaftor for patients with cystic fibrosis.

To assess the implementation of Individualized Medicine approaches into medical research and clinical practice requires more than an analysis of whether IM will live up to its clinical expectations. Since the late 1990s, IM has also raised ethical concerns. The following two parts deal with these ethical issues, including the conceptual layers of IM (Part IV) and applied research ethics (Part V).

The task of the philosophical project of GANI_MED is to produce a conceptual and critical reconstruction of the approach of IM. Adopting the reconstructive theory of science, **Konrad Ott** and **Tobias Fischer** (chap. 8) deal with constitutive momenta of IM (stratification, diagnosis, prediction, prevention and risk) and four different medical approaches (lifeworld, traditional-conventional medicine, human-ecological and "alternative" ways of healing, and the molecular-genetic approach). Over and above these conceptual issues, the contribution proposes an ethics of IM based on three pillars (informed consent, "cura sui" and solidarity). The authors argue that IM cannot be attributed to the molecular-genetic approach alone. Instead they opt for an understanding and framing of IM as an integrative health science in contemporary societies. Rather than deploring the vagueness of the term, Ott and Fischer emphasize the flexibility and the openness of the "epistemic grammar" of IM enabling the design of new scopes of medical practice.

In the second chapter of the part on concept-based issues (chap. 9), **Johann-Christian Põder** and **Heinrich Assel** address criticism according to which IM contributes to an extension of the concept of disease and a pathologization of life. In the near future, biomarker-based predictive medicine and the rapidly growing amount of health-related information might considerably increase the accuracy in predicting the onset of a disease leading to new personal and social patterns of identification (the "healthy ill persons"), decision-making and responsibility. To answer the question of whether IM will contribute to a pathologization of life, three disease theories—the naturalistic theory by Christopher Boorse, the reconstructive theory by Peter Hucklenbroich and the "practical" theory by Dirk Lanzerath—are discussed. Against this background, the authors argue that medical terms and concepts should not only be elaborated from a medical (or political and economic) but also from a philosophical perspective as an ongoing task and challenge.

In addition to conceptual questions, Part V focuses on application-oriented ethical issues. One of the peculiarities of GANI_MED is that the ethical requirements and regulatory demands of clinical epidemiological research have been investigated within the collaborative project. To assure ethically appropriate ways of dealing with the recruitment of patient cohorts and the establishment of a biobank- and IT-infrastructure, bioethicists and clinical researchers have collaborated closely. This interdisciplinary cooperation between life sciences and the humanities has led to pragmatic, patient-centered solutions and procedural improvements that meet legal standards and establish new ethical standards that can serve as a model on how to integrate ethical research into clinical epidemiological settings for future research consortia.

In the first chapter of Part V **Martin Langanke, Jakob Fasold, Pia Erdmann, Roberto Lorbeer and Wenke Liedtke** present the informed consent process of the GANI_MED project including the consent documents and the response pattern of the patients participating (chap. 10). The challenge is that clinical epidemiological studies typically try to pursue several scientific goals at once. Data from different sources including those from third parties have to be collected to form a sample to generate hypotheses in the field of IM. The challenge then is that the complexity of epidemiological research designs has to be incorporated in the consent form in an easily comprehensible way. The ethicists of the GANI_MED joint project have designed a consent form for each cohort of patients and with several sections allowing the patient to consent or to refuse his or her full participation or to rule out certain aspects of the study. The analysis of the empirical data demonstrates that patients actually have seized this possibility to deselect and that the patients of different morbidity cohorts answer differently.

Once the patient has signed the informed consent documents, the next challenge is to store this information to guarantee that the contents of the agreements and the corresponding data can easily be connected, and made available. This next step of the management of informed consent—the processing of data items that belong to persons who have agreed to take part in medical research—requires both technical and ethico-legal considerations. In the following contribution (chap. 11) **Thomas Bahls, Wenke Liedtke, Lars Geidel** and **Martin Langanke** present a high-level architecture of an IT platform that guarantees quality and ethical validity of the informed consent documents and their automatic application during a data use and access process.

One of the distinctive features of the GANI_MED project is the integration of empirical-ethical studies. Whereas in research ethics, studies have focused on abstract ethical requirements, only few studies explore the actual expectations of participants before consenting and their stress during and after the examinations. Particularly the whole-body MRI examination leads to incidental findings constituting a challenge for both researchers and participants on how to handle them. **Pia Erdmann** takes up this issue by carrying out quantitative and qualitative surveys with participants of the "Study of Health in Pomerania" (SHIP) who underwent a whole-body MRI examination (chap. 12). The analysis of the questionnaires and interviews provide important insights into the subjective perception of health and the shifting risk-benefit evaluation as well as potential misjudgments of the conditions for participating. To conduct MRI examinations in an ethically appropriate way, researchers should avoid Diagnostic Misconception, as well as factors causing stress, by adjusting the mode of communication.

Part VI is devoted to different aspects of assessing IM from a health economic perspective. At first, **Steffen Flessa** and **Paul Marschall** examine in their contribution "Individualized Medicine: From Potential to Macro-Innovation" (chap. 13) the possibilities of IM to initiate a paradigm shift in the German health care provision. Economically the relevance of a novel idea or a new product can be captured by the concept of innovation and the associated penetration of society or a market. This can be linked to the question of whether IM has the potential to transform thinking and behavior within the health system. Different stages of innovation can be distinguished which express the adoption level within the system. Stakeholders and their attitude play a crucial role in the corresponding process. The authors analyze whether IM has the potential to change the relationship between doctors, patients and other agents, to alter the rules, institutions and regulations of the health care sector and even to influence societal values. In addition major barriers preventing the key stakeholders adopting this new approach to medicine must be considered. Flessa and Marschall firstly analyze these barriers and whether IM has the potential to become a macro-innovation, thus the most comprehensive level. Based on a characterization of the attributes of the current status of IM the authors study what has to be fulfilled so this approach can become the new standard solution for the health care system for allocating scarce resources.

In the medical section of this book **Marcus Dörr**, **Uwe Völker** and **Stephan B. Felix** (chap. 6) exemplify the use of biomarkers for the prediction of treatment response within the context of widespread diseases. **Paul Marschall**, **Timm Laslo**, **Wolfgang Hoffmann**, **Kerstin Weitmann** and **Steffen Flessa** address themselves to the same clinical example. In: "Assessing Individualized Medicine—the Example of Immunoadsorption" (chap. 14) they provide the corresponding picture from the economic perspective. Based on time studies and investigations of used resources at the University Medicine Greifswald they provide preliminary results for the costs of Immunoadsorption therapy with subsequent IgG (IA/IgG) substitution and the corresponding gene expression analysis. Under the current setting the latter can be regarded as diagnostics for deciding whether IA/IgG is appropriate. Currently, both parts of the IM tandem are not implemented in combination in clinical routine. The authors show that the reimbursement system has a critical role for providing incentives for health care providers to translate research into routine. For assessing whether a new therapy approach is useful, a full economic evaluation of costs and consequences is necessary. Marschall et al. also present some preliminary results of outcome evaluation based on disease-related quality of life.

Heart failure is currently one of the most cost-intensive diseases in Germany. It also represents one of the most common reasons for hospitalization. By now it is evident that due to demographic change prevalence and incidence of heart failure is enormously increasing. **Timm Laslo**, **Paul Marschall** and **Steffen Flessa** finally cover this issue in: "How individualized is medicine today? The case of heart failure in the G-DRG system" (chap. 15). Currently the remuneration of hospitals in the German DRG system (G-DRG) is carried out through the payment of a lump sum for each case of inpatient treatment. The system must be able to map complex cases. Based on a comprehensive data set from the University Medicine Greifswald the authors analyze how a pathway for clinical care is apparent for heart failure using the example of the base DRG F62. In this context the concept of cost homogeneity is of high relevance. In addition, criteria for the clinical course of symptoms of heart failure are determined by multivariate analysis. Furthermore, the authors discuss whether the G-DRG system is ready for the implementation of IM approaches into the clinical routine.

Finally in chapter 16 conclusions and recommendations based on the first 5 years of GANI_MED with respect to socio-cultural, ethical and health-economic issues are presented. Thereby we draw heavily on the experience of the studies pointed out in Parts I-VI.

Part I
Definition and Concept
of Individualized Medicine

Chapter 2
The Meaning of "Individualized Medicine": A Terminological Adjustment of a Perplexing Term

Martin Langanke, Wolfgang Lieb, Pia Erdmann, Marcus Dörr, Tobias Fischer, Heyo K. Kroemer, Steffen Flessa and Heinrich Assel

Abstract This chapter introduces "Individualized Medicine" as a technical term. In order to do this the chapter first gives a precise, logical and conceptual analysis of relevant explanations and definitions from English and German speaking areas. It secondly presents a definition according to which the term "Individualized Medicine" should be used for describing research approaches and health care practices, when the

M. Langanke (✉) · P. Erdmann · H. Assel
Theologische Fakultät, Lehrstuhl für Systematische Theologie, Ernst-Moritz-Arndt-Universität Greifswald, Am Rubenowplatz 2–3, 17487 Greifswald, Germany
e-mail: langanke@uni-greifswald.de

P. Erdmann
e-mail: pia.erdmann@uni-greifswald.de

H. Assel
e-mail: assel@uni-greifswald.de

W. Lieb
Institut für Epidemiologie, Christian-Albrechts-Universität zu Kiel, Campus UKSH, Niemannsweg 11, 24105 Kiel, Germany
e-mail: wolfgang.lieb@epi.uni-kiel.de

M. Dörr
Klinik und Poliklinik für Innere Medizin B, Universitätsmedizin Greifswald, F.-Sauerbruch-Straße, 17475 Greifswald, Germany
e-mail: mdoerr@uni-greifswald.de

T. Fischer
Department für Ethik, Theorie und Geschichte der Lebenswissenschaften, Universitätsmedizin Greifswald, Walther-Rathenau-Str. 48, 17475 Greifswald, Germany
e-mail: tobias.fischer@uni-greifswald.de

H. K. Kroemer
Universitätsmedizin Göttingen, Grete-Henry-Str. 9, 37085 Göttingen, Germany
e-mail: heyo.kroemer@med.uni-goettingen.de

S. Flessa
Rechts- und Staatswissenschaftliche Fakultät, Lehrstuhl für Allgemeine Betriebswirtschaftslehre und Gesundheitsmanagement, Ernst-Moritz-Arndt-Universität Greifswald, Friedrich-Loeffler-Str. 70, 17487 Greifswald, Germany
e-mail: steffen.flessa@uni-greifswald.de

© Springer International Publishing Switzerland 2015   11
T. Fischer et al. (eds.), *Individualized Medicine,* Advances in Predictive,
Preventive and Personalised Medicine 7, DOI 10.1007/978-3-319-11719-5_2

biomarker-based prediction of (a) diseases and/or (b) the effectiveness of therapies by stratification is central. The relevant terms "research approach", "health care practice", "biomarker", "prediction" and "stratification" will be discussed in detail. Finally the term "Individualized Medicine" will be examined regarding its extension and be compared to "Personalized Medicine", which is also understood terminologically.

Keywords Definition · Aristotelian concept of definition · Individualized Medicine · Personalized Medicine · Medical research · Health care · Biomarker · Stratification

2.1   Background

"Individualized Medicine is fraudulent labeling and fiction." This provocative statement of Prof. Wolf-Dieter Ludwig, chairman of the Drug Commission of German Physicians, is quoted in an article from March 2011 which can be still found in the archive of the website of the Association of German Internists (Individualisierte Medizin: Etikettenschwindel 2011).

Such criticism of Individualized Medicine (IM), suggesting that the term is misleading, cannot be ignored by scientists who understand their research as a contribution to the establishment of IM. Moreover, such critical voices are not only part of the non-medical "accompanying discourse" about IM, but they come also—as shown by the initial quote—from researching physicians. The accusation made is serious and massively affects the integrity and academic respectability of any work in the field of IM. If this criticism is justified, IM would be only a label, which is, at best, useful for the acquisition of funding but it would not seriously describe a current branch of medicine.

To respond to this accusation is also in the interest of those research groups which are part of the joint project Greifswald Approach to Individualized Medicine (GANI_MED) of the Ernst-Moritz-Arndt-University Greifswald (Grabe et al. 2014; Langanke et al. 2011; Langanke et al. 2012a). There is a risk that their activities, which are within one of the most extensive projects concerning IM in Germany, will be discredited through the accusation of fraudulent labeling. In the light of this, this paper secures the result of the discussions within the interdisciplinary GANI_MED working group, which was established in order to do the terminological demands justice with regard to a refined "IM" term. Experts from the field of medicine, health economics, ethics and theory of sciences were part of this working group.

2.2   Preliminary Methodological Considerations

The spectrum of what is called IM today includes

a. medicine which is based on the use of unique therapeutic measures i.e. in the course of Tissue Engineering or cell therapy

b. pharmacogenetics and
c. other lines of research which aim at the improvement of the prediction of diseases and/or courses of diseases with the help of so-called biomarkers (Costigliola 2009; Hüsing et al. 2008; Kollek and Lemke 2008; Niederlag et al. 2010; Schleidgen et al. 2013).

If one reflects on possible introductory strategies for the term "IM", with regard to the differences concerning this term, one could at first consider the option to provide "IM" as a simple collective term and to list all the relevant trends which are understood as "IM".

The advantage of such a "collective term", created in an enumerative way, is that everything which praises itself as "IM" can be accepted as IM. However, this leads to the disadvantage of lacking a depth of focus. In particular, such an "IM" term leaves the question open of whether the different "IMs" match methodologically to one feature, or to a group of features, which is valid as a specific group characteristic in the sense that it is common for exactly all —"IMs", but not for comparable fields of action within medicine.

This disadvantage is particularly crucial in the present case: Within IM there are two different concepts of individualization circulating, as Hüsing et al. 2008 detected. These two concepts can methodologically not be reduced to a common concept. Whereas unique therapeutic measures are therapeutic interventions

> for the individual patient [...] where the "individualization" is based on the manufacturing process of the custom-made item and the resulting product (Hüsing et al. 2008, p. 9).

"individualization" in the light of concepts like pharmocogenetics and/or biomarker-based IM means

> a division of the patient population into clinical relevant subgroups (so-called stratification) [...] which goes beyond the status quo. Leading factors are the presumption that diagnostics, specification of risks and interventions can be more accurate if more criteria, including specific criteria, can be used for the group division (Hüsing et al. 2008, p. 9).

There are lines of research within IM which aim at the development of therapeutic options for only and exactly one individual patient, as well as lines of research which "just" aim at a *more* individual treatment of all patients who belong to a certain group. Therefore the validity of the definitions which eliminate these significant differences has to be questioned. This problem becomes greater if one assumes that methodologically the approach of stratification depends on statistic procedures. Hüsing et al. 2008 are able to bring both lines into coexistence because they introduce "IM" by using a typology of five individualization concepts. Behind them one can presume the same three "drivers" (Hüsing et al. 2008, p. 7).

The methodological problem mentioned above becomes quite clear wherever a definition of "IM", following the "Aristotelian" scheme of genus and specific differences, is aimed at, or at least used as heuristic orientation. According to the tradition of the Aristotelian philosophy of science, a term can be defined on the one hand by putting it under a generic term which includes all the phenomena which can be asked for, if they fall under the concept defined and on the other hand by indicating certain characteristics (specific differences) which have only and exactly the phenomena which fall under the concept defined.

Not every term can be introduced by the scheme of genus and specific difference. (On the level of our everyday experience, colors like "green" or "red" are such a problem case and much debated because they are included in a generic term "color" but cannot be defined with regard to specific differences between the single colors. Thus—on the level of our everyday experience—colors can only be introduced by giving examples and counter examples.) However, the Aristotelian scheme embodies an ideal of definition theory.

If one takes up this ideal for "IM", methodological decisions have to be made at two points specifically:

1. In order to introduce "IM" under the use of a relation between a generic and subsumable concept a decision has to be made with regard to the genus of "IM". Thus, it has to be determined which phenomena of which kind are candidates for proving whether they are included in the term of "IM" or not.
2. Following the Aristotelian strategy of definition, one has to make criteria based decisions within the field, which can be outlined by listing a "collective term" such as "IM" in the broadest sense, in favor or to the disadvantage of some approaches. An "IM" term which is defined in the Aristotelian way cannot be as tolerant as a solely enumerative term.

2.3   The Question of Genus

It is characteristic for the German discussion that "Individualized Medicine" and "Personalized Medicine" are used equally. It is common that both terms appear next to each other in one article without any reflection on this alternative use (e.g. Fricke 2011). "Preventive Medicine" is another term which is commonly used to refer to, at least, certain approaches within the large field of medical lines of research and health care practices, which can be called "IM" by using the "IM" term of Hüsing et al. 2008.

This result needs a more specific classification in the frame of this paper by answering the question of whether the existence of different terms should be used for an objective difference as well. However, we do not want to artificially narrow down the discussion here. In the following, the question of genus will be raised with regard to firstly explanations which explicitly refer to the term "IM", and secondly, by using relevant text passages which use the terms "Personalized Medicine" or "Predictive Medicine". This "Babylonian language confusion" can be tolerated methodologically as long as the language use is only described but not standardized in the sense of a definition.

2.3.1   "IM" as Health Care

Hüsing et al. introduce their typology of "Individualized Medicine," saying that "Individualized Medicine" means "a possible future health care" (Hüsing et al. 2008, p. 7).

The indefinite article suggests that Hüsing et al. (2008) understand "IM" as a subsumable concept of "health care" or the way around "health care" as a genus or generic term for "IM". One could generally compare "IMs" to other forms of health care, according to this suggestion. If one follows the terminological suggestions which were established in the course of the three pillar model of health care, which were put up for discussion by Pfaff (2006) for the field of health care research, all *activities of health care institutions and personnel* are included in the term health care, which aim at

a. *the prevention or health promotion (preventive health care) and/or*
b. *measures for acute care in acute care clinics and family doctor or specialist practice (curative health care) and/or*
c. *a reintegration of the patient into society (rehabilitative health care)*

The use of the term "health care" in Hüsing et al. (2008) can be logically referred to the terminologically regulated discourse about "health care" in health care research according to Pfaff (2006): It should be clear that "IM", in the sense of Hüsing et al. (2008), is *not* a fourth pillar beside preventive, curative and rehabilitative health care and is therefore *no* fourth "health care type" *sensu* Pfaff (2006). It rather shows a possible manner of how health care can be designed on the one hand type-independent and on the other hand in all three fields. Moreover, it has to be noticed that the health care concept described by Pfaff (2006) covers the required logical possibility claimed by Hüsing et al. (2008), i.e. that several designs of health care, thus health care*s*, can be distinguished. Although every medical field has dependent on the indication a range of different methods, it can be indicated fairly precisely for a certain point in time t_1 and for a certain indication X which procedure in the context of the so-called conventional medicine is the standard way of health care. However, the decision concerning what a standard method is depends on the medical state of knowledge and temporal changes.

Generally, it is possible that within the three different health care types described by Pfaff (2006), procedures will be established as standard methods which are based on the use of unique therapeutic measures or the preventive use of biomarkers in the future. In the sense of Hüsing et al. (2008) this future health care could be characterized as "IM" and is to be separated from current health care, in which such procedures play only a subordinate role. From today's perspective one can say: "IM" will remain only a "possible future health care" until procedures like *Tissue Engineering* or biomarker-based prediction are used in the clinical practice significantly more often. If this is the case in the future, IM will have "become reality". By its constitutive embedding of "IM" in the genus "health care", the explanation by Hüsing et al. (2008) is one of the most sophisticated approaches of standardizing the "IM" term in the German language area. With regard to its underlying genus decision, prominently published publications followed Hüsing et al. (2008) in the German language area (e.g. Niederlag et al. 2010).

In this chapter we cannot list all explications which understand "IM", "Personalized Medicine" or "Predictive Medicine" as health care phenomenon. This is also due to many borderline cases which are linguistically so loose that competitive reading is possible which can be distinguished only with disproportionate hermeneutical effort.

2.3.2   "IM" Between Health Care and Research

The situation in the English language area is quite similar. One can find an existing range of explanations of the "IM" or "PM" term here also which linguistically and logically cannot reach the severity of a scientific definition. Only two examples shall be provided here:

The Council of Advisors on Science and Technology, which is advising the US president, discusses the term "Personalized Medicine" in its report "Priorities for Personalized Medicine". Although the health care perspective is the main focus here under the keyword "tailoring of medical treatment", a wording is used which is broadening, if not softening, regarding the genus problem:

> "Personalized medicine" refers to the tailoring of medical treatment to the individual characteristics of each patient. It does not literally mean the creation of drugs or medical devices that are unique to a patient but rather the ability to classify individuals into subpopulations that differ in their susceptibility to a particular disease or their response to a specific treatment (President's Council 2008).

When talking about the "tailoring of medical treatment" the cited explanation refers to a metaphoric expression at a logically crucial point. This is blurring with regard to the genus question in respect that one could ask if "tailoring" of medical treatment in view of individual characteristics of patients is understood as part of health care or if it is rather situated in the field of medical research.

Another explanation given by Costigliola et al. (2009) lies on the border between research and health. This explicative "border-crossing" is linguistically also caused by a figurative phrase. Costigliola et al. (2009) write in order to introduce the expression "Predictive Medicine":

> Predictive Medicine is a new philosophy in healthcare and an attractive subject for currently initiated research activities aimed at a potential application of innovative biotechnologies in the prediction of human pathologies, a development of well-timed prevention and individual therapy-planning. The issue has several aspects which allow the expectations of great advantages for predictive diagnostics and personalized treatment as the medicine of future (Costigliola et al. 2009, p. 1).

According to this explanation, "Predictive" Medicine is not simply "a possible future health care", as in Hüsing et al. (2008), but rather—much less clear—a "philosophy in health care". "Philosophy" does not mean the academic subject, of course, but rather—in a figurative sense—a "(basic) orientation", "background concept", "guiding principle", "trend", "development" (the authors use the expression "development" themselves for reasons of explanation), maybe "vision" also, in any case something abstract, which begins to shape and change health care on the level of terms, concepts and ideas.

The explanation by Costigliola et al. (2009) aims at or tolerates a certain openness of the term "Predictive Medicine" towards not solely health care aspects. However, it does not provide a suggestion which meets the *scientific* demands of a definition.

2.3.3 "IM" as Medical Line of Research

The two English explanations are examples of how figurative phrases can lead to the problem that "IM" or "PM" cannot be assigned to the genus of health care, as clearly as in Hüsing et al. (2008). However, there are also explanations which are really alternative to the understanding of "IM" (or "PM") as new health care insofar as the term "IM" (or "PM") is decidedly used for describing a phenomenon which is situated in the field of medical *research*.

One definition which develops this genus alternative with a demand of linguistic exactness is the following "working definition" suggested by Marckmann (2011).

> Working definition: personalized (or individualized) medicine tries to identify individual (especially biological) factors which allow a better prediction of the risks of diseases and effects of therapies. [The] aim [of personalized or individualized medicine is]: a better prevention, diagnosis, prognosis and therapy tailored for the individual (Marckmann 2011, additions within the quotation by the authors).

Although this definition does not have an "Aristotelian" structure externally, the special wording "tries to identify individual factors" shows that "PM"/"IM" is seen as a line of research which is characterized and united by a certain τέλος, i.e. the "individualization" of health care by identifying predictors. Thus, "PM"/"IM"— both expressions are used explicitly as synonyms by Marckmann 2011—fall under another genus than the "PM"/"IM" terms which have been discussed so far.

Accordingly, PM was qualified as a medical research approach in a no longer accessible article on the German website of the pharmaceutical company Merck-Serano, which is literally quoted and thereby partially conserved in Langanke et al. (2012b):

> Personalized (stratified) medicine is a new and important approach of the modern medical research. In the course of this concept, medical solutions shall be found which are specifically tailored for the need and circumstances of individual patients or certain patient groups (Langanke et al. 2012b, p. 302).

If "IM" or "PM" is included in the generic terms "medical line of research" or "medial research approach" as in the cited explanations, it has to be clarified what "line of research" or "research approach" means. This task is demanding in the view of theory of science. However, for our purposes it is sufficient to adjust both terms in the following way: "Line of research" (in the following synonymous: "research approach") means a system of scientific assumptions and procedures which are especially suitable for reaching a certain aim within a certain science or scientific branch. A line of research or a research approach does not have to have the form of a proper research program *sensu* Lakatos (1978). But it is crucial that representatives of the same research approach are like-minded when it comes to pursuing a certain scientific target in a certain way. This way is described by a certain spectrum of methods and the (theoretical and/or empirical) assumptions which support these methods.

If one accepts this explanation, further specification is needed: aims within sciences can be hierarchically graded, i.e. they can be more or less general or specific. The "primary" or "highest" aim in medicine, for instance, could be to heal or to reduce suffering. However, such a general aim is not suitable for constituting a line of research because it cannot be operationalized by using certain methods. Aims which constitute fertile and promising lines of research are defined on a lower level in that sense that they are formulated in the light of specific methods and in the knowledge about what these methods can do. In these operationalized aims working instructions regarding their pursuit are roughly "included". Such a "well-formed" aim is, for instance, given when pharmacogenetics, as a medical line of research is aiming at a better prediction of the effects or side effects of pharmaceuticals through the systematic use of biological information about metabolizing dispositions of certain patient groups. This aim can be pursued with specific procedures.

2.4   Is Everything IM? The Question of Specific Difference

The leading aspect of evaluating the explanations and definitions in paragraph 2.3 was the genus problem. To which class of phenomena—we asked—belong IM/PM according to certain definitions? It was noticed that beside many loose and therefore not clear definitions, there are two scientifically valid suggestions: on the one hand, the IM and PM terms aim at activities which can be summarized under the generic term "health care". On the other hand, the IM and PM terms open up a category under the generic term "medical line of research".

If one asks, whether one of the two alternatives should be preferred or if a synthesis of both is desirable, it must be noticed, that by clarifying the genus questions, only half of the demands of the Aristotelian concept of definition will have been met. According to this concept a term is defined by providing specific characteristics. Then based on these specific characteristics a certain subgroup of phenomena can be distinguished from similar phenomena within the same genus. If applied to the IM or PM term, further investigation is needed with regard to how and by which means IM/PM can be distinguished from other health care or other medical lines of research.

2.4.1 Individual or More Individual Medicine?

Hüsing et al. (2008) distinguish five "concepts of individualization" within PM/IM which can be assigned to two strikingly different scientific and technological branches:

> The first branch includes scientific and technological developments which aim at tailoring therapeutic and preventive intervention for the individual patient in the sense of unique measures. [...] The second [...] branch is based on the knowledge that dispositions for certain diseases determine the development of the disease and its course through a complex interaction of genes, environmental factors, lifestyle and social status as well as intervention [...]. The manifestation of these factors is different for every individual. With the help of sufficient biomarkers a better stratification of the patient population shall be achieved regarding the actual clinical problem which goes beyond the status quo (Hüsing et al. 2008, p. 37).

If applied to the genus decision by Hüsing et al. (2008), this means that IM/PM is a possible future health care that

a. is orientated at using therapeutic and preventive measures which are adjusted for the individual in a strict sense *and also*
b. could be characterized by using medical means which address patient *collectives*, but in a more efficient way because the *means* are adapted more precisely to a certain subgroup of patients by making use of biomarker based stratification.

That this "and also" is "covering up" an important methodical difference on the level of the specific difference which was already mentioned in paragraph 2.2. Thus, it is sufficient here to emphasize that from our point of view the so-called "first strain" and the stratification concept of the "second strain" by Hüsing et al. 2008 are methodologically incommensurable approaches.

2.4.2 Stratification—A New Paradigm Within Health Science?

We have methodological concerns about combining the two approaches mentioned above, namely the specific patient approach and the concept of biomarker based stratification. If these concerns are met with approval, explanations such as the explanations of the US Council of Advisors on Science and Technology, which refer to one of these concepts of individualization become more plausible.

> It does not literally mean the creation of drugs or medical devices that are unique to a patient but rather the ability to classify individuals into subpopulations that differ in their susceptibility to a particular disease or their response to a specific treatment (President's Council 2008).

If one chooses the stratification concept in order to understand the specific aspect of IM/PM, one has to face another problem, which cannot be ignored here: "Stratification". In the German term "*Feinerstratifizierung*" one can see, by the use of the comparative form for the prefix, that stratification is not a *categorically*, but only a *gradually* definable phenomenon: patient populations have for a long time already been stratified in the field of medicine. If IM/PM is now claiming that it uses the procedure of classification (or: taxonomical) division, which is the logical key principle of stratification of patient populations, in a more sophisticated, i.e. more precise way, it means indeed that within IM/PM nothing fundamentally new is done, rather a well-known instrument is being used in a refined way only. In the light of this it must be noted that wherever

a. on the one hand, IM/PM stands for an approach, which is based on the stratification of externally uniformed patient populations with the help of biomarkers, and whereas
b. on the other hand, this procedure is declared "new" or "innovative" or even— with regard to the very demanding discourse about the philosophy of science—a new medical "paradigm" (Kuhn 1970),

one has to be cautious. Certain medical fields have been using the underlying principle for a long time in the frame of their diagnostic and therapeutic routine. Thus, it is already in use as standard in the field of oncology where (1) endocrinological or genetic parameters are used for the molecular characterization of tumors and can therefore be used for a more individual and more effective coordination of therapy or (2) where certain medications are only prescribed when a genetic test was done beforehand which proved the patient's genetic disposition of metabolism or that certain molecular structures (surface molecules), where certain therapeutics (e.g. monoclonal antibodies against these surface molecules) target, are expressed or over-expressed (Hasler-Strub 2009).

2.5   Individualized Medicine—Definition and Explanations

The explanation by Hüsing et al. (2008), which we have referred to several times, is suitable to show the semantic "irritations" which caused the authors of this chapter to work out a term of its own for GANI_MED.

a. The "IM" term in Hüsing et al. (2008) is semantically narrowed on the genus level. This constriction occurs because the IM term is solely ascribed to the generic term of health care. Thus, a linguistic standardization is made for the term "IM" which is not suitable to define certain activities in the field of medical *research* as "IM". For the authors of this chapter, this seems counter-intuitive. Moreover, this terminological decision leads to linguistic problems within the discourse about current scientific research projects regarding IM.
b. On the other hand, the opinion of Hüsing et al. (2008) seems to be too tolerant when it comes to the question of specific difference. By combining the two methodologically incommensurable concepts of individualization, the term of "Individualized Medicine" is not provided in a way that it defines a methodologically joined and in this sense homogenous class of phenomena. It rather defines a heterogeneous field of different medical activities in a loose way.

c. Speaking of "one health care" is not specific enough when considering that in health care research one speaks of *types* of health care terminologically and that Individualized Medicine, according to the current explanations and definitions, is not a fourth type of health care (Pfaff 2006), alongside preventive, curative and rehabilitative health care.

d. From our point of view, the wording used by Hüsing et al. (2008) "possible future" health care is, after all, criteriologically blurred. If one reconstructs this manner of speaking in a way that paragraph 2.3.1 suggested, one could ask for a threshold: To what quantitative extent do IM-based procedures have to be integrated into diagnostic and therapeutic routine in order for health care to be subsumed under the IM term? Is it even possible to determine a valid quantitative "saturation level" which allows an intersubjective communicable decision in this case? The only thing which is clear *prima facie,* is the opinion of Hüsing et al. (2008), that the existing biomarker-based applications are not yet sufficient in order to understand IM as already existing. Otherwise, the use of the modal-logical and future-charged attributes "possible" and "future" will remain incomprehensible in their explanation.

As a correction for these problems, we suggest the following definition:

"Individualized Medicine" ("IM" in short) stands for current medical fields of research on the one hand, and for health care practices on the other hand. Medical fields of research can be subsumed under the term Individualized Medicine if they aim at identifying, validating and integrating biomarkers into the clinical routine which allow to predict the outbreak or the course of diseases and/or the effect of therapies or their unwanted effects for certain patient groups in a better way. Preventive, therapeutic or rehabilitative health care practices, which can be included in the term of Individualized Medicine, are characterized by the fact that they use biomarkers for a systematic prediction of risks or courses of diseases and/or for the prediction of the effect of therapies or their unwanted effects.

The term "Stratifying Medicine" can be used as a synonym for "Individualized Medicine" in the sense stated above.

This entire definition needs further explanations:

1. Although the Aristotelian concept of definition was used as methodical heuristic for developing this definition, the suggested definition goes beyond the scope of this frame because it places the term "Individualized Medicine" into two different genera so that the relation of the generic and subsumable terms, relevant for understanding the term, cannot be displayed with *one* taxonomy.

2. *Within* the two genera, i.e. on the level of the particular difference, the logical option of adjunction is used by speaking of "disease prediction *and/or* of prediction of therapy success". An adjunction (logical link with the help of an *including* "or"—"and/or") between two statements is fulfilled when either one of the statements is true or both statements are true. A marker which can predict the risk for a certain disease can at the same time be a marker which gives predictive information regarding the success chances of a certain therapy, as the example of the endocrinologically or genetically based stratification of the mammary carcinoma shows.

3. According to the definition above "Individualized Medicine" deals with patient *collectives*. But "individualization" obviously is counted among the same word family as "individual". This circumstance supports the criticism cited at the beginning of this chapter that Individual Medicine is "fraudulent labeling". The authors want to respond to this criticism as follows: If "individual" is understood in the sense of everyday language or even of logical understanding, one can only speak of "more individual" medicine regarding biomarker-based "IM". Thus, one cannot simply praise the approaches of biomarker-based stratification as "Individual Medicine". The word "individualization" formed through the nominalization of the verb "individualize" seems suitable for showing that a precisely stratifying medicine, which uses the chances of biomarker-based prediction of diseases and therapy success by creating smaller subgroups, can contribute to the development that diagnostics and therapy can be tailored to the individual patient in a better way in the future. Whoever worries that the expression "individualization" may create unrealistic expectations regarding the establishment of an "individual case medicine" is free to use the more technical term "Stratifying Medicine", which was introduced as a synonym for "Individualized Medicine" in the definition above.

4. Crucial for the definition above is the concept of a biomarker. A "biomarker" is a molecular measurand (Vasan 2006), which allows by itself or together with other measurands to subdivide a patient population with regard to a relevant predictive feature or a set of such features. Relevant for diagnostics and therapy in the sense of IM, are predictive characteristics, which can be combined with probabilistic statements about whether and how its carriers develop a certain disease, and/or how its carriers will respond to a certain therapy. The connection between the quantitative tangible characteristic or set of characteristics, i.e. collected during the measuring, and the probabilistic predictive statement, is methodologically provided by the so-called association analyses. Here, it is examined whether the possession of a certain molecular feature, or a set of them, is connected in a statistically significant way, i.e. associated with the appearance of a certain course of disease (disease variant) or a certain (e.g. particularly good or bad) therapy response. If this is the case, a subpopulation can be distinguished with reference to the particular characteristic or set of characteristics. But such a characteristic or a set of such characteristics is relevant for clinical practical diagnostics and/or therapy, only if it is sufficiently sensitive and specific. This precondition is not self-understandingly fulfilled in a statistical association.

5. Although the program of "Individualized Medicine" had initially a very strong boost through innovations in the field of molecular *genetic* therapy success prediction in the light of pharmacogenetics, and although it is still today put on the same level as molecular *genetic* prediction, the definition above does not narrow down the term biomarker to the characteristics tangible by measurement and testing methods of molecular genetics. Rather, the possibility is logically granted, that (a) also on the level of organismic processes biomarkers can be found, which are accessible with the help of other molecular methods, more

specifically: with other -omics methods (proteomics and metabolomics) and
(b) combinations ("mixed scores") from non-molecular and -omics based characteristics are used for prediction.

6. In accordance with Schleidgen et al. (2013) we do *not* recommend, to subsume
medical research and healthcare activities related to predictive scores which
use *purely* non-molecular measurands (e.g. blood pressure, data from anthropometry, data from imaging and so on), under the term "biomarker based". In
consequence such activities should not be subsumed under the term "IM". This
decision makes the "IM" term presented in this chapter extensionally less tolerant than the elder version of this term in Langanke et al. (2012b).

7. With regard to the definition above one can say that IM leads to the establishment of new subgroups within a population which was considered homogenous
before: Where IM shows success in the field of predicting therapy success, it
allows, for instance, to subdivide groups of patients with regard to relevant
characteristics in different responder classes (e.g. Non-Metabolizer, Low
Metabolizer, High Metabolizer within pharmacogenetics). Moreover, where IM
can—this branch of IM is particularly successful in the field of oncology—
mark out groups of people with a higher or lower risk of a certain disease, or a
variant of it, or a certain course of disease (e.g. on the basis of endocrinological
and/or genetic characterization of tumors), IM leads to the establishment of
new subclasses within a disease phenotype, which was considered homogenous
before. Such subgroups are important for practical therapy to the extent that
patients, who are likely not to respond to a certain therapy, will be prevented
from this therapy, whereas patients who respond to it particularly well will be
offered this therapy.

8. The term "healthcare practice" used in the definition above needs to be terminologically distinguished from the term health care type *sensu* Pfaff (2006).
Health care practices, which are examples of IM in the sense of the definition
above, can occur within all three health care types, i.e. prevention, treatment
and rehabilitation. However, they do not constitute a fourth type.

9. The *same* health care practice is represented by different preventive, curative
and/or rehabilitative individual procedures if these procedures are based on the
same methods. Thus, different surgeries, for instance, which use the same protocol, represent the same health care practice.

10. The general expression "therapy" was chosen in the definition above on purpose. In particular, it should be avoided to only speak restrictively of drugs. On
the one hand, not all therapies are based on medication, on the other hand—
as shown in the example of immunoadsorption as a novel therapy option for
patients with dilated cardiomyopathy (Ameling et al. 2013; Felix and Staudt
2008; Herda et al. 2010; Staudt et al. 2010; Staudt et al. 2007; Staudt et al.
2006; Trimpert et al. 2010)—current research results justify the thesis that also
the response of patients to certain non-drug-based therapies can be predicted on
the basis of biomarkers (for this example see also Dörr et al. chap. 6, pp. 81–92

and Marschall et al. chap. 14). Thus, the choice of the generic term "therapy" is displayed methodologically here.

11. The term "Individualized Medicine", as developed in this section, remains orientated to the established understanding of this term and familiar phrases like "Personalized Medicine". It seems to be extensionally congruent with the term "Personalized Medicine", as suggested by Schleidgen et al. (2013). Furthermore it is very close to the definition given by the President's Council of Advisors on Science and Technology with its clear vote for the concept of individualization in the sense of stratification. Having said this, it should be noted that the metaphorical expression of "tailoring" chosen for the explication of the Council still needs more interpretation on the genus level. This is determined by the definition above because by it "Individualized Medicine" is anchored in two different genera, "medical research approach" and "health care practices". The "IM" term, which was provided by the discussed definition by Marckmann (2011) and the term of Personalized Medicine, which results from the explication by MerckSerano (Langanke et al. 2012b, p. 302), are extensionally fully included in the "IM" term suggested here through their narrow genus determination, so that the following applies: everything that is included in "IM" *sensu* Marckmann (2011) or "PM" *sensu* MerckSerano, is also included in the IM term which is provided here, but not the other way around.

2.6   Demarcation—Marking off the boundaries

How scientific a definition is, can be proven particularly by the fact that by the definition clear lanes become possible in the sense of terminological demarcation and exclusion, in an insufficiently structured field. Therefore it is crucial for a definition theory to give account for what does *not* or does *not anymore* fall under a defined term according to the underlying definition:

1. Approaches like cell therapy or *tissue engineering*, which are based on giving unique therapy to a specific patient (*specific patient approach*) cannot be subsumed under our IM term. This informed terminological decision is open to the critique of the *Scientific Community*. In particular, it should be noted that one could find it more appropriate to use the expression "Individualized Medicine" for labeling the *specific patient approach*, whereas "Personalized Medicine" should be used for describing the biomarker-based approach. But the terminological choice used by us is different here from this option, in terms of our definition. With regards to our decision it is *de facto* surely relevant that GANI_MED stands especially for "Greifswald Approach to *Individualized Medicine*". However, more important than decisions, which only concern the choice of words, are the reasons for the terminological differentiation of the *specific patient approach* and the *stratification concept*. There are serious methodological differences

between both concepts. In the light of this one should not be tempted, from the authors' point of view, to talk about the two concepts in a similar way.

2. Complementary medicine approaches and medical-technical motivated approaches like Ambient Assisted Living (AAL, for this approach see Manzeschke 2011, Manzeschke and Oehmichen 2010), which are treated like Individualized Medicine or as a contribution to it in certain contexts, are not included in "IM" according to the suggested definition. It is explicitly admitted that all these approaches can contribute to the realization of a better health care tailored for the individual. However, these approaches are *not* based on the biomarker concept in the sense of the definition above and thus do not meet the suggested criteria of subsumption.

3. With regard to the complementary medical approaches one could expand the demarcation by adding "conventional medicine" to the definition; but this is logically not necessary because the biomarker term is sufficiently discriminative. After all, complementary medical approaches like homeopathy or traditional Chinese medicine do not use the measurand concept in the same way as conventional medicine.

4. Particularly the concerns of a "communicative medicine", which rightly emphasizes the communicative aspects of a doctor-patient-relation regarding its importance for therapy and, in the light of this, criticizes the one-sided "technical" orientation of medicine, are taken seriously by the authors. Yet, it cannot be recommended terminologically to address such positions also as "Individualized Medicine" *at the same time* with those approaches based on the biomarker concept.

2.7   Does Individualized Medicine Already Exist?

If "existence" in the question above is understood as a quantifier, i.e. the prerequisites for existence are fulfilled when, on the level of the particular objects, at least one object can be identified, which falls under the defined term, one can say: the question concerning the existence of Individualized Medicine can be answered methodologically with regard to the definition in paragraph 2.5 by "screening" the genera "medical research approach" and/or "health care practice" towards the question of whether there are any activity fields within the two genera which satisfy the criteria laid down in the definition.

The result for the genus "medical line of research" is then very clear: Medical research approaches like pharmacogenetics, tumor genetics or genetics of multifactorial diseases are examples of "Individualized Medicine" in the sense of the definition above. This is due to the fact that they are lines of research which aim at the development of a biomarker-based predictive medicine in the sense explained above.

The picture in the field of health care practices is not less clear when looking at the definition above. Particularly in the field of oncology the use of certain test-drug-combinations has been a part of health care routine for a longer time. Relevant examples, picked up several times in the literature about Individualized or Personalized Medicine, are Herceptin or Tamoxifen, two drugs which are successfully being used against mamma carcinoma. In the case of Tamoxifen e.g., it is already standard that the drug is only prescribed, if a test showed beforehand that the tumor has certain molecular biological characteristics, which hint at a successful response to the drug. In the light of such cases, it can be said that: The oncological treatment of certain tumor diseases is a case of IM in the sense of the definition above.

2.8   Prospects—"Individualized Medicine" in the Field of Personalization Strategies

Cell therapy, Tissue engineering, but also medical developments like AAL are not included in the term IM according to the definition given in paragraph 2.5 because the biomarker concept and the attached stratification approach, as explained above, do not play a role here.

However, it was already emphasized that the IM defined here is congruent with these approaches by aiming at a better health care tailored for the individual. Medical and medical technical approaches, which do converge with this aim, should be called "Personalized Medicine". Thus, we use the existence of both different word groups "individualized"/"individualization" and "personalized"/"personalization" in order to show the *objective* difference.

The aim to tailor health care for the individual cannot be achieved directly, so that we—unlike the case with IM—plead for *not* using "Personalized Medicine" to label a certain research approach in the sense above, but as an expression for describing a methodological very heterogeneous field of different approaches which are "only" similar to each other concerning a very general aim. One of the approaches within this field is IM as in the sense above.

The fact that so many highly heterogeneous research activities within medicine or medical technology claim to contribute to a Personalized Medicine may be due to the current trend of "individualization" (the term is used here in a sociological sense) in Western societies, which does not spare the field of medicine, and also shapes and influences the expectations of patients. The patients, in the Western countries do not simply want the "health care standard", rather they increasingly expect that their treatment is oriented at personal needs.

The question of what kind of new ethical and health economic challenges could possibly arise from this trend will not be discussed here (for this topic see Ott and Fischer chap. 8, pp. 115–163 as well as Põder and Assel chap. 9, pp. 165–180). This discussion goes beyond the scope of this chapter. Only one question has to be raised here: Does Personalized Medicine not always need a counterweight in the form of general standardized health care, so that medicine, in the course of personalization, does not develop into luxurious medicine, for wealthy and well informed people, which will then fail to reach the relevant population groups? From our point of view, great political challenges concerning health care are apparent here, which will only be solved on a middle course, which avoids extremes and fits in with "communicative medicine". If such a political middle course in health care is found and followed, a modern medicine is possible, which will not simply be Individualized Medicine, but to which the different personalization approaches make important contributions.

References

Ameling S, Herda LR, Hammer E et al (2013) Myocardial gene expression profiles and cardiode-pressant autoantibodies predict response of patients with dilated cardiomyopathy to immuno-adsorption therapy. Eur Heart J 34(9):666–675

Costigliola V, Gahan P, Golubnitschaja O (2009) Predictive medicine as the new philosophy in healthcare. Introduction of the general concept. In: Golubnitschaja O (ed) Predicitive diagnostics and personalized treatment. Nova Publishers, New York, p 1–3

Felix SB, Staudt A (2008) Immunoadsorption as treatment option in dilated cardiomyopathy. Autoimmunity 41(6):484–489

Fricke A (2011) Individualisierte Medizin: Hoffnung und Risiko zugleich. Deutsche Ärzte Zeitung 22.02.2011. http://www.aerztezeitung.de/politik_gesellschaft/arzneimittelpolitik/article/642186/individualisierte-medizin-hoffnung-risiko-zugleich.html. Accessed 27 June 2014

Grabe HJ, Assel H, Bahls T et al (2014) Cohort profile: Greifswald approach to individualized medicine (GANI_MED). J Transl Med 12:144

Hasler-Strub U (2009) Aktuelle Therapien beim metastasierten Brustkrebs. Eine Übersicht über evidenzbasierte Optionen. Schweizer Zeitschrift für Onkologie 5:20–22

Herda LR, Trimpert C, Nauke U et al (2010) Effects of immunoadsorption and subsequent immunoglobulin G substitution on cardiopulmonary exercise capacity in patients with dilated cardiomyopathy. Am Heart J 159(5):809–816

Hüsing B, Hartig J, Bührlen B et al (2008) Individualisierte Medizin und Gesundheitssystem, TAB-Arbeitsbericht Nr. 126. http://www.tab-beim-bundestag.de/de/publikationen/berichte/ab126.html. Accessed 27 June 2014

Individualisierte Medizin: Etikettenschwindel (2011). http://www.bdi.de/allgemeine-infos/aktuelle-meldungen/ansicht/article/individualisierte-medizin-etikettenschwindel.html. Accessed 27 June 2014

Kollek R, Lemke T (2008) Der medizinische Blick in die Zukunft. Gesellschaftliche Implikationen prädiktiver Gentests. Campus-Verlag, Frankfurt a. M.

Kuhn TS (1970) The structure of scientific revolutions, 2 edn. University of Chicago Press, Chicago

Lakatos I (1978) The methodology of scientific research programmes, philosophical papers Vol. 1. Cambridge University Press, Cambridge

Langanke M, Brothers KB, Erdmann P et al (2011) Comparing different scientific approaches to personalized medicine: research ethics and privacy protection. Per Med 8(4):437–444

Langanke M, Erdmann P, Dörr M et al (2012a) Gesundheitsökonomische Forschung im Kontext Individualisierter Medizin: Forschungsethische und datenschutzrechtliche Aspekte am Beispiel des GANI_MED-Projekts. Pharmacoecon Ger Res Artic 10(2):105–121

Langanke M, Lieb W, Erdmann P et al (2012b) Was ist Individualisierte Medizin? Zur terminologischen Justierung eines schillernden Begriffs. Zeitschrift für medizinische Ethik 58:295–314

Manzeschke A (2011) Technische Assistenzsysteme. Eine Antwort auf die Herausforderung des demographischen Wandels? Pro Alter 5:36–40

Manzeschke A, Oehmichen F (2010) Tilgung des Zufälligen. Ethische Aspekte der Verantwortung in Ambient-Assisted-Living-Kontexten. Jahrbuch für Wissenschaft und Ethik 15:121–138

Marckmann G (2011) BMBF-Verbundprojekt „Individualisierte Gesundheitsversorgung: Ethische, ökonomische und rechtliche Implikationen für das deutsche Gesundheitswesen". 1. Expertenworkshop, München, 20. September 2011. http://www.igv-ethik.de/vortraege/projektvorstellung.pdf. Accessed 27 June 2014

Niederlag W, Lemke HU, Golubnitschaja O et al (eds) (2010) Personalisierte Medizin. General Hospital, Dresden

Pfaff H (2006) Präventive Versorgung. Begriffsbestimmung und theoretisches Konzept. Prävention und Gesundheitsförderung 1:17–23

President's Council of Advisors on Science and Technology (2008) Priorities for personalized medicine. Report of President's Council of Advisors on Science and Technology, September 2008. http://www.whitehouse.gov/files/documents/ostp/PCAST/pcast_report_v2.pdf. Accessed 27 June 2014

Schleidgen S, Klingler C, Bertram T et al (2013) What is personalized medicine: sharpening a vague term based on a systematic literature review. BMC Medical Ethics 14:55

Staudt A, Herda LR, Trimpert C et al (2010) Fcgamma-receptor IIa polymorphism and the role of immunoadsorption in cardiac dysfunction in patients with dilated cardiomyopathy. Clin Pharmacol Ther 87(4):452–458

Staudt A, Eichler P, Trimpert C et al (2007) Fc(gamma) receptors IIa on cardiomyocytes and their potential functional relevance in dilated cardiomyopathy. J Am Coll Cardiol 49(16):1684–1692

Staudt A, Hummel A, Ruppert J et al (2006) Immunoadsorption in dilated cardiomyopathy: 6-month results from a randomized study. Am Heart J 152(4):712.e711–712.e716

Trimpert C, Herda LR, Eckerle LG et al (2010) Immunoadsorption in dilated cardiomyopathy: long-term reduction of cardiodepressant antibodies. Eur J Clin Invest 40(8):685–691

Vasan RS (2006) Biomarkers of cardiovascular disease. Molecular basis and practical considerations. Circulation 113:2335–2362

Chapter 3
Individualized Medicine Within the GANI_MED Project

Hans-Jörgen Grabe and Henri Wallaschofski

Abstract In 2009 the University Medicine Greifswald, Germany, has launched the "Greifswald Approach to Individualized Medicine" (GANI_MED) project to address major challenges of Individualized Medicine. The major goals of Individualized Medicine require excellent clinical stratification of patients as well as the availability of genomic data and biomarkers as prerequisites for the development of novel diagnostic tools and therapeutic strategies. In this chapter, we describe the implementation of the scientific and clinical infrastructure. Patient cohorts with an emphasis on metabolic and cardiovascular diseases are being established aiming at the inclusion of more than 5,000 patients. Standardized protocols for the assessment of medical history, laboratory biomarkers, and the collection of various biosamples for biobanking have been developed. A multi-omics based biomarker assessment is applied to identify new biomarkers or biomarker-signatures that complement disease-specific biomarkers at an individual or stratified level. A central data management structure has been implemented to integrate all relevant clinical data for research and translational purposes. Ethical research projects on informed consent procedures and the reporting of incidental findings were initiated as well as health economic evaluations.

Keywords Personalized Medicine · Individualized Medicine · Epidemiology

3.1 Introduction

Recently developed high throughput -omics technologies are thought to enable more targeted diagnostic and treatment approaches. The introduction of these -omics technologies into everyday clinical routine will enable an increase in

H.-J. Grabe (✉)
Klinik und Poliklinik für Psychiatrie und Psychotherapie, Universitätsmedizin Greifswald und HELIOS Klinik Stralsund, Rostocker Chaussee 70, 18437 Stralsund, Germany
e-mail: grabeh@uni-greifswald.de

H. Wallaschofski
Facharzt für Innere Medizin Endokrinologie Diabetologie Andrologie,
Krampferstr. 6, 99084 Erfurt, Germany
e-mail: praxis@endokrinologie-erfurt.de

© Springer International Publishing Switzerland 2015   29
T. Fischer et al. (eds.), *Individualized Medicine*, Advances in Predictive,
Preventive and Personalised Medicine 7, DOI 10.1007/978-3-319-11719-5_3

individualized diagnostic and therapeutic strategies. Further, treatment efficacy and safety will be improved as well as better individual outcome prediction and risk assessment. A more efficient allocation of economic resources is pursued. Still, many aspects of Individualized Medicine remain controversial. A questionable clinical utility of most novel biomarkers is raised amongst the most common criticisms (Völzke et al. 2013). Furthermore, potential societal and ethical consequences of Individualized Medicine are not adequately addressed or even considered. However, great advances in pharmacogenomic testing enable a much better management of non-responses due to increased or decreased metabolism of drugs as well as the prediction of serious side effects (e.g. statins). The genetic testing and the informative link to large data bases are already commercially available. Other important examples of the great potential of Individualized Medicine come from predictive research for type I and type II diabetes (Wang et al. 2013; Zhang et al. 2013) and Alzheimer's Disease (Mapstone et al. 2014).

In the research project "Greifswald Approach to Individualized Medicine" (GANI_MED) we implement individualized diagnostic and therapeutic strategies in a university hospital. Primary focus is on cardiovascular, cerebrovascular and metabolic diseases as well as on ethical and economic aspects of Individualized Medicine.

The main objectives of GANI_MED are as follows:

- Recruitment of large patient cohorts for common vascular and metabolic diseases and conditions (heart failure, renal failure, stroke, metabolic syndrome, fatty liver disease) at the university hospital Greifswald and the thorough examination of these cohorts with standardized clinical measures, imaging, and molecular analyses.
- Special efforts ensure high quality data even in the clinical context and facilitate the use of these data for scientific purposes.
- Establishment of a fully automated biobank.
- Further development of modern bio-analytical methods, focusing on pharmacogenomics, metabolomics, and proteomics in order to enhance discovery of biomarkers for disease conditions of interest.
- Generation of complementing multi-omics data for association with clinical and subclinical phenotypes in case-control designs using data generated for the Study of Health in Pomerania (SHIP, (Völzke et al. 2011)) as control.
- Identification of novel biomarkers and bio-signatures that enable individualized diagnostics, prediction of risks and outcomes as well as individualized therapeutic approaches.
- Assessment of potential ethical and economic implications of Individualized Medicine.

3.2 Methodological Hallmarks of GANI_MED

3.2.1 Examination Standard Across all Patient Cohorts

Key elements of GANI_MED are extensive quality control and standardization in the data acquisition. We have implemented an extended array of quality control measures regarding clinical phenotyping of the GANI_MED patients. These include computer-assisted standardized interviews to obtain the medical history and the medication taken by the patient. Standard operating procedures have been defined for each clinical examination, including the measurements of blood pressure, height, weight, hip circumference, waist circumference, and different ultrasonographic measures of the carotid arteries, kidney, liver, pancreas, and uterus. In contrast to established procedures in clinical practice, the medical staff have not only been trained for performing examinations, but also certified according to SHIP standards (Völzke et al. 2011). In addition, questionnaires are used to further assess the patients' individual characteristics. For example, in all GANI_MED cohorts childhood traumata (CTS, (Grabe et al. 2012)) and current depressive symptoms (PHQ-9, (Kroenke et al. 2001)) are measured to enable an integrative approach to the impact of early traumatization and depression on the course and treatment response in somatically ill patients.

Moreover, data collection is constantly monitored as well as examiner variation and time trends. We have implemented an automated monitoring system allowing for online recruitment monitoring and weekly feedbacks to the examiners.

3.2.2 Standardized Medical History

To avoid the unsystematic taking of medical history, which would limit the scientific use of the data, we have developed a standardized taking of medical history across the GANI_MED cohorts. A basic set of obligatory, standardized questions is applied extended by optional cohort-specific special questions as part of a hierarchical structure of the interview. Answers to structured questions are directly coded into a portable computer. Each patient recruited for GANI_MED is interviewed as part of the taking of medical history. The main areas covered by the interview are hypertension, cerebrovascular diseases, cardiac disorders, thrombosis, pulmonary embolism, peripheral vascular diseases, and gastrointestinal, hepatic, biliary, pancreatic as well as kidney diseases. Furthermore the history of metabolic diseases, stroke, cancer, mental disorders, surgery, current complaints, and current therapies as well as the social history and the history of health behaviors and lifestyle factors are recorded. The IT solution for the medical history covers a total of approximately 8,000 electronic Case Report Form (eCRF) variables.

3.2.3   Laboratory Measures and Biosampling

We have defined a specific set of laboratory parameters that is routinely measured for every patient in GANI_MED. These parameters include blood count, electrolytes, renal function parameters, liver function tests, blood lipids, inflammatory markers, and urinary markers. A comprehensive set of biomaterials including EDTA-plasma, serum, urine, saliva as well as buccal and tongue smears are collected, assayed (in 850 µl cryo tubes) and stored in an automated biobank (STC12k-ULT KiWi Store, FA Liconic, Liechtenstein) for future laboratory analyses.

3.2.4   Cohort Specific Examinations

Besides the core examination program, which includes the computer-assisted basic medical history and a comprehensive medication assessment, blood pressure and anthropometric measurements, a basic dental examination, and the common laboratory measurements, each cohort is characterized by additional examination procedures. For further information see (Grabe et al. 2014).

3.3   Patient Cohorts in GANI_MED

Six main cohorts with common cardiovascular, cerebrovascular or metabolic conditions: heart failure (expected $n=1200$), stroke (expected $n=600$), periodontal disease (expected $n=800$), renal insufficiency (expected $n=400$), metabolic syndrome (expected $n=1600$), and fatty liver disease (expected $n=400$) are being recruited. The official start of the patient recruitment was July 7, 2011. Three further cohorts (patients with sepsis, pulmonary diseases, and adverse medication effects) have been launched in the meantime. GANI_MED was approved by the local ethics committee. After a complete description of the study to the patients, written informed consent is obtained.

3.3.1   Heart Failure Cohort

Patients with suspected or known chronic heart failure (HF) due to systolic (HF with reduced ejection fraction = HFREF) or diastolic left ventricular dysfunction (HF with preserved ejection fraction = HFPEF) are recruited for this cohort. Within the HFREF cohort a specific focus is on patients with dilated cardiomyopathy (Dickstein et al. 2008). HFPEF is diagnosed in concordance with current recommendations (Paulus et al. 2007).

3.3.2   Cerebrovascular Disease Cohort

This cohort enrolls patients with acute ischemic stroke or transitory ischemic attacks (TIA) (Sacco et al. 2013) who are admitted to the Stroke Unit of the Department of Neurology. Stroke subtypes are classified according to TOAST and ASCO criteria (Adams et al. 1993; Amarenco et al. 2009). Cognitive function is assessed with the mini-mental state examination (MMSE) (Folstein et al. 1975).

3.3.3   Periodontal Disease Cohort

Periodontal disease is an inflammatory disease caused by an infection of the supporting tissue around the teeth which may lead to tooth loss if left untreated. Besides being a major cause of tooth loss, periodontal disease may be a risk factor for various systemic conditions and diseases (Chapple and Genco 2013; Ramirez et al. 2014). We recruit patients, who have been on maintenance therapy for a long time, patients who were incompliant with maintenance and dropped out, and untreated newly admitted patients. Biosamples include subgingival plaque, tongue smear, and stimulated saliva. Comprehensive dental examinations are performed.

3.3.4   Renal and Renovascular Disease Cohort

About two thirds of the cohort is represented by patients with end-stage kidney disease (ESKD) on dialysis. CKD is assessed by estimating the glomerular filtration rate from serum creatinine levels using the Modification of Diet in Renal Disease (MDRD) formula (Levey et al. 1999), and by measuring the albumin-to-creatinine ratio in spot urine (KDIGO 2013).

3.3.5   Metabolic Syndrome Risk Cohort

Subjects who have an increased risk of developing, or who actually suffer from, metabolic syndrome are eligible for this cohort. Briefly, the metabolic syndrome is defined as fulfilling three or more of the following five criteria: (1) Abdominal obesity; (2) elevated blood pressure; (3) elevated non-fasting glucose; (4) elevated triglycerides; (5) reduced high-density lipoprotein (HDL). All patients fulfilling the criteria for metabolic syndrome without meeting the diagnosis of diabetes mellitus receive an oral glucose tolerance test. In order to assess possible hormonal disturbances related to distinct comorbid conditions associated with the metabolic syndrome, this cohort consists of three sub-cohorts:

- Patients with the primary diagnosis of mental disorders are recruited from the day clinic of the Department of Psychiatry and Psychotherapy. The main diagnoses of these patients are depressive disorders (90 %), anxiety and somatoform disorders, trauma-related disorders and personality disorders.
- Women with polycystic ovary syndrome (PCOS) are recruited at the Department of Obstetrics and Gynecology. PCOS is a highly prevalent heterogeneous disease affecting 1 in 5 women in reproductive age. PCOS is characterized by clinical or biochemical androgen excess and/or anovulation, and polycystic ovaries on ultrasound. PCOS increases the risk of insulin resistance, type 2 diabetes mellitus, visceral obesity, cardiovascular disease, infertility, and depression.
- Patients with the metabolic syndrome and prevalent cardiovascular diseases (e.g., coronary artery disease, myocardial infarction, hypertensive heart disease) and at least three metabolic syndrome criteria are recruited from the Department of Internal Medicine B/Cardiology.

3.3.6   Fatty Liver Disease Cohort

This cohort includes patients with non-alcoholic fatty liver disease (NAFLD). Screening of patients is performed by transabdominal ultrasound. In the case of a hyperechogenic pattern of the liver parenchyma in comparison to the kidney, patients are recruited for the study. Patients with at least one of the following conditions are excluded: Liver cirrhosis, alcohol abuse or consumption > 30 g alcohol per day, pulmonary hypertension, dilated cardiomyopathy with ejection fraction < 30 %, autoimmune hepatitis, viral hepatitis, Wilson disease (hepatolenticular degeneration), hemochromatosis, α-1-antitrypsin deficiency, amyloidosis, malignoma of the liver, primary biliary cirrhosis or primary sclerosing cholangitis. In a fraction of patients liver biopsy is obtained, provided that it is clinically indicated.

3.3.7   Cohort of Adverse Medication Effects

This cohort includes all patients who are admitted to the Departments of Internal Medicine due to severe adverse drug reactions (ADR). In patients with suspected ADR, a detailed history is obtained and results of clinical and laboratory tests as well as ADR description (e.g., severity, course and outcome) are gathered. Patients receiving cancer chemotherapy are excluded.

Outcomes of ADRs are classified according to International Conference of Harmonization (ICH) guidelines. The assessment of ADR severity is based on the adverse drug reaction reporting system described by Hartwig et al. (Hartwig et al. 1992). Classification of ADR types is performed according to Edwards and Aronson (Edwards and Aronson 2000). A standardized causality assessment of each drug taken before the hospital admission is made using the Begaud algorithm (Bégaud et al. 1985). All ADRs are reviewed in a quality assurance and clinical plausibility check by a second clinical pharmacologist.

3.3.8 Cohort of Pulmonary Diseases

The aim of this cohort is a thorough characterization of the individual cardiovascular risk factor profile and the co-morbidities in patients with chronic obstructive pulmonary disease (COPD). The cohort specific phenotyping includes a six-minute walk test (6-MWT) (Enright 2003), lung function analysis, blood gas analysis, and polysomnography. In a selected subgroup, additional measurements will be performed including MRI scanning of thorax, heart, and hamstring muscle. Examinations are complemented by specific questionnaires regarding sleep-associated symptoms (e.g. Berlin Questionnaire (Netzer et al. 1999)).

3.3.9 Cohort of Sepsis

Sepsis and septic shock still contribute significantly to in-hospital mortality, in particular in intensive care units, despite major progress in hygiene and antibiotic treatment (Stevenson et al. 2014). It remains unclear which combination of patient characteristics may determine the clinical course of the condition, and therefore the survival of the patients (McEvoy and Kollef 2013). Patients are enrolled in the cohort, based on established sepsis criteria. All patients are followed up for three consecutive time points. The major aim of this cohort is to identify clinical and omics-markers that allow for prediction of the sepsis course and may guide individual treatment decisions.

3.4 Infrastructure

3.4.1 Medical Informatics and Data Management

Individualized Medicine requires complex solutions for data sets, data handling and data storage. A high degree of automation is necessary to efficiently collect and manage data as well as to facilitate the output of selected data sets for research purposes (Meyer et al. 2012). The integration of very different data sources like eCRF applications, medical devices (e.g., MRI, sonography, ECG), and hospital information systems is a challenging task. The software architecture has to protect data privacy in compliance with current data protection regulations of the Medical Device Directive (MDD) of the EU. Further technical aspects like IT security, scalability or high availability need to be addressed.

The informed consent is a basic requirement for the scientific use of patient data. GANI_MED adopts a modular consent design which allows patients to agree or disagree to all or only to particular uses of the data for research purposes. Areas of specific informed consent include re-contacting of the patient, use of clinical data, use of biomaterial, or data application of secondary data at the level of health insurance companies. An independent Trusted Third Party manages all patients' identifying data. The Trusted Third Party also generates pseudonyms for the research data sets.

The cohort management is done using Java-based offline-capable eCRF software on mobile clients, so-called Mobile Clinical Assistants (Meyer et al. 2013). These are used to record patients' personal data, document the informed consent, and generate a cohort-specific number of dynamic eCRF-based data, including lab order IDs and similar links to external sources. All data are synchronized with a server application which functions as one of the data sources. Currently, the cohort management is used for nine GANI_MED cohorts, implementing 225 electronic forms, more than 8,000 variables and 200,000 data sets. The Extraction Layer of the platform accounts for separating personal (i.e., identifying) data from medical data and performs source-specific transformations into a platform-internal unified data format for further processing.

3.4.2 Functional Genomics Facilities

Deep phenotyping of patients with genomics and functional genomics approaches provides molecular information that will allow much more in-depth description of the (patho)physiological status of a patient. The -omics data generated at different levels (genomics, transcriptomics, proteomics, and metabolomics) will likely allow for better descriptions of health and disease states (Chen et al. 2012). Currently, genomic variants are analyzed for 1,800 patients with the aid of the HumanCoreExome + v1.1-Psych Array. For a subset of 1,200 patients, whole blood expression data are generated using Illumina's HT-12 bead chips. These data will be complemented with proteomics and metabolomics data using established workflows. Integrated analyses of these data for associations with clinical and subclinical phenotypes will be performed by the bioinformatics groups of the GANI_MED consortium.

3.5 Economic and Ethical Aspects of Individualized Medicine

3.5.1 Economic Research

The health economic analyses contribute to the comprehensive assessment of advantages and disadvantages of Individualized Medicine by comparing costs of innovative diagnostics with their effectiveness. One focus lies on the performance of an economic evaluation of the predictive value of genetic and non-genetic biomarkers to forecast health care costs. The analysis, which is based on SHIP data, correlates genetic and non-genetic biomarkers with the health seeking behavior and/or health care expenditure (Haring et al. 2010). Secondly, cost analyses of severe drug-related side effects are performed by the economic research group as the side effects are a major cause of hospital admissions and cause a large fraction of the global health care expenditure (Schwebe et al. 2012). Thirdly, economic analyses evaluates the cost-effectiveness of specific therapeutic interventions of Individualized Medicine (Flessa and Marschall 2012).

3.5.2 Ethical Research

One major challenge for the ethical working group was to develop an appropriate information concept to enable patients to give their informed consent. This informed consent procedure covers the multilevel assessment of personal, clinical and biological data as well as the storage and use of data and the biobanking. All informed consent procedures are carried out according to standards of law, data protection, and research ethics. This includes the preparation of the information and consent forms, an ongoing training for the GANI_MED staff as well as the design and implementation of workflows and standard operation procedures regarding all ethically sensitive processes in GANI_MED (Langanke et al. 2011; Langanke et al. 2012).

Some GANI_MED cohorts also include non-treatment related study examinations. Ethical challenges emerge, if incidental findings, that might be relevant for the health of the participants, occur in these cases. Their management does not follow clinical routine care. Briefly, when the participant does not suffer from any symptoms of illness, researchers need to weight the potential harm against the potential benefit of disclosure of medical findings for the participant. To address this issue, one of our ethical research projects specifically focuses on the disclosure of incidental findings (Völzke et al. 2011) with a special focus on the whole body MRI. The view of the participants is inferred from their answers to questionnaires and interviews with SHIP participants who were informed about their incidental MRI findings (Erdmann et al. 2013; Rudnik-Schöneborn et al. 2013; Schmidt et al. 2013).

When IM propagates a paradigm shift in clinical research and practice, this also means an epistemological turning point for the concepts of the individual and responsibility. Therefore, Individualized Medicine needs to be understood in its cultural and historical setting and its socio-cultural implications need to be historically contextualized (Gadebusch Bondio and Michl 2010). In addition, a medical-philosophical approach explores the concept of IM concerning epistemological prerequisites and normative-ethical implications. In order to justify the estimation of the consequences of Individualized Medicine a lifeworld reconstruction of a discourse about the concepts of health and disease; a presuppositional and analytical examination of IM's impact on medical ethical principles and problem areas; as well as an analyses of the skeptical objections of IM are focused on (Ott and Fischer 2013).

3.6   Research Strategies and Topics Within GANI_MED

3.6.1   Reference Groups for Case-Control Comparisons

For the evaluation of biomarker findings in patients well characterized control samples are necessary. At Greifswald, the large prospective and representative cohort study SHIP was implemented in 1997 (SHIP-0) to investigate a broad spectrum of health-related parameters in West Pomerania, a rural area in the north-east of Germany with a relatively short life expectancy as compared to that in the average German population (Völzke et al. 2011). As the university hospital Greifswald is the only major hospital in the area the SHIP sample is derived from the same geographic area, ensuring that patient cohorts and control groups originate from the same population. Over the past 15 years, the population-based study SHIP has evolved as a cohort study with one of the most comprehensive phenotype assessments worldwide covering the follow-up waves SHIP-1, SHIP-LEGEND and SHIP-2. From 2008 to 2012 a new, independent sample called SHIP-TREND was recruited ($n=4420$).

3.6.2   Translational Strategies

Associations primarily identified in clinical setting can be tested for their relevance in the general population (SHIP). Identified or suspected biomarkers can be tested for their predictive value for incidental diseases in the longitudinal waves of SHIP. On the other hand, the translational strategies also include the testing of associations in SHIP-0 and the replication of findings in SHIP-TREND. In a further step, those associations can be transferred to the clinical samples of GANI_MED to check for their validity in severely ill patients. Clinical samples from GANI_MED can be stratified based on clinical and biological markers and investigated for differential outcomes, courses, and treatment responses. Biological mechanisms will be addressed by biological systems approaches and other experimental approaches such as cell culture and animal studies, which are performed in collaboration with our local, national or international collaborators.

3.6.3   Research Within the Field of Individualized Medicine

As data collection was still ongoing in GANI_MED and -omics data only became available in summer 2014, research to date has focused on the analyses of SHIP and laboratory data and on smaller patient samples. New hypotheses have been generated for validation and translation into clinical samples. Research in the field of hormones like testosterone, prolactin and IGF-1 as individual biomarkers and putative predictors of clinical outcomes (e.g., hypertension, depression, inflammation) has been especially active (Friedrich et al. 2011; Ohlsson et al. 2011; Friedrich et al. 2012b;

Haring et al. 2012; Ziemens et al. 2013; Sievers et al. 2014). Novel biomarkers like angiopoietin-2, Tie-2 (Lorbeer et al. 2013), homoarginine and asymmetric dimethyl-arginine (ADMA) (Atzler et al. 2013) are under investigation. Metabolites have been analyzed by NMR and mass spectroscopy and are currently being tested for their association and predictive properties in various diseases (Friedrich et al. 2012a). Strong associations between metabolites and genetic variations have been discovered (Suhre et al. 2011). Currently, metabolic data are being used to develop a "metabolic age score" which indicates the biological age versus the chronological age. Myocardial gene expression profiles have been used to successfully predict the treatment response to immunoadsorption therapy in patients with dilated cardiomyopathy (Ameling et al. 2013). The important interface between oral and systemic health is increasingly addressed in GANI_MED (Schwahn et al. 2013), especially the inflammatory effects of periodontitis (Holtfreter et al. 2013). Growth factor receptor-mediated signaling is meanwhile recognized as a complex signaling network. The interactome of the epidermal growth factor receptor (EGFR) was identified and quantified. The newly developed Cytoscape plugin ModuleGraph facilitated the extension and functional investigation of this network (Foerster et al. 2013). A short 5-item questionnaire for routine clinical assessment of childhood abuse and neglect has been developed and validated (Grabe et al. 2012; Glaesmer et al. 2013). Thus, the effects of gene-environment interactions in somatically burdened patients can be investigated (Grabe et al. 2010; Grabe et al. 2011). A standardized protocol for fat quantification in fatty liver disease on liver MRI has been developed and validated within this cohort (Kuhn et al. 2011; Kuhn et al. 2013a; Kuhn et al. 2013b).

Acknowledgement The contribution to data collection performed by study nurses, study physicians, ultrasound technicians, interviewers, and laboratory workers is gratefully acknowledged. We are also appreciative of the important support of IT- and computer scientists, medical documentarists, and administration staff. We also thank all study participants whose personal dedication and commitment make this project possible.

References

Adams HP Jr, Bendixen BH, Kappelle LJ et al (1993) Classification of subtype of acute ischemic stroke. Definitions for use in a multicenter clinical trial. TOAST. Trial of Org 10172 in acute stroke treatment. Stroke 24(1):35–41

Amarenco P, Bogousslavsky J, Caplan LR et al (2009) New approach to stroke subtyping: the A-S-C-O (phenotypic) classification of stroke. Cerebrovasc Dis 27(5):502–508

Ameling S, Herda LR, Hammer E et al (2013) Myocardial gene expression profiles and cardiodepressant autoantibodies predict response of patients with dilated cardiomyopathy to immunoadsorption therapy. Eur Heart J 34(9):666–675

Atzler D, Rosenberg M, Anderssohn M et al (2013) Homoarginine–an independent marker of mortality in heart failure. Int J Cardiol 168(5):4907–4909

Bégaud B, Evreux JC, Jouglard J et al (1985) Imputation of the unexpected or toxic effects of drugs. Actualization of the method used in France. Therapie 40(2):111–118

Chapple IL, Genco R (2013) Diabetes and periodontal diseases: consensus report of the joint EFP/AAP workshop on periodontitis and systemic diseases. J Periodontol 84(4 Suppl):106–112

Chen R, Mias GI, Li-Pook-Than J et al (2012) Personal omics profiling reveals dynamic molecular and medical phenotypes. Cell 148(6):1293–1307

Dickstein K, Cohen-Solal A, Filippatos G et al (2008) ESC guidelines for the diagnosis and treatment of acute and chronic heart failure 2008: the task force for the diagnosis and treatment of acute and chronic heart failure 2008 of the European Society of Cardiology. Developed in collaboration with the Heart Failure Association of the ESC (HFA) and endorsed by the European Society of Intensive Care Medicine (ESICM). Eur J Heart Fail 10(10):933–989

Edwards IR, Aronson JK (2000) Adverse drug reactions: definitions, diagnosis, and management. Lancet 356(9237):1255–1259

Enright PL (2003) The six-minute walk test. Respir Care 48(8):783–785

Erdmann P, Langanke M, Assel H (2013) Zufallsbefunde, Risikobewusstsein von Probandinnen und forschungsethische Konsequenzen. Medizin und Technik, Risiken und Folgen technologischen Fortschritts. Mentis, Münster, pp 15–47

Flessa S, Marschall P (2012) Individualisierte Medizin: vom Innovationskeimling zur Makroinnovation. PharmacoEconomics—German Research Articles 10(2):53–67

Foerster S, Kacprowski T, Dhople VM et al (2013) Characterization of the EGFR interactome reveals associated protein complex networks and intracellular receptor dynamics. Proteomics 13(21):3131–3144

Folstein MF, Folstein SE, McHugh PR (1975) "Mini-mental state". A practical method for grading the cognitive state of patients for the clinician. J Psychiatr Res 12(3):189–198

Friedrich N, Schneider HJ, Spielhagen C et al (2011) The association of serum prolactin concentration with inflammatory biomarkers—cross-sectional findings from the population-based study of health in pomerania. Clin Endocrinol (Oxf) 75(4):561–566

Friedrich N, Budde K, Wolf T et al (2012a) Short-term changes of the urine metabolome after bariatric surgery. OMICS 16(11):612–620

Friedrich N, Schneider HJ, Haring R et al (2012b) Improved prediction of all-cause mortality by a combination of serum total testosterone and insulin-like growth factor I in adult men. Steroids 77(1–2):52–58

Gadebusch Bondio M, Michl S (2010) Individualisierte Medizin. Die neue Medizin und ihre Versprechen. Dtsch Arztebl 107(21):A-1062/B-934/C–922

Glaesmer H, Schulz A, Häuser W et al (2013) The childhood trauma screener (CTS)—development and validation of cut-off-scores for classificatory diagnostics. Psychiatr Prax 40(4):220–226

Grabe HJ, Schwahn C, Appel K et al (2010) Childhood maltreatment, the corticotropin-releasing hormone receptor gene and adult depression in the general population. Am J Med Genet B Neuropsychiatr Genet 153B(8):1483–1493

Grabe HJ, Schwahn C, Appel K et al (2011) Update on the 2005 paper: moderation of mental and physical distress by polymorphisms in the 5-HT transporter gene by interacting with social stressors and chronic disease burden. Mol Psychiatry 16(4):354–356

Grabe HJ, Schulz A, Schmidt CO et al (2012) A brief instrument for the assessment of childhood abuse and neglect: the childhood trauma screener (CTS). Psychiatr Prax 39(3):109–115

Grabe HJ, Assel H, Bahls T et al (2014) Cohort profile: Greifswald Approach to Individualized Medicine (GANI_MED). J Transl Med 12(1):144

Haring R, Baumeister SE, Völzke H et al (2010) Prospective association of low serum total testosterone levels with health care utilization and costs in a population-based cohort of men. Int J Androl 33(6):800–809

Haring R, Baumeister SE, Völzke H et al (2012) Prospective inverse associations of sex hormone concentrations in men with biomarkers of inflammation and oxidative stress. J Androl 33(5):944–950

Hartwig SC, Siegel J, Schneider PJ (1992) Preventability and severity assessment in reporting adverse drug reactions. Am J Hosp Pharm 49(9):2229–2232

Holtfreter B, Empen K, Gläser S et al (2013) Periodontitis is associated with endothelial dysfunction in a general population: a cross-sectional study. PLoS One 8(12):e84603

KDIGO (2013) KDIGO 2012 clinical practice guideline for the evaluation and management of chronic kidney disease. Kidney Int 3(Suppl):1–150

Kroenke K, Spitzer RL, Williams JB (2001) The PHQ-9: validity of a brief depression severity measure. J Gen Intern Med 16(9):606–613

Kuhn JP, Evert M, Friedrich N et al (2011) Noninvasive quantification of hepatic fat content using three-echo dixon magnetic resonance imaging with correction for T2* relaxation effects. Invest Radiol 46(12):783–789

Kuhn JP, Hernando D, Mensel B et al (2013a) Quantitative chemical shift-encoded MRI is an accurate method to quantify hepatic steatosis. J Magn Reson Imaging 39(6):1494–1501

Kuhn JP, Jahn C, Hernando D et al (2013b) T1 bias in chemical shift-encoded liver fat-fraction: role of the flip angle. J Magn Reson Imaging. doi:10.1002/jmri.24457

Langanke M, Brothers KB, Erdmann P et al (2011) Comparing different scientific approaches to personalized medicine: research ethics and privacy protection. Per Med 8(4):437–444

Langanke M, Erdmann P, Dörr M et al (2012) Gesundheitsökonomische Forschung im Kontext Individualisierter Medizin—Forschungsethische und datenschutzrechtliche Aspekte am Beispiel des GANI_MED-Projekts. PharmacoEconomics German Research Article 10(2):105–121

Levey AS, Bosch JP, Lewis JB et al (1999) A more accurate method to estimate glomerular filtration rate from serum creatinine: a new prediction equation. Modification of diet in renal disease study group. Ann Intern Med 130(6):461–470

Lorbeer R, Baumeister SE, Dörr M et al (2013) Circulating angiopoietin-2, its soluble receptor Tie-2, and mortality in the general population. Eur J Heart Fail 15(12):1327–1334

Mapstone M, Cheema AK, Fiandaca MS et al (2014) Plasma phospholipids identify antecedent memory impairment in older adults. Nat Med 20(4):415–418

McEvoy C, Kollef MH (2013) Determinants of hospital mortality among patients with sepsis or septic shock receiving appropriate antibiotic treatment. Curr Infect Dis Rep 15(5):400–406

Meyer J, Ostrzinski S, Fredrich D et al (2012) Efficient data management in a large-scale epidemiology research project. Comput Methods Programs Biomed 107(3):425–435

Meyer J, Fredrich D, Piegsa J et al (2013) A mobile and asynchronous electronic data capture system for epidemiologic studies. Comput Methods Programs Biomed 110(3):369–379

Netzer NC, Stoohs RA, Netzer CM et al (1999) Using the berlin questionnaire to identify patients at risk for the sleep apnea syndrome. Ann Intern Med 131(7):485–491

Ohlsson C, Wallaschofski H, Lunetta KL et al (2011) Genetic determinants of serum testosterone concentrations in men. PLoS Genet 7(10):e1002313

Ott K, Fischer T (2013) Can objections to Individualized Medicine be justified? In: Dabrock P, Braun M, Ried J (eds) Individualized Medicine between hype and hope. Exploring ethical and societal challenges for healthcare. LIT, Berlin, pp 173–200

Paulus WJ, Tschöpe C, Sanderson JE et al (2007) How to diagnose diastolic heart failure: a consensus statement on the diagnosis of heart failure with normal left ventricular ejection fraction by the Heart Failure and Echocardiography Associations of the European Society of Cardiology. Eur Heart J 28(20):2539–2550

Ramirez JH, Parra B, Gutierrez S et al (2014) Biomarkers of cardiovascular disease are increased in untreated chronic periodontitis: a case control study. Aust Dent J 59(1):29–36

Rudnik-Schöneborn S, Langanke M, Erdmann P et al (2013) Ethische und rechtliche Aspekte im Umgang mit genetischen Zufallsbefunden, Herausforderungen und Lösungsansätze. Ethik in der Medizin 26(2):105–119

Sacco RL, Kasner SE, Broderick JP et al (2013) An updated definition of stroke for the 21st century: a statement for healthcare professionals from the American Heart Association/American Stroke Association. Stroke 44(7):2064–2089

Schmidt CO, Hegenscheid K, Erdmann P et al (2013) Psychosocial consequences and severity of disclosed incidental findings from whole-body MRI in a general population study. Eur Radiol 23(5):1343–1351

Schwahn C, Polzer I, Harring R et al (2013) Missing, unreplaced teeth and risk of all-cause and cardiovascular mortality. Int J Cardiol 167(4):1430–1437

Schwebe M, Marschall P, Flessa S (2012) Schwerwiegende Arzneimittelnebenwirkungen von Phenprocoumon: stationäre Kosten der gastrointestinalen Blutungen und Einsparungspotentiale. PharmacoEconomics German Research Articles 10(1):17–28

Sievers C, Auer MK, Klotsche J et al (2014) IGF-I levels and depressive disorders: results from the Study of Health in Pomerania (SHIP). Eur Neuropsychopharmacol 24(6):890–896

Stevenson EK, Rubenstein AR, Radin GT et al (2014) Two decades of mortality trends among patients with severe sepsis: a comparative meta-analysis. Crit Care Med 42(3):625–631

Suhre K, Wallaschofski H, Raffler J et al (2011) A genome-wide association study of metabolic traits in human urine. Nat Genet 43(6):565–569

Völzke H, Alte D, Schmidt CO et al (2011) Cohort profile: the study of health in pomerania. Int J Epidemiol 40(2):294–307

Völzke H, Schmidt CO, Baumeister SE et al (2013) Personalized cardiovascular medicine: concepts and methodological considerations. Nat Rev Cardiol 10(6):308–316

Wang TJ, Ngo D, Psychogios N et al (2013) 2-Aminoadipic acid is a biomarker for diabetes risk. J Clin Invest 123(10):4309–4317

Zhang Q, Fillmore TL, Schepmoes AA et al (2013) Serum proteomics reveals systemic dysregulation of innate immunity in type 1 diabetes. J Exp Med 210(1):191–203

Ziemens B, Wallaschofski H, Völzke H et al (2013) Positive association between testosterone, blood pressure, and hypertension in women: longitudinal findings from the study of health in pomerania. J Hypertens 31(6):1106–1113

Part II
Perspectives of Socio-Cultural and Historical Studies

Chapter 4
Inventing Traditions, Raising Expectations. Recent Debates on "Personalized Medicine"

Susanne Michl

Abstract Since the late 1990s, the term "Personalized Medicine" has been coined to enable collaborations between different stakeholders in and outside research units. As a concept, it constitutes an imaginary framework of expectations and claims for a better, patient-centered and efficient health care system. Rather than deciding whether such trends represent "hype" or "hope", scholars from the social studies of technology and science emphasize that the expectations revolving around new technology are not only accessory parts of scientific inventions or innovation networks. Instead, they regard them essential in shaping these technologies. The aim of the following chapters 4 and 5 is twofold: (4) analyzing the semantic and socio-cultural contexts in which new technologies could come into being or be implemented on a larger scale (5) analyzing the continuing significance of epistemological key categories (e.g. the focus on the biological individuality) in the field of medical research and practice, and their influence on past visions of medical future.

Chapter 4 analyzes writings about "Personalized Medicine" addressed to a scientific and a popular public. They represent different promoting strategies while sharing normative assumptions that are rarely articulated: "Personalized Medicine" is made to appear to be part and parcel of a venerable tradition of past medical advances, an "invented tradition" that seems to herald a brighter and more democratic future for (Western) societies. Debates about the implications of new trends need to render the normativity of such claims explicit to allow for more informed judgments, rational critique and a more careful choice of research priorities.

Keywords Personalized Medicine · Sociology of expectation · Medical history · Epistemology · Popularization

S. Michl (✉)
Institut für Geschichte, Theorie und Ethik der Medizin, Universitätsmedizin der Johannes Gutenberg-Universität Mainz, Am Pulverturm 13, 55131 Mainz, Germany
e-mail: susmichl@uni-mainz.de

© Springer International Publishing Switzerland 2015
T. Fischer et al. (eds.), *Individualized Medicine,* Advances in Predictive, Preventive and Personalised Medicine 7, DOI 10.1007/978-3-319-11719-5_4

4.1   Introduction

On April 16th 1999, two science journalists of the "Wall Street Journal" proclaimed a 'new era' of medicine, "the era of Personalized Medicine". Robert Langreth and Michael Waldholz described the formation of a huge single nucleotide polymorphism (SNP) consortium comprising of ten major drug companies, the Wellcome Trust and five academic research centers that aimed to draw up a comprehensive SNP map of the human genome. Surprisingly enough, only a few months later, a medical journal, "The Oncologist", republished the article in identical terms (Langreth and Waldholz 1999).

Langreth and Waldholz in the "Wall Street Journal" and "The Oncologist" were not the first to use the phrase "personalized medicine" in the way we understand it today. Two years before, in 1997, Andrew Marshall had already coined the term in an article entitled "Laying the foundations of personalized medicines", albeit in a more modest terminology and significantly writing of "medicines" in the plural (Marshall 1997a, also quoted by Tutton 2012, p. 1721). Instead of postulating an homogenous entity of "Personalized Medicine" (PM) as a nascent technology in itself and therefore using the singular, Marshall took into account the underlying idea of the new advances of genomic research, that is drafting new or more efficient therapies for patients (in the plural). In his following article on "Getting the right drug into the right person" Marshall abandoned the notion of PM replacing it entirely by "pharmacogenomics" (Marshall 1997b). He foregrounded a research field whose origins date back to the late 1950s, rather than making the case for a vague entity that invoked scenarios of how medical research and practice might develop in the future. Though no less enthusiastic about the advances in genomic-based and information technologies than the two science journalists, Marshall's idea of personalized medicines could more be seen as sequentially emerging in a continuing process evolving from an already well-known research field of pharmacogenetics, and -genomics. Whereas Langreth and Waldholz's report revolved around an economically and scientifically powerful and future-oriented merging research field.

By reporting on developments within drug and biotech companies, the two science journalists contributed to a rhetorical framing of what became known as "Personalized Medicine" (in the singular). Since then PM has become a powerful language through which significant change in medical practice has been imagined and in which the interests of various actors in politics, economics, science and patient organization seem to converge.

How did we get from the plural "personalized medicines" or "individualized therapies" (at which physicians of past times had taken great pains in both their daily practice as well as in their research efforts of finding new and better diagnostic and therapeutic measures) to "Personalized Medicine" in the singular, a powerful vision as a single set of concepts, practices and technologies, capable of bringing together actors from various fields, professions and disciplines? Many observers wonder whether the "Personalized Medicine" approach marks a beginning of a "new era" (Langreth and Waldholz 1999; Jørgensen 2009), a "revolution in medicine" (Collins 2010a; Collins 2010b), a "paradigm shift" (Stock and Sydow 2013; Vanfleteren et al. 2014) or on the contrary if it is rather a "misnomer" and has always been within the scope of medical practice and research (see for the general discussion on "hype or hope" Dabrock et al. 2012; Pray 2008). Instead, this chapter takes a step back and views PM as a specific set of meanings, claims and visions in a broader social-cultural context. To consider PM as a social phenomenon is to take into account that proponents and opponents alike contribute to the shape of PM whether it will be implemented in future health care systems or not. This chapter also takes a step back concerning the specific "Greifswald Approach to Individualized Medicine" by considering PM as a broader movement uniting different biomedical initiatives, supported not only by big research consortia but also by other promoting and monitoring parties. Hence this chapter considers itself as a contribution to what has been called in literature the "Politics of Personalized Medicine" (Hedgecoe 2004) or "social practice of Personalized Medicine" (Brown et al. 2000; Tutton 2012). It is also for this reason that this chapter analyzes the term "Personalized Medicine" instead of "Individualized Medicine". This choice refers to the fact that it is the more common term in Anglo-American and in other European countries. Within the sample this study is based upon, scientists and popular authors mostly used the term "Personalized Medicine". Since the 1980s, oncologists often refer to "individualized therapies" (a fact that lies beyond the scope of this chapter). In contrast to the preceding chapter 2 that responds to the need for a clear definition of "Individualized Medicine," Part II focuses on the fluid and contradictory uses of the term "Personalized Medicine" in medical and popular literature, its function in different contexts and its normative claims.

How, then, can we most fruitfully approach PM in a perspective of social and historical studies of medicine? Certainly, PM does not emerge out of nothing and it is possible to write a history in a genealogical sense tracing back in time the technological, scientific or societal developments in the last decades leading to particular configurations and networks that enabled the emergence and the success of the claim to individualize medicine. PM as a set of visions and claims is deeply entrenched in a social and cultural context including researchers, drug companies, patient groups, consumer movements or political agendas. Furthermore, there is a narrower disciplinary history of pharmacogenetics or population genetics and other core research fields of PM. The interviews with clinical researchers conducted by the sociologist Adam Hedgecoe, whose book provides fruitful insights into the "Politics of Personalized Medicine", documents the hesitations and even resistance to these new technologies of genetic testing. Yet, this group often is excluded from the debate surrounding the future of PM (Hedgecoe 2004). The sociologist conducted the interviews between April 2001 and July 2003. At that time the rhetorical framing of PM had become very influential in medical and popular literature, by raising hopes and expectations of an economically more efficient, better and more individualized health care system.

At the same time, the marketing of consumer genomics companies has started to focus on the participatory aspects of PM as one of its main "promissory virtues"

(Juengst et al. 2012a). Social networks and patient or consumer groups fostered this idea of PM as a joint venture in which providers and consumers, as well as clinical researchers and patients could participate. Tracking one's own health information, making them public and sharing them with others in order to receive useful advice about one's behavior and have personal control over one's health, seem to attract more and more people. Meanwhile, there is an emerging advice literature concerning better health-related behavior available, empowering patients and consumers, particularly within the rhetorical framework of "Personalized Medicine" or personalized health care (Davies 2010; Goetz 2010; Collins 2010a). It seems as if there are diametrically opposed developments: wide-spread doubts in parts of the researchers' community directly involved with PM (Hedegecoe 2004) about the usefulness of some of these technologies declared as "Personalized Medicine" stand in contrast with multiple forms and co-operations in society as well as the agendas from drug companies and political parties.

As a social phenomenon in modern societies, PM is still largely unexplored. In this chapter it is not possible to address all the relevant aspects and multiple stakeholders in order to explore the complex dynamics that we could observe in the field of PM (for a more comprehensive approach see McGowan et al. 2014; Hedgecoe 2004). Instead, the sample on which the analysis draws consists on the one hand in the medical and popular literature that began to sprout up soon after the emergence of the rhetorical framing of "Personalized Medicine" (chap. 4). Commentaries and editorials at the beginning outweighed by far the number of papers presenting original works (Hedgecoe 2004). According to Hedgecoe, it is a particular kind of science writing that opens up a "discursive space within which to speculate about particular technologies and to create expectations about how pharmacogenetics may impact on healthcare provision and practice in the future" (Hedgecoe 2004, p. 18). Greg Myers makes a similar argument in his analysis of the role of review articles for the construction of knowledge (Myers 1990). On the other hand, chapter 5 emphasizes the particular role of the main pioneers in the young field of pharmacogenetics and –genomics throughout the twentieth century as well as around the turn of the millennium and the rhetorical invention of PM.

The aim is twofold: (1) analyzing the semantic and socio-cultural contexts in which new technologies could come into being or be implemented on a larger scale (2) analyzing the continuing significance of epistemological key categories (e.g. the focus on the biological individuality) in the field of medical research and practice, and their influence on past visions of medical future.

Chapter 4 analyzes the rhetorical framing of PM today. Within the field of sociology a number of scholars have already adopted this approach by focusing on its different champions, critics and stakeholders (Hedgecoe and Martin 2003; Hedgecoe 2004; McGowan et al. 2014) as well as the role of the "promissory virtues" of PM (Juengst et al. 2012a). Based on the findings of these works the emphasis will be on the specific ways different proponents of PM have connected past, present and future. PM is first and foremost about the future. While much has been written about its promises and perils, we know little about the many medical and popular writers who championed this new approach to medicine, invented traditions (Hobsbawm and Ranger 1992) and placed PM in a larger medical tradition.

Whereas chapter 4 considers how the past is constructed and used to explain and to legitimitate the emergence of the new field of PM, chapter 5 explores past visions of the future. It focuses on the past developments of some core fields of what would later be subsumed under the overall term of PM, especially the history of pharmacogenetics/-genomics. This chapter is not a history of the continuous discoveries and the trajectory of this scientific field. The emphasis will rather be on the self-conceptions of several spokespersons. For a better understanding on how a single set of visions, claims and meanings comes into being, it is of special interest to analyze when, how and why these actors considered themselves as part of PM or accordingly what have been alternative terms and prospects to it before the rise of "Personalized Medicine" as a powerful language.

4.2 Fashioning Continuities: Connecting Past, Present and Future

In the last two decades a growing number of studies in science and technology research provided new insights about the role of expectations, imaginings and visions in the emergence of new technologies (Van Lente 1993; Martin 1999; Brown et al. 2000). These expectations consist of both hopes and fears and not only describe a desirable or rejectable future, therefore being simply accessory, irrelevant parts in the emergence of new technologies. They are also part of the way this future is shaped because they actively contribute to the success or failure of nascent technologies. This new interest presumably accounts for a historical shift in the technology development itself. To the extent that in advanced industrial modernity, the intensity and pace of technology development has increased, research has become more strategic and has, thus, given rise to a rhetorical space for visions, imagining and promises (Borup et al. 2006, p. 268). Especially in the field of biomedicine, we have witnessed in the past decades several emergences of new technologies, their either failing or successful implementation into research strategies and clinical contexts accompanied by a powerful rhetoric of inexhaustible promises. Several of these biomedical research fields have been analyzed by scholars of the studies of science and technology, for instance xenotransplantation (Brown and Michael 2003), stem cell research (Martin et al. 2008), regenerative medicine (Wainwright et al. 2008), nanotechnolgy (Selin 2006) as well as pharmacogenetics (Hedgecoe and Martin 2003) to name only a few. PM is, though, a special case. As a rhetorical framing it is a reality what is more, unfolding an astonishing capacity of creating new coalitions, merges, institutional forms and funding sources. That is to say, there are materialities that begin to structure what PM is. But then again and in contrast to the new technological fields mentioned above, PM could hardly be (yet) understood as a new technology in itself. The use of the term is vague and fluid and its critics consider it even as a false and misleading labeling and a misnomer. The different networks, coalitions, movements or stakeholders are equally varying from context to context in a more or less stable manner. In their insightful contribution on the "Rise of Membrane Technology: From Rhetorics to Social Reality" which provided the methodological and theoretical framework to a study of "Sociology of expectation", Harro van Lente and Arie Rip took another stance and analytical framework. The two sociologists analyzed the function of such an umbrella term within the nascent field of membrane technology and came to this conclusion: "Together with the introduction and establishment of the umbrella term, a *history* is created, the history of membrane technology, which in a sense creates the field henceforth to be covered by that term" (Van Lente and Rip 1998, p. 227). In this reading an umbrella term has the capacity to merge different histories into one. Analogically, by launching the admittedly appealing umbrella term of PM, one history is created. This essay analyzes the history, or more precisely, the construction of traditions of PM. The focus will be on the one hand on medical literature, especially editorials or commentaries keen to shed light on what we should understand by the rather cloudy and fuzzy term of PM (sect. 4.2.1). In order to take into account that PM is not only a new approach within scientific and research context, but moreover a joint venture between politics, science, economy and society, the focus will be on the other hand on the popular discourse that has developed over the last years (sect. 4.2.2). Indeed several recent books aim to explain the prospects and goals of PM to a wider public by emphasizing its participatory potential. Based upon this sample of writings, which addresses both academics and a wider public, the aim of the following section is to explore and to identify strategic issues in the promoting of PM.

4.2.1 Medical Writings: The Timeline of "Personalized Medicine"

In an introductory article to a special issue on the sociology of expectation Borup et al. point out that "expectation of technology is also seen to foster a kind of historical amnesia—hype is about the future and the new—rarely about the past—so the disjunctive aspects of technological change are often emphasized and continuities with the past are erased from promissory memory" (Borup et al. 2006, p. 290). Social Studies of Science and Technology have focused on how the future of new technologies is designed by promoting or opposing parties (Guice 1999). Rarely, the shaping of the past has been considered as constituting a strategic way to foster technologies even if they are oriented towards the future (Tutton 2012; Selin 2006). In regards to nanotechnology, Cynthia Selin has identified different "timescapes"—a term borrowed from Barbara Adam (Adam 1998)—in discourses of the future technology. By this, she means that "time is built into stories" (Selin 2006, p. 126) and distinguishes a more linear model of trajectories and path dependency, discontinuous and disruptive timescapes as well as timelines marked by immediacy and indetermination. Each of these timescapes corresponds to specific temporal codings and innovation strategies. The coordination and negotiation of these different timescapes by different stakeholders, she argues, constitutes a great part in building up stable networks or contributes to their failings. Either way they are constitutive in the shaping of nascent technology.

Similarly, the promises revolving around PM could be analyzed within a framework of different timescapes. A first approach would be to look at specific historical narratives sketched out in medical writings. PM is actually neither defined as a clearly outlined research field (by comparison to pharmacogenetics or -genomics for example) nor could it be narrowed down to a simple technological, material invention (by comparison to SNP array analysis). Even if the highly performative sequencing technology prepared the grounds to develop its full potential, PM is primarily emerging at the crossroad of past developments of multiple methods, instruments and disciplines on the one hand and overlapping scientific, economic and political interests of different actors on the other hand. In this sense the corresponding historical narrative is a multiplicity of histories, all of which converge at one point in time and under specific circumstances giving rise to what is called PM, and which is accordingly the inevitable end point of a development. This linear vision of PM is clearly expressed by Charles Cantor, one of its fervent promoters in the early years. Having accounted for the genome-technological advances in the last 20 years, Cantor drew the following conclusion:

> Like it or not, we will soon have the ability to predict the variation in drug responsiveness of large numbers of individuals in large numbers of therapeutic settings. Once this ability is established and its usefulness is validated, it should become standard medical practice. The tools for gathering, analyzing, and explaining such data to physicians and patients will become mainstream and not rare curiosities. A dramatic change in the practice of medicine over what we know today is certain. Each patient will in practice have to be treated as the individual we already know he or she is (Cantor 1999, p. 288).

Instead of historical amnesia, I would like to argue that medical literature establishes a specific connection between past, present and future marked by the following features: a presumable culmination of logic and irreversibility ("like it or not"), the temporalities of an innovation life-cycle, which includes the prophecy of profound change ("we will see soon") followed by large-scale standardization ("should become mainstream"), as well as the ability to finally and adequately respond to a long-lived knowledge drawing upon the individuality of each patient ("we already know") and going along with strong normative claims ("each patient will have to be treated").

This connection between past, present and future in a teleological perspective makes a strong argument for promoting PM as an unavoidable and irreversible development. This timeline becomes even more important when its beginning is situated in the early stages of medical history. Jain in his handbook on Personalized Medicine, even starts the account of the landmarks in the development of "Personalized Medicine" with primitive medicine 10,000 years ago, passing by nearly all medical discoveries and ending up with the "Personalized Medicine Act" of the later President Obama in 1996 and the "Genetic Information Nondiscrimination Act" in the USA in 1998 (Jain 2009, p. 4). In the construction of this impressing genealogy, PM is not a revolution, as Jain explicitly claims, but more than that: PM is deeply rooted in the evolution of a medical thinking in its historical dimension and another step in the medical progress we have been witnessing in the last thousands of years. In contrast to the promotion strategies of other new technologies, it is noteworthy that the timeline of PM is generally marked by continuities and not by disruptions. The revolutionary rhetoric suggesting a sudden change ostensibly stands in sharp contrast to this continuous and slowly developing timeline. However, what counts as revolutionary or "dramatic change" prompted by PM, is exactly its capacity to be the solution to a centuries-old problem of medical history, or in other words to be at the core of medical history.

Besides this example of a comprehensive genealogy of PM, different spokespersons or pioneers are being granted of having introduced the concept of PM long before the term had been coined. Most frequently medical commentaries quote Hippocrates which is tantamount to invoking the mythical origins of medicine as a science. Hippocrates' statement "It is more important to know what sort of person has a disease than to know what sort of disease a person has" not only inscribes PM in an medical long-lasting, venerable tradition addressing each patient in his or her individuality. The reference also suggests that genomic-based instruments provide present medicine with powerful tools in order to implement into scientific practice what physicians in ancient times only were able to perform intuitively and arbitrarily. The same is true with the "father of modern medicine" William Osler: "If it were not for the great variability among individuals, medicine might as well be a science and not an art." Osler, known and admired for his clinical skills, embodies a perfect figurehead for the claim to bridge the gap between clinic and research or, in other words, art and science. Here again, the strength of this historical argumentation stems from the suggestion to consider PM as a systematic, scientific-based solution to a centuries-old problem of clinical practice (Kennedy 2007; Hamburg and Collins 2010; Horwitz et al. 2013). What was the art of medicine—the intuitive, therapeutic decision-making of physicians based on their clinical experience—has now become a science—a systematic, data-driven and more precise way of patient care. PM, as Allen Roses puts it, is about to reverse the logic of Osler's statement: "Sir William Osler, if he were alive today, would be re-considering his view of medicine as an art not a science" (Roses 2000, p. 857).

Not surprisingly the debate surrounding the promises of PM is shaped by these classical divides. As several scholars have convincingly shown, "science" and "art" are often invoked when new technologies are implemented into the clinic leading to a redistribution of power within the realm of medical practice (Schlich 2007; Gordon 1988). In an account about the historical dimensions to contemporary expectations of Personalized Medicine, Richard Tutton shows how different historical forms of medical "universalism and specificity" have marked Western medicine by opposing individualized efforts on the part of clinicians to the production of universal knowledge in research since the middle of the nineteenth century (Tutton 2012). It is questionable if this historical account does not simply relocate the ongoing debate into the past. Nonetheless, Tutton's brief historical overview reminds us that the tension between "universalism and specificity" was in medical history and is still a very active tension, giving rise to a contest of older systems and new forms and innovative approaches to medicine. Against this historical backdrop, conjuring this classical divide between medicine as an art and medicine as a science leads us to a powerful strategy for promoting PM today. It is not the purpose of this chapter to answer the question whether this promise of PM is well-founded or not. For the purpose of this analysis, the emphasis should be rather on the recurring discursive pattern in these medical writings, stressing the convergence of art and science—a strategy that accompanies the professionalization of medicine since the nineteenth century (Bynum 1994). In this powerful but also contested line of argument, the promises of PM mean individualized care by means of scientific-based instruments: "Listening, evaluating, and responding to the patient to promote his or her interests is the foundation of clinical medicine. Using genetics to customize care, in whatever form that may take, could be a powerful tool to further refine the healing process" (Yurkiewicz 2010, p. 16).

What makes the strength in promoting PM as a promissory technology is at the same time its weakness. The criticism that PM is a misnomer and only refers to what physicians already do in their daily practice is following the same logic, but stresses the down sides of a complete scientization of medicine. In this reading, a scientific-based instrument could never replace the art of medicine, this means clinical experience and doctors knowing his or her patient best. Whether PM will fully realize its potentials in the future, as part of the transformation of the health care system, is still unclear. However what can be said is that the debate surrounding PM is following the opposition lines that were drawn during the transformation of medical practice and research since the nineteenth century.

As noted above, in medical writings PM seems to follow its own internal, historical logic, deeply rooted in medical thinking. This common historical narrative comes frequently along with evaluative conclusions. The moral imperative precisely consists in pushing forward the developments of PM in order to realize what in previous times physicians could only have done intuitively and had never fully achieved. From an ethical point of view, the legitimacy of such normative claims could surely be questioned. If the consideration of the patient's individuality should be without doubt a strong moral requirement and directive for physicians in their daily encounter with patients, it does not mean that this argument alone leads to a decision about the allocation of resources pouring into different research fields. Even if the likely benefit for one single patient or one single patient group is very high, in a broader ethical discussion one has to weigh the effect of different research strategies which are likely to lead to improvements of health. Besides ethical considerations the historical narrative in itself is flawed. Clinical practice and even more medical science was often centered on a better understanding of the pathological in terms of common, universal traits, although this was by no means an effort to deny one's individuality or neglect the range of variability in therapeutic response. Hence, it is a fallacy to assume that the best way for improving health care is uniquely based on the individual level and on research strategies finding out what makes us different rather than what we all have in common. The normative claims going along with historical argumentation, as implicit as they are, have to be made explicit and reconsidered in a broader ethical and historically informed discussion.

4.2.2 Popular Writings: "Personalized Medicine" as Democratic Empowerment?

Editorials and commentaries in medical journals articulate the visionary, but nonetheless past-oriented, character of PM. Despite its highly technological character, PM is regarded as deeply rooted in venerable medical traditions. In contrast to medical writings that commented on PM research projects and consortia, popular writings not only address a wider public, they also refer to another phenomenon that emerged at about the same time: social networks on internet platforms, as well as commercial genetic testing companies sprouted up since the beginning of the twenty-first century as a Personalized Medicine movement "from below" (Gollust et al. 2002; Prainsack 2005; Prainsack 2014; Paul et al. 2014). In the last couple of years several books have been published for a wider audience of readers with the aim to instruct how to join this movement (Collins 2010a; Topol 2012; Goetz 2010; Angrist 2010; Frank 2009). As fashionable as these writings may appear, this literary genre adds to the tradition of the nineteenth century advice literature on health related issues. Especially, the idea of prevention and the assumption of a predictable and manageable future led to new forms of activism and thus to a booming market of health related advice literature and other forms of health education initiatives (Lengwiler and Madarász 2010). Since the late nineteenth century, new preventive and predictive possibilities have been accompanied by an appeal to individuals to stay healthy for the sake of one's own well-being or for the sake of the community. The connection of the four "Ps"—personalized, preventive, predictive and participatory—is certainly not new. At least in Western societies, preventive measures have been and are still centered upon individual responsibility and activism. What is more remarkable is that today, ever more importance has been attached to the participatory aspects of the PM venture, even though it is not yet sure whether PM will fulfill the three other promises of "personalized", "preventive" and "predictive". On the contrary, one of the main messages of these books is that only by joining the movement, PM could realize its full potential. In this reading, the four "Ps" are conceptually not on the same level. Only an expanding public participation could guarantee that more and more information is gathered and stored on digital platforms from which personal information could be extracted. Participation is seen as a means and necessary condition to achieve "personalization", "prevention" and "prediction".

That is to say that these writings are not only an explanatory commentary on current research efforts, they are considerably broadening the scope of PM. Eric Topol, for example, describes how digital advancements in the "science of individuality" will create more knowledge about diseases or other medical problems of a patient. In this perspective, he suggests that "the need for activism" (Topol 2012, p. 218) of a patient in the digital revolution might even lead to a "democratization of medicine" (Topol 2012, p. 12). In this reading, the promise of PM concerns not only medical improvements but political issues on justice and fairness. Henceforth, PM is not simply and exclusively a scientific research program, but beyond a narrow and hermetic circle of specialists, it is a joint venture in which every person could participate. These new forms and platforms are blurring the boundaries of people carrying out scientific research and those benefiting from it. While sharing its health information with others and gathering more and more data, the promise is to use this information sampling, and even biobanks, for potential research issues and to increase the awareness of individuals about their own health which could lead to better decisions on how to improve it. The production of knowledge and health-related behavior are no longer separate courses of actions but closely linked and coordinated by one platform.

The authors of these writings range from renown health planners and scientists such as Francis S. Collins, the former director of the National Human Genome Research Institute and now director of the National Institute of Health, or Eric Topol, cardiologist and professor at the Scribbs Institute to science journalists, the Wired-Magazine editor Thomas Goetz or the Danish journalist and biologist Lone Frank. Accordingly, there are no clear-cut expert roles within these writings. Collins and Topol as researchers aim to share their expert knowledge with a wider public, whereas Goetz reports on his own experience with social networks or consumer testing. These different forms of PM, "top down" or "bottom up", are understood as partly cooperative, partly competitive. Especially doctors are accused of ignoring the revolution going on under their eyes. As Francis Collins puts it: "From reading this book, you almost certainly know more than your doctor about personalized medicine" (Collins 2010a, p. 277). Expert knowledge does not arise from professionals with a special training, but from participating in the revolution, and could easily bypass the doctor-patient-relationship. This transformation implies that market forces are structuring henceforth what generally is seen as constitutive for a trustful medical encounter. In contrast to past medical developments, PM in popular writings is propagated by genetic consumer companies such as "deCODEme", "23andMe" or "Navigenics". Most of the authors outline their interviews and meetings with the companies' CEOs and write about their experience in having their personal genomes decoded. Hereby, they give the impression that these companies and joint ventures are considered to be the relevant future stakeholders, or "research revolutionaries" as Lone Frank labels them. Eric Topol's text about the "Creative Destruction of Medicine" is the most outspoken in this category of books. The title obviously alludes to Joseph Schumpeter's thesis of innovation as a form of creative destruction (Schumpeter 1942) of outmoded structures. Accordingly, journalist Scott Gottlieb praised Topol's oeuvre in a book review of the Wall Street Journal as a "venture capitalists delight" (Gottlieb 2012).

According to most of the writers, destruction and creation concern firstly patient identities that are shifting from being the medical patient, who either suffers from a disease, carries it or is at risk of becoming ill, to the medical consumer who is defined by his or her personal data and acts accordingly. Start-up direct-to-consumer testing companies first circumvented the personal encounter between patients and doctors. Despite state regulations, for example in California, where genetic testing could formerly only be ordered by a physician, the genetic consumer industry established an internet ordering service and a massive lobbying effort to have these regulations abolished. In the age of information technology, individual data sampling and transferring, it might not even be necessary to have eye-to-eye contact with an examining physician. Although many companies sell their products without a physician or a geneticist as intermediaries today, some of them have successfully applied for FDA approval and turned to a model, which includes a health care professional (Howard and Borry 2012). It remains to be seen in how far this modulates the current effects of direct-to-consumer marketing of genetic testing or if there is an economic pull for physicians to participate in and contribute to this market.

Secondly, as the title of Topol's book indicates, the status of medicine itself is changing. As seen in the precedent analysis on medical writing, the boundaries between medicine as an "art" and medicine as a "science" are voluntarily blurred by celebrating the triumph of PM as another step or even the last step to overcome this traditional distinction. In the case of the popular discourse on PM, this becomes even more important. The greater the emphasis on the collection of huge amounts of data and individual health (care) information, the more marginal the art of medical practice becomes (Tutton 2012). Even though doctors collect personal health information in diagnostic tests on the basis of which diagnosis they make, the vast amount of information seems to render medical practice as clinical experience obsolete. For this reason, the popular writings stress the participatory aspect presenting PM as a widespread digital movement (see also Juengst et al. 2012b). For example, after each chapter Collins gives a summary and recommendations under the topic "What can you do to join the personalized medicine revolution", often with links to websites where one could register, for example, to document the family's history of breast cancer (Collins 2010a). Wired-Magazine editor Thomas Goetz aims to provide the reader of his book "The Decision Tree" with a complete set of guidelines on how to "make better choices and take control of your health" in the context of PM developments (Goetz 2010).

As in medical writings, this popular discourse about PM invokes invented traditions and nurtures expectations about a brighter future. In general, popular writings are less reluctant to postulate a turning point for the history of medicine and more broadly for mankind. So, for example, Collins refers to the historical dimension of the new scientific findings. We could live in an "astounding time of our history", he explains, and the historical impact would be so significant that everybody "will remember this" and the participants will tell their future grandchildren (Collins 2010a, p. 2). Therefore, "revolution" as a keyword is used at times in an inflationary manner. A fragmentary search in Kevin Davies "The $ 1,000 Genome: The Revolution in DNA Sequencing and the New Era of Personalized Medicine" indicates that there are more than 100 hits for the word in his text. For the Danish science journalist and biologist Lone Frank, the new research findings will answer the essential questions on one's future health "because a revolution is under way" (Frank 2011, p. 8). Regardless of that, Frank championed in 2009, in a precedent book how brain science also will change the world (Frank 2009).

What are the histories that are written in these books? Mostly the promise of progress refers to the eradication of diseases within a historically classical account on the continuing medical progress witnessed in the last decades: "Medical science has made stupendous progress over the past 100 years" writes Goetz (Goetz 2010, p. xi). He compares the situation of today with the one 100 years ago in respect to diabetes, polio and smallpox. Once deadly and incurable, they are now regarded as harmless due to medical progress. The next step concerns the eradication of other common diseases such as cancer, diabetes or Alzheimer diseases. For another writer, medical progress is tantamount to technological advancements. Kevin Davies describes in his book "The $ 1,000 Genome" how technological developments, and therefore cheaper genome sequencing, will offer the breakthrough in personal and genetic medicine. "Will history repeat itself?", he asks, and he also alludes to the once deadly ages of polio and smallpox (Davies 2010, p. 266). Davies presents his readers the obviously widely shared opinion that a cheap genome might even "eradicate most kinds of genetic diseases". However he mentions that there are also geneticists who believe this prospect is too optimistic (Davies 2010, p. 266).

Francis Collins even compares the decoding of the genome sequence, and thus the possibilities of PM, to quasi-mythical events in American history: "We have been engaged in a historic adventure. Whether your metaphor is Neil Armstrong or Lewis and Clark, your metaphor is at risk of falling short", he cites his keynote speech at a Cold Spring Harbor Laboratory meeting (Collins 2010a, p. 2). Collins alludes here to the frontier-myth of American history, to astronaut Neil Armstrong, the first man on the moon, or to explorers like Meriwether Lewis and William Clark who discovered and mapped the American West at the beginning of the nineteenth century. What they have in common is that they were considered pioneers in the discovery of formerly unknown territory. They pushed the boundary further whether on land or in space thereby shaping the future of American society. According to Collins, the scope of PM, once it will have realized its full democratic and participatory potential, will be beneficial well beyond clinical issues or medical practice.

4.3 Conclusion

To summarize the first chapter of Part II on recent debates on PM it can be said that the expectations raised in writings addressing a medical and a broader public differ. They represent different strategies to promote PM, even if both types of writings share evaluative assumptions. The rhetoric of PM emerging since the late 1990s is not a neutral one, but goes along with far-reaching normative claims neatly linked to a historical argument. According to medical writers, PM is the logical culmination, deeply entrenched in past medical thinking, bridging formerly irreconcilable opposition lines between medicine as an "art" and medicine as a "science". In contrast to this presumably old tradition in medical thinking of which PM becomes a part, the popular discourse presents PM as a modern phenomenon, integral to advances in modern medicine and to a democratic project changing our societies. It is equally imbued with normative requirements presuming that the way to deepen the understanding of one's individuality and act in the purpose of bettering one's health is by joining PM as a digital revolution. This synergy between commercial interests, normative claims and promises for individual and societal well-being makes this discourse a powerful means to promote PM. The historical narrative is here (information) technological and medical progress and accordingly the myth to push the (last) frontier further into unexplored territory. The appeal of this framing stems from the fact that scientific and technological progress is considered as being carried out by a joint venture societal movement, rather than in a more conventional form by a small, hermetically sealed community of researchers and funding parties.

PM certainly provides a powerful rhetorical framing and an astonishing integrative potential. Nevertheless, current medical research and practice would be well advised to pay careful attention not only to what is included but also what is excluded in current promoting strategies. In both medical and popular writings, the meaning of medicine as professional practice becomes blurry. Either "science" grows at the expense of "art" or the participatory power of PM overshadows or even eclipses the relationship between professional health care providers and their patients.

References

Adam B (1998) Timescapes of modernity. The environment and invisible hazards. Routledge, London

Angrist M (2010) Here is a human being: at the dawn of personal genomics. HarperCollins, New York

Borup M, Brown N, Konrad K et al (2006) The sociology of expectation in science and technology. Technol Anal Strateg 18(3/4):285–298. doi:10.1080/09537320600777002

Brown N, Rappert B, Webster A (eds) (2000) Contested futures: a sociology of prospective technoscience. Ashgate, Aldershot

Brown N, Michael M (2003) A sociology of expectations: retrospecting prospects and prospecting retrospects. Technol Anal Strateg 15(1):3–18

Bynum WF (1994) Science and the practice of medicine in the nineteenth century. Cambridge University Press, Cambridge

Cantor C (1999) Pharmacogenetics becomes pharmacogenomics: wake up and get ready. Mol Diagn 4(4):287–288

Collins FS (2010a) The language of life: dna and the revolution in personalized medicine. HarperCollins, New York

Collins FS (2010b) Has the revolution arrived? Nature 464:674–675. doi:10.1038/464674a

Dabrock P, Braun M, Ried J (eds) (2012) Individualized medicine between hype and hope: exploring ethical and societal challenges for healthcare. LIT Verlag, Wien

Davies K (2010) The $ 1,000 Genome: the revolution in DNA sequencing and the new era of personalized medicine. Simon & Schuster, New York

Frank L (2009) Mindfield: how brain science is changing our world. Oneworld Publications, London

Frank L (2011) My beautiful genome: exposing our genetic future, one quirk at a time. Oneworld Publications, London

Goetz T (2010) The decision tree: taking control of your health in the era of personalized medicine. Rodale, New York

Gollust SE, Chandros S, Wilfond BS (2002) Limitations of direct-to-consumer advertising for clinical genetic testing. JAMA 288:1762–1767

Gordon D (1988) Clinical science and clinical expertise: changing boundaries between art and science in medicine. In: Lock M, Gordon D (eds) Biomedicine examined. Kluwer, London, pp 257–298

Gottlieb S (2012) Digital doctoring. Wall Street Journal. http://online.wsj.com/news/articles/SB10 00142405297020474090457719319107711 7530. Accessed 8 July 2014

Guice J (1999) Designing the future: the culture of new trends in science and technology. Res Policy 28(1):81–98

Hamburg MA, Collins FS (2010) The path to personalized medicine. N Engl J Med 363:301–304

Hedgecoe A, Martin P (2003) The drugs don't work: expectations and the shaping of pharmacogenetics. Soc Stud Sci 33(3):327–364

Hedgecoe A (2004) The politics of personalised medicine—Pharmacogenetics in the clinic. Cambridge University Press, Cambridge

Hobsbawm E, Ranger T (1992) The invention of tradition. Cambridge University Press, Cambridge

Horwitz RI, Cullen MR, Abell J et al (2013) (De)Personalized medicine. Science 339:1155–1156

Howard H, Borry P (2012) To ban or not to ban? Clinical geneticists' views on the regulation of direct-to-consumer genetic testing. EMBO reports 13(9):791–794

Jain KK (2009) Textbook on personalized medicine. Springer, Dordrecht

Jørgensen JT (2009) New era of personalized medicine: a 10-Year anniversary. Oncologist 14(5):557–558. doi:10.1634/theoncologist.2009-0047

Juengst ET, Settersten RA, Fishman JR et al (2012a) After the revolution? Ethical and social challenges in "personalized genomic medicine". Per Med 9:429–439

Juengst ET, Flatt MA, Settersten RA (2012b) Personalized genomic medicine and the rhetoric of empowerment. Hastings Cent Rep 42(5):34–40. doi:10.1002/hast.65

Kennedy D (2007) Breakthrough of the year. Science 318(5858):1833. doi:10.1126/science.1154158

Langreth R, Waldholz M (1999) New era of personalized medicine: targeting drugs for each unique genetic profile. Oncologist 4(5):426–427. doi:10.1634/theoncologist.2009-0047

Lengwiler M, Madarász J (eds) (2010) Das präventive Selbst. Eine Kulture-schichte moderner Gesundheitspolitik. transcript, Bielefeld

Marshall A (1997a) Laying the foundations for personalized medicines. Nat Biotechnol 15:954–957. doi:10.1038/nbt1097-954

Marshall A (1997b) Getting the right drug into the right patient. Nat Biotechnol 15(12):1249–1252

Martin P (1999) Genes as drugs: the social shaping of gene therapy and the reconstruction of genetic disease. Sociol Health Illness 21(5):517–538

Martin P, Brown N, Kraft A (2008) From bedside to bench? Communities of promise, translational research and the making of blood stem cells. Sci Cult 17(1):29–41. doi:10.1080/09505430701872921

McGowan ML, Settersten RA, Juengst ET et al (2014) Integrating genomics into clinical oncology: ethical and social challenges from proponents of personalized medicine. Urol Oncol 32(2):187–192. doi:10.1016/j.urolonc.2013.10.009

Myers G (1990) Writing biology: texts in the social construction of scientific knowledge. University of Wisconsin Press, Madison

Paul NW, Banerjee M, Michl S (2014) Captious certainties: makings, meanings and misreadings of consumer-oriented genetic testing. J Comm Genet 5:81–87

Prainsack B (2005) Personalized medicine in times of "global genes":making sense of the "hype". Per Med 2(2):173–174

Prainsack B (2014) The powers of participatory medicine. PLoS Biol 12(4):e1001837. doi:10.1371/journal.pbio.1001837

Pray L (2008) Personalized medicine: hope or hype? Nat Educ 1(1):72

Roses AD (2000) Pharmacogenetics and the practice of medicine. Nature 405:857–865. doi:10.1038/35015728

Schlich T (2007) The art and science of surgery: innovation and concepts of medical practice in operative fracture care, 1960s-1970s. Sci Technol Hum Val 32(1):65–87

Schumpeter JA (1942) Capitalism, socialism and democracy. Harper, New York

Selin C (2006) Time matters: temporal harmony and dissonance in nanotechnology networks. Time Soc 15(1):121–139. doi:10.1177/0961463X06061786

Stock G, Sydow S (2013) Personalisierte Medizin. Bundesgesundheitsblatt—Gesundheitsforschung—Gesundheitsschutz 56(11):1495–1501

Topol E (2012) The creative destruction of medicine: how the digital revolution will create better health care. Basic Books, New York

Tutton R (2012) Personalizing medicine: futures present and past. Soc Sci Med 75(10):1721–1728

Vanfleteren LE, Kocks JW, Stone IS et al (2014) Moving from the Oslerian paradigm to the postgenomic era: are asthma and COPD outdated terms? Thorax 69(1):72–79. doi:10.1136/thoraxjnl-2013-203602

Van Lente H (1993) Promising Technology: the dynamics of expectations in technological development. PhD Thesis, Department of Philosophy of Science and Technology University of Twente, Enschede

Van Lente H, Rip A (1998) The rise of membrane technology: from rhetorics to social reality. Soc Studies Sci 28(2):221–254. doi:10.1177/030631298028002002

Wainwright SP, Michael M, Williams C (2008) Shifting paradigms? Reflections on regenerative medicine, embryonic stem cells and pharmaceutical. Sociol Health Ill 30(6):959–974. doi:10.1111/j.1467-9566.2008.01118.x

Yurkiewicz S (2010) The prospects for personalized medicine. Hastings Cent Rep 40(5):14–16

Chapter 5
The Epistemics of "Personalized Medicine". Rebranding Pharmacogenetics

Susanne Michl

Abstract Whereas chapter 4 focuses on uses and normative claims of the rhetorical frame "Personalized Medicine" in medical and popular writings, chapter 5 analyzes the intellectual formation of pharmacogenetic, -genomics as a disciplinary field. It explores when, how, and why the leading journals in the field present themselves as part of the overall phenomenon labelled "Personalized Medicine". Pharmacogenetic journals founded at the beginning of the twenty-first century, not only adopted the rhetorical framing of PM, but also branded pharmacogenomics as a milestone in medical history. Their vision extended to a large societal context that included not only pharmacology and genetics, but also broad movements of multiple stakeholders within society.

An analysis of past framings of pharmacological and genetic research centered on the epistemics of "individuality" or "variability" illustrates alternative conceptualizations of scientific progress in medicine. Three scholars exemplify these multifaceted developments: the clinician Archibald E. Garrod, the pharmacologist Werner Kalow and the biochemist Roger J. Williams. It is a history of survival or oblivion of scientific byways, detours and dead-ends that draws our attention to the fact that there was not a linear technological and scientific development that led to contemporary "Personalized Medicine". Such an analysis sheds light not only on the question of what we gain but also on what we lose by framing specific research projects and traditions as "Personalized Medicine".

Keywords Personalized Medicine · Pharmacogenetics · Pharmacogenomics · Epistemology · Medical history · Individuality · Variability

S. Michl (✉)
Institut für Geschichte, Theorie und Ethik der Medizin, Universitätsmedizin der Johannes Gutenberg-Universität Mainz, Am Pulverturm 13, 55131 Mainz, Germany
e-mail: susmichl@uni-mainz.de

© Springer International Publishing Switzerland 2015
T. Fischer et al. (eds.), *Individualized Medicine,* Advances in Predictive,
Preventive and Personalised Medicine 7, DOI 10.1007/978-3-319-11719-5_5

5.1   Introduction

As noted in chapter 4, medical and popular writers did not suffer from "historical amnesia" (Borup et al. 2006). Instead, they connected past unmet desires, present performances and future promises, thereby paving the way to a specific understanding and politics of "Personalized Medicine" (PM). With this in mind, one may ask, if PM has a history (or several histories) at all. Or isn't it more adequate to approach PM as a phenomenon deeply entrenched in current seemingly prolific convergences between political, economical, scientific, and in a broader sense, civic and societal interests? In short, may PM be considered as a singular phenomenon that could only occur at this moment in time and location (on a global perspective PM certainly is a phenomenon of advanced industrialized countries)?

Doubtlessly there are several histories of PM to write: the history of pharmacogenetics and pharmacogenomics (Jones 2013; Tutton 2012), the history of information technologies (Bud 1993), the history of population genetics and the emergence of epidemiological, large-scale population based studies (Dawber 1980; Rothstein 2003) and the history of drug companies and their shifting commercial investment agendas (Gaudillère and Hess 2013), to mention only a few examples. One history is more or less legitimate than the other. As seen above, the multiplicity of histories is an integral part of the concept of PM. Therefore PM has the potential to integrate the interests of various actors and to bridge the gap between past performances and future promises.

Chapter 5 follows another approach by analyzing the point of view of the actors carrying out and bringing forward the research fields which could be considered as preparing the grounds for a PM to come. The main question is when the actors in these fields considered themselves as part of the overall phenomenon called PM. Another important question is how did this shifting self-positioning take place, and were there alternative terms framing the research goals of these fields before the emergence of the rhetorical entity of PM. To answer these questions, I focus on pharmacogenetics/-genomics as one of the core research fields of PM.

5.2   Pharmacogenetics Gets Personalized

Compared to human genetics and genomics, which despite several turning points have a long-standing and progressive history throughout the twentieth century, its subset pharmacogenetics/-genomics has developed rather slowly (for a more detailed historical account see Kalow 2001a; Nebert et al. 2008; Paul and Roses 2003). This is surprising considering the fact that soon after William Bateson's rediscovery of Mendel's law of heredity in 1900, trained clinician and pediatric Archibald Garrod applied Mendel's new groundbreaking findings to biochemical properties of the human body (Garrod 1902, 1923). The value of Garrod's contributions for the emerging research field of pharmacogenetics was, however, only recognized in the 1950s. It was at that time that the term "pharmacogenetics" was coined by Friedrich Vogel (Vogel 1959, p. 117). First discoveries such as the Glucose-6-Phosphate dehydrogenase (G6PD) deficiency by Arno Motulsky in 1956 (Motulsky 1957) or the genetic variation in ethanol metabolism by Werner Kalow in 1964 (Kalow 1965) formed the basis for further developments. Nevertheless from the late 1950s until the late 1980s the interest in this new research field, and in inter-individual variations in general, was fairly marginal. After the rise of molecular biology and genomic-based research in the late 1980s however, pharmacogenetics, then soon to be called pharmacogenomics, began to rapidly accelerate in its development. In 1988, for example, Frank Gonzalez and his colleagues succeeded in isolating several alleles of the human CYP2D6 gene, part of the P450 cytochrome complex of genes and responsible for metabolism in the liver (Gonzalez et al. 1988). This was first and foremost a technological success and launched the era of high performative sequencing and the use of information technology, the latter of which set the stage for processing the data and turning it into potentially clinically relevant information.

The shift from pharmacogenetics to –genomics, albeit two disciplines with overlapping aims, has certainly raised the promises, fears and hopes revolved around PM by broadening considerably the foci of research and developing new sequencing technologies and methods of investigation. The pharmacologist Werner Kalow, one of the pioneers of this research field since the 1950s, historically and conceptually distinguished pharmacogenetics, pharmacogenomics and "Personalized Medicine" by referring to the launching of PM as a deliberate, rhetorical act: "The term 'personalized medicine' was coined in the hope of creating collaborative uses of drug therapy and genetic knowledge. It represents the seemingly straightforward thought that the choice of drug for treating a given subject may be improved by considering the patient's genetic make-up; however, the frightening complexities on the way to that aim will not allow us to reach it quickly" (Kalow 2006, p. 164). What Kalow highlights in his skeptical statement is the creative power and performativity of the rhetorical entity PM, allegedly coined on purpose to bring research fields and the various actors within these fields together.

The question could then be raised when "Personalized Medicine" as an ultimate goal became the overall framing of research conducted in pharmacogenomics. It is less surprising that in 1991, the year of the launch of the first issue of the journal "Pharmacogenetics", terms as "Personalized Medicine" or even "individualized drug therapies" are absent from the editorial's mission statement: "The need for a new journal was quite simple: the molecular biology of drug metabolism has now come of age and in so doing has also stimulated renewed interest in the original biochemical and clinical studies of the inherited phenomena of drug and chemical responsiveness which has kept the beacon alight for over thirty years" (Idle and Gonzalez 1991, p. 1). Reference has been made to former achievements since the late 1950s, the technological and scientific progress the field had encountered since then and the prospect of heralding "new and exciting findings" in the future. The editorial of the first issue of "Pharmacogenetics" anticipated "new data and insights into chemical disease pathophysiology and its individual diathesis".

Diathesis, in this context understood as the constitution of a disease, is a very old term that made its first appearance in the Hippocratic writings. The medical historian Erwin H. Ackerknecht has analyzed the term's rise and fall as well as its philosophical and theoretical meanings (Ackerknecht 1982). According to him, Archibald Garrod himself made reference to the term when he described his discovery of uric acid in the blood of the gouty as a confirmation of the gouty diathesis (Ackerknecht 1982, p. 324). Ackerknecht's history of the term mirrors the struggle between "the ontological approach, which sees the disease, and the individual approach, which sees the patient" (Ackerknecht 1982, p. 325). The reason why the term "diathesis" has more or less disappeared from contemporary research issues is explained by the shift of medicine from philosophical ambitions to a more pragmatic approach. A diathesis could not have been treated so easily as the corresponding disease. Nevertheless the vagueness and the many-sided character make it a "prominent member of the august family of verbal constructs", appropriate to fill the gaps of knowledge in the "riddles of pathology" (Ackerknecht 1982, p. 325). It remains to be seen if "diathesis" will make its reappearance in future research. Besides this brief intermezzo in the editorial of the first issue of "Pharmacogenetics", diathesis as an epistemological category for a deeper understanding of bodies, diseases and persons has not been taken up by researchers in the field of pharmacogenetics in particular or "Personalized Medicine" in general. Other terms like "markers" have been increasingly used since the late 1980s. Their theoretical and conceptual foundations and implications as well as their lexical history still await thorough research.

Whereas in 1991 the term "Personalized Medicine" had not at that point been coined, it is more remarkable that even after the turn to pharmacogenomics (the journal adopted the new extended title "Pharmacogenetics and Pharmacogenomics" in 2007) "Personalized Medicine" did not become the new rhetorical framing for the numerous contributions to the journal. "Personalized Medicine" appeared as a key word only five times, the first time in an article published in 2007 concerning a drug-response marker in the glatiramer acetate therapy for multiple sclerosis: "This study provides an additional step towards addressing the challenge of adequately tailoring drugs to the individual patients, according to their own genetic profile. Hence "Personalized Medicine" forms seem within reach" (Grossman et al. 2007). The optimism of the authors regarding the promises of PM may well be justified in the case of multiple sclerosis therapies, it is rather a unique example for the use of "Personalized Medicine" as an overall term of pharmacogenetic research with "promissory virtues" (Juengst et al. 2012) in the journal.

In the literature, though, the shift from pharmacogenetics to –genomics has been usually accompanied by a new framing in terms of "Personalized Medicine" or "Individualized Therapies" (Licinio and Wong 2002). Soon after the arrival of the term "pharmacogenomics" and the completion of the Human Genome Project in 2000, the handbook "Pharmacogenomics" edited amongst others by Werner Kalow (Kalow et al. 2001), could not omit to make reference to "Personalized Medicine" as the ultimate goal of pharmacogenomics research efforts. Kalow mentioned the term in his introduction as well as in the concluding remarks written together with another pioneering pharmacologist Arno Motulsky: "The expectations which have been aroused by pharmacogenomics are substantial and involve the gradual development of Personalized Medicine or therapeutics, leaving behind the present statistically based medicine" (Kalow and Motulsky 2001, p. 389). In this statement, PM was clearly considered as the clinical impact of pharmacogenomics, opposed to research based on statistical rather than individual levels. The future-oriented direction of the research field has then become neatly linked to the notion of "individualized therapy for the patient" (Kalow and Motulsky 2001, p 390). The promises of such an approach have been articulated in B. Michael Silver's contribution to the volume. Though realistic about the limited clinical impact, he even ventured precise predictions on what future impacts on pharmacogenomics and "Personalized Medicine" could look like, by scheduling a timetable of the next four decades ahead. His predictions for 2040 reached from the most desirable prognoses of "comprehensive genomics-based healthcare" and "disease predisposition determined, possibly at birth" to disconcerting consequences such as "worldwide inequities remain, contributing to international tensions" and "serious debate is underway about humans possibly taking charge of their own evolution" (Silber 2001, p. 26). The societal impacts of this new approach, both hopes and fears, have existed from the start of the reasoning surrounding PM.

"The Pharmacogenomics Journal" was founded in 2001. Remarkably, allusions to past performances in this journal were not directed towards achievements in the decades preceding the launch of the journal and linked directly to the narrow field of pharmacology and genetics, as had been the case with the journal "Pharmacogenetics". The editor Julio Licinio raised broader issues by asking "What will define medicine in the twenty-first century?" Against this background he invoked what he considered major hallmarks of twentieth-century medical history "like the discovery of antibiotics, vaccines, chemotherapy, progress in surgery and anesthesiology" and treatment of chronic diseases such as heart disease, diabetes and mental illness. He concluded by asking: "Are we poised for the next step in medicine, the utilization of DNA sequences to identify new treatment targets and the introduction of more effective—and personalized—therapies?" (Licinio 2001, p. 1). Pharmacogenomics has since found its place within the timeline of past performances in medical history in general and not only in the narrower field of pharmacology. An integral part of the politics of PM was to take into consideration the research field within its societal context by introducing the rubric concerning ethical, economical, legal and societal implications of pharmacogenomics. To finish the analysis of the journal landscape, in the journal "Pharmacogenomics and Personalized Medicine" the mission statement announced in the editorial of "Pharmacogenomics" is made even more visible by transferring it into the title. Considering the fact that journals play an important role in the institutionalization of new research areas and technologies, PM has become increasingly institutionalized in the last two decades. From a very circumscribed research field in pharmacology linked to genetics, it has grown into one that is compared to past milestones in medical history and seen as a scientific progress transforming society.

5.3   Pharmacogenetics and its Predecessors

This brief overview of the main journals leads us to the historical issues on alternative terms and framings that had been used by researchers to describe their aims and future visions before the advent of PM at the turning of the century. In order to investigate this issue, the last section of this chapter will be dedicated to the works of the "founding fathers", most of them influential researchers contributing in some way or other to the development of pharmacogenetics as a discipline. Hence, by examining some works of prominent figures, the aim is to explore the evolving intellectual formation of a new discipline which emerged in the late 1950s at the crossroad of genetics and pharmacology but may be traced back still earlier to the beginning of the twentieth century with the surprising findings of the clinician Archibald Garrod. Since the late 1950s pharmacologists Arno Motulsky, Werner Kalow and the geneticist Friedrich Vogel maintained their advocacy for pharmacogenetics for four decades. Throughout this period their writings give us rich insights on the formation of pharmacogenetics as a discipline; the shifting framings of the research directions; and the aims which it has taken. In addition to this more disciplinary framework, the work of the biochemist Roger J. Williams is significant. Connecting genetics, nutritional science and biochemistry, Williams is less well known for his contribution to the development of the field. He nevertheless chose a different way to shape this new research field, to sketch out future prospects by addressing a wider public.

5.3.1   Archibald E. Garrod (1857–1936)

The work of Archibald E. Garrod "Inborn Errors of metabolism", resulting from his Croonian Lectures delivered at the Royal College of Physicians in June 1908 and first published in series in "The Lancet", is generally cited as one of the germinal works in the field, considered as well ahead of his time (Garrod 1909,1923). Sir Archibald Edward Garrod was a trained pediatric and general physician practicing in different London hospitals at the time of his research into human biology (Bearn and Miller 1979). For this reason, he was passionately interested in linking physiological chemistry to clinical problems. Whereas the few geneticists at that time were enthusiastic about the studies on drosophila or other animals, Garrod applied the Mendelian laws of inheritance soon after their re-discovery by William Bateson to human beings and their specific metabolic properties.

Garrod studied four biochemical individual particularities as albinism, alkaptonuria, cystinuria and pentosuria. Before coining the notion of "inborn errors of metabolism", Garrod developed the concept of "chemical individuality" in an article published in "The Lancet" in 1902 (Garrod 1902) and republished around the turning of the century (Garrod 1996, 2002). Hence, the rediscovery of the findings of Garrod in the late 1950s (Knox 1958) seems to be mirrored by a newly revived interest in these issues starting in the late 1990s.

Especially alkaptonuria, the "black urine disease", served as a case study of "merely extreme examples of variations of chemical behavior which are probably everywhere present in minor degrees and that just as no two individuals of a species are absolutely identical in bodily structure neither are their chemical processes carried out on exactly the same lines (Garrod 1902, p. 1620). Garrod's reasoning followed two lines of argument. Firstly by observing the appearance of black urine in numerous families, Garrod plausibly argued that alkaptonuria was a congenital condition following a Mendelian pattern of recessive inheritance. Secondly, Garrod referred to the notion of variability between species derived from evolutionary theory by transferring it to the idea of variability within one species: "If, then, the several genera and species thus differ in their chemistry we can hardly imagine that within the species, when once it is established, a rigid chemical uniformity exists" (Garrod 1902, p. 1620). While the concept of biochemical inter-individual difference sounds quite familiar to us, at the turning of the last century, Garrod's findings might have casted substantial doubt on the persistently and broadly accepted concept of a fixed internal environment or "milieu intérieur" developed by Claude Bernard in the middle of the nineteenth century (Bernard 1878, p. 113). It was only a few years after Garrod had published his book on the inborn errors of metabolism that Walter Cannon expanded Bernard's concept on milieu intérieur to the idea of "homeostasis", the capacity of the human body to maintain steady levels of vital conditions (Cannon 1932). Whereas in anthropometry and external body morphology the search for inter-individual differences has been a longstanding undertaking, physiological and even more biochemical properties were often considered as displaying a narrower range of variability.

From his publication in "The Lancet" in 1902 to the publication of the book and former Croonian Lectures, Garrod tried to render his clinical findings based on family histories plausible to geneticists, biochemists and clinicians alike. As he was not able to demonstrate but only postulate an enzymatic deficiency, he was reliant on a coherent line of argument. In addition, Garrod struggled even more to find an adequate rhetorical framing for his findings. The very titles—"chemical individuality" and "inborn errors" are not only shifting the focus from biochemistry to genetics, but also from a semantic of individual differences to a more common clinical divide between the normal and the pathological. There is indeed a wavering between these different kinds of rhetorical framings throughout his writings. In order to describe alkaptonuria, he mostly used neutral terms as "conditions", "modes of incidence", "particularities" or "peculiarities", which alternate with pathological notions of "abnormalities" or "disease"– though only a few times he specified his findings by using the term "individualities". This search for an adequate semantic refers to Garrod's own questioning on the status of his research findings. Could alkaptonuria be classified within the system of pathological findings and treated like so many other symptoms as a deviation from the normal, even though the clinical relevance was not manifest? Alkaptonuria as the majority of the phenomena studied by Garrod was not a morbid condition and it was not even sure that it was a condition predisposing for diseases like diabetes. In this respect the reference to the traditional normal-pathological divide, though widely common and easily comprehensible, was not quite appropriate, and Garrod clearly stated that alkaptonuria was not a disease although "those who exhibit the phenomenon are in a state of unstable equilibrium" (Garrod 1902, p. 1619). Furthermore, Garrod was strongly concerned to open his research findings to more generalized patterns of the human biology linked to a clinical impact. Therefore he described the studied phenomena as "a 'sport' or an alternate mode of metabolism" (Garrod 1902, p. 1619), speculating that it might not be an isolated example. Limited by the analytical instruments of his time, Garrod anticipated broader application fields regarding "drugs and the various degrees of natural immunity against infections" (Garrod 1902, p. 1620).

Hence, besides remaining in the realm of pathology, Garrod intuitively chose a rhetorical framing open to a much broader field of investigation and an alternate grasp on the metabolism of human biology. The term of "chemical individuality" or "individualities of metabolism" used by Garrod captured this heuristically different approach to an exploration of inter-individual differences beyond the classical normal-pathological divide.

It was only in the late 1950s that researchers working in the young field of pharmacogenetics mentioned Garrod as one of the "founding fathers" of biochemical genetics. In an overview of the advancements of the last century, Arno Motulsky paid respect to two pioneering researchers bridging the gap between biochemistry and genetics: Archibald Garrod and Linus Pauling (Motulsky 1967). Given the prestige and reputation of the two-time Nobel Prize laureate Pauling, the reference made to Garrod's work was tantamount to ensuring Garrod a place in medical history. Nevertheless, whereas Garrod's "inborn errors of metabolism" was often quoted, "chemical individuality" generally received less attention. The same holds true for the overview article of Motulsky who quoted the "Inborn errors of metabolism" but not the article on "chemical individuality" in "The Lancet" issue of 1902. The first generation of researchers in the young field of pharmacogenetics generally dismissed the term and respectively the concept sketched out by Garrod. Even if one could argue that this is only a rhetorical or semantic choice without substance, this obvious neglect refers to the slightly shifting interests and framings of the research programs.

5.3.2   Werner Kalow (1917–2008)

In the first edition of the first handbook on pharmacogenetics, Werner Kalow defined pharmacogenetics as "pharmacological responses and their modification by hereditary influences" (Kalow 1962, p. 1). The pharmacologist offered a huge data collection of research findings published in other medical literature. Despite his focus on the existence of hereditary influences, Kalow points out that one should not "minimize the importance of the effects of internal or external environment" (Kalow 1962, p. 2). Total variation, he concludes, might be caused both by hereditary and nonhereditary factors. Indeed, sketching out a new research field and calling it "pharmacogenetic" which means to consider oneself within the broader field of human genetics, was not a matter of course by that time. The debate on

hereditarian and environmentalist explanation and thus on inter-individual variation concerning human biology was far from being solved (Jones 2013, p. 39).
Biochemical pharmacology was constantly referring to an external factor such as
nutrition or toxic substances. It was also by that time that studies showed high
ranges of intra-individual variability, throughout a life course or even within one
day due to changing life habits (see Schreider (1966) who summarizes the state of
research in the 1950s). Despite Kalow's concession of the limited explanatory reach
of hereditarian factors, he took up the term coined a few years earlier by Friedrich
Vogel "Pharmacogenetik" (Vogel 1959, p. 117) and thus steered this new approach
in pharmacology in the direction of genetics.

Kalow's interest lied in an explication of human and animal variability deeply
entrenched in evolutionary thinking and the idea of natural selection. To demonstrate and explain the increasing complexity of life forms was one of the main
objectives of his textbook, beginning with unicellular organisms and ending with
multicellular organisms. Accordingly, the first chapter deals with drug resistance
of bacteriological microorganisms, the second with tissue cultures, then insects,
vertebrates and finally "man". In the chapters that deal with humans, Kalow differed between heritable factors recognized by the use of drugs. For example, he
mentioned the prolonged action of succinylcholine and atypical cholinesterase, the
person-to-person differences in the metabolic conversion of isoniazid or the ability
to taste phenylthiourea. In addition, Kalow focused on human hereditary defects
with altered drug response. He mentioned in this context glucuronide formation and
familial nonhemolytic jaundice, the glycogen storage disease or phenylketonuria
(Kalow 1962, p. 146).

Thus, the first textbook on pharmacogenetics thoroughly followed the analytical
framework and statistical tools of natural or evolutionary history. For addressing
these issues and finding an explanation, Kalow referred to the problem of "discontinuous variation", a notion also borrowed from evolutionary theory which points to
the overwhelming complexity and variation found in the animal and human world
(Bateson 1894). Kalow argued that pharmacologists generally would recognize
only the commonly known distribution of continuous variation, the imperceptible
degrees of differences clustered around a mean value and given graphic expression
in Gauss' distribution curve. In genetics continuous variations might correspond to
the pattern of multifactorial inheritance. According to Kalow, the continuous variation was fostered by the standardizing effort in pharmacology to define the given
potency, namely "by that dose which produces a given effect in 50 per cent of a
group of individuals" (Kalow 1962, p. 1). Widely neglected is the other type of variation, that is discontinuous variations which is caused by monofactorial inheritance
and which facilitate the arrangement in groups differing from each other by specific
varieties. By bringing pharmacology, genetics and mathematical and statistical procedures together, Kalow expressed the hope "that pharmacogenetics will advance,
helping individuals and groups of human beings to understand their own and each
other's responses to drugs and toxic agents" (Kalow 1962, p. 209). "Variations"
or "variability" was the more common term within the framework of evolutionary

theory precisely dealing with biological variations of all kind and assessing it by mathematical and statistical means.

Kalow was influenced by another preoccupation that accompanied the beginnings and the further development of pharmacogenetics. Interestingly, he finished the book with a chapter on racial differences in the response to drugs. Though there might be some observations on racial differences, he noticed, these findings had never been "substantiated by measurement" (Kalow 1962, p. 209, see also 1982, 2001b). The medical historian David S. Jones illustrates the "deep roots of the genetic and racial preoccupations in pharmacology" since the late 1950s (Jones 2013, p. 1). Especially Kalow's work shows "how race emerged as a central concern at the origins of the field and has remained an irresistible attraction for pharmacologists" (Jones 2013, p. 5). In a historical overview of the advances in pharmacogenetics/-genomics, Kalow described his emergence into the field of pharmacogenetics as neatly linked to the discovery of a significant difference in Canadian immigration society in the 1970s. While conducting a study with the aim to investigate the metabolism of the barbiturate drug Amobarbital, he was himself surprised to find out that the seven subjects who presented anomalies were all students from China (Kalow et al. 1979). Individualities, one could argue, are in the early writings of pharmacogenetics, basically bound to racial differences; much later sex or age differences have been taken into account. In an article published in 2006, while the genomic turn has shifted the coordinates of the debate, Kalow discussed the possibilities of the classification of person according to the well-known, and for each physician easily accessible divides of age, sex, ethnicity, or race (Kalow 2006, p. 164), amongst which race appears to be the most promissory profiling method. The marked medical benefit among Afro-Americans of the new drug BiDil, used to treat heart failure, led Kalow to speculate whether racial differences trump individual differences. It is still an ongoing debate whether individualized health care, despite its potential to classify individuals along other divides than ethnicity and race, in fact contributes to racial profiling by doctors and scientists and gives rise to race-tailored therapies for patients (Fullwiley 2007; Foster et al. 2001). If there might be clear benefits for certain groups from more knowledge on specific susceptibilities or drug responses, the "molecularization of race" (Fullwiley 2007) gets ethically controversial if individual genetic profiles are still lacking and racial categories serve as proxies for potential genetic differences. A historical analysis suggests that the close ties between pharmacogenetics and racial issues in the second half of the twentieth century may be one of the reasons why individual differences in metabolism have been discussed within the framework of evolutionary theory and population genetics, rather than stressing "biochemical individualities", which would blur these boundaries.

5.3.3   Roger J. Williams (1893–1988)

The question could therefore be raised whether an emphasis on biochemical individualities represents a research strategy distinct from population genetics. By the time that pharmacogenetic research was about to take off, a biochemist at the University of Texas at Austin, known to be the first to synthesize vitamin B1, proposed a completely different framing than his colleagues from pharmacology. In his book published in 1956 on "Biochemical Individuality: The Basis for the Genetotrophic Concept" Roger Williams upheld a uniform terminology by speaking consequently of individualities neither considering the classical normal-pathological divide nor common categories of social identities like ethnicity or race. Drawing on evidence that "each human being possesses a highly distinctive body chemistry" (Williams 1956, p. 166), Williams combined biochemical, genetic approaches and environmental, especially nutritional needs to the human metabolism. The choice of the term "individuality" is significant. Williams deliberately dismissed the idea of dividing humankind into two groups: a vast majority with a "normal" range of properties and a small minority that shows deviations or "abnormalities". According to Williams, this widespread view could only be upheld while considering one attribute at a time, but would get more complicated when numerous measurable items would be included into the analysis: "The existence in every human being of a vaster array of attributes which are potentially measurable (whether by present methods or not), and probably often uncorrelated mathematically, makes quite tenable the hypothesis that *practically every human being is a deviate in some respects*" (Williams 1956, p. 3). Whereas Garrod had based his theory on the normal/abnormal divide, Williams, albeit paying his respect to Garrod, refused the rhetorical framing of "errors" or "abnormalities" of metabolism (Williams 1956, p. 97) and preferred the terms of individual peculiarities or "individuality". Consequently, throughout his book, he marked this different kind of approach by putting "normal" in quotation marks.

If Williams' account for the range of variability and individuality basically is grounded in genetic patterns, his interest in individual nutritional needs blurred the boundaries between hereditarian and environmentalist approaches. He claimed that without heredity *and* environmental influences "we are nothing" (Williams 1956, p. 168). These two decisive influences interplay in the sense that we can try to alter the basic conditions and requirements for the well-being of each individual in either direction: "*Understanding and appreciating what heredity distinctively does for an individual may make it possible to cope environmentally with his difficulties*" (Williams 1956, p. 169). Williams called this his "genetotrophic concept".

Methodologically this concept had several implications: Accordingly, the biochemist based his method not on the study of representative population but on repeated observations on the same individuals or smaller group samples. He voluntarily reversed the scientific proceeding: To develop a "science of man", the traditional strategy has been to start with the study of individual human specimens and to arrive at generalizations. In contrast, an emphasis on variability renders the extraction of generalized patterns impossible. For this reason, Williams proposed henceforth the opposite way that is to select a specific unresolved problem and then "*investigate how individual human differences enter into these specific problems*" (Williams 1956, p. 183). More precisely Williams suggested a three-step-approach: "(1) Select a disease the etiology of which is obscure. (2) Explore the known metabolic peculiarities, and look for new ones, which may be associated with the disease or susceptibility to it. (3) Seek to correct the metabolic failures by applying the genetotrophic principle in whatever manner seems most appropriate for the disease in question" (Williams 1956, p. 185). Williams was aware that this was a draft for future research programs within biochemistry, genetics and medicine. Starting with the Framingham Heart Study in 1948, the twentieth century witnessed the emergence of large-scale population based clinical trials with the aim to build up a data base which could lead to more individualized health care. In the face of Williams' ideas, these undertakings would have certainly shed light on generalizations and a deepening of our understanding of the average man, but not on the particular individualities of metabolism and on individual carriers that deviate from this mean value. Population based trials are pursued in order to stratify individuals into groups that ultimately respond to an intervention. The specific characteristics that serve as useful information for future interventions are identified in order to establish what works best for single patients. Williams started with the opposite idea of observing one single or a small group of individuals in the course of time. In a more sophisticated way and informed by statistical knowledge drawn out of the huge population based trials, the emergence of the n-of-1 or one single subject clinical trial considering the individual as the sole unit of observation seems to mirror Williams' approach. These n-of-1 clinical trials apply the strict methodology of the statistical techniques of standard population based clinical trials, including randomization, washout and crossover periods, as well as placebo controls, to a single subject clinical trial (Lillie et al. 2011). By combining the outcomes of these single subject trials, distinguishing characteristics of the single patients could be recorded and used as a data collection towards insights to inform for future intervention for a much larger group of patients. It is not the purpose of this chapter to decide the best way to come to generalizable findings that would benefit patients the most or to discuss and weigh up the advantages and pitfalls of the top-down or the bottom-up approach. And even if Williams' concept could not be paralleled with current research methods, it is noteworthy to highlight alternative ways, both in present and past times, to implement the challenging claim to individualize medicine.

Williams' influence within the scientific community, and for the development of a particular approach centered upon the individual, was nearly marginal. Contemporaries such as Friedrich Vogel qualified some of Williams' thoughts as fuzzy and exaggerated, but "certainly fundamentally correct" (Vogel 1959, p. 123). Likewise Motulsky paid his respect to Williams by identifying him as one of the researchers paving the way to the intellectual formation of pharmacogenetics, even though he noted his minimal influence (Motulsky 2002).

Williams himself considered his approach centered on biochemical individuality as an exception and deplored the lack of interest in variation and individuality among researchers. This interest, he stated, "has often been considered a hobby and has not led to serious publication." And he added: "This field of interest has not gained the respectability that it deserved"(Williams 1956, p. xvi). This statement, certainly not free from rhetorical over-statement, points nevertheless to the shifting interests that occurred in the second half of the twentieth century. Needless to say, the focus on individualities is far from being a pastime for a handful of researchers.

In light of the emergence of the rhetorical framing of "Personalized Medicine", the research field has rather become worth of investigating, large-scale funding and political support.

The emergence of pharmacogenetics in the late 1950s certainly did not attract as much political and public attention as the rise of "Personalized Medicine" does today. The textbook of Kalow in 1962 and his presentation of his work in the "New York Academy of Sciences" by October of this year, though, have not been passed unnoticed. The "New York Times" ran an article and an editorial about his findings (Schmeck 1962; Anonymous 1962). In contrast to his colleagues from the nascent field of pharmacogenetics, Williams clearly followed a political agenda projecting not only a new research field but also broader social and political implications of his new approach. The problem that had to be solved in the late 1950s was the continuous growth of the mega-cities and its overcrowding population. In this situation a new concern to individualize and to organize efficiently the living together, according to specific needs in nutrition and education emerged. In Williams' attempt to promote the individualizing approach, he addressed politicians, educators and clinicians, whose interest lied in urban planning, education or the health care system. Moreover, far from being confined within the disciplinary boundaries, Williams contributed to debates embracing scholars from different disciplines, including natural science, economics, history, literature, philosophy, politics, rhetoric and sociology. For example, Williams was invited to attend a "Symposium on Individuality and Personality" sponsored by Princeton University in New Jersey in September 1956, given by the Foundation for American Studies, which expounded the problems "of man's freedom in the face of modern society's seemingly irresistible urge to socialize and regiment the thought and action of the individual" (Morley 1958, p. 5; Williams 1958). Far from being confined to the realm of biosciences, the interest in individuality as a scientific and multidisciplinary research area was closely linked to the modernization of our societies in the 1950s.

Williams, as a biochemist, emphasized the leading and educating role of the biosciences in this matter. He thought that the recognition of biochemical individualities should play an important role in avoiding suffering not only in medicine, but also in the realm of politics, crime prevention, family relation, race problems, education, philosophy, fine arts and religious beliefs. Furthermore, in his book "You are extraordinary", he invented another labeling to underpin and popularize his research to a much wider public (Williams 1967). His visions should come true in a modern society in which people knew their constitution and could act accordingly. Even if the contexts had completely changed since the late 1950s, Williams anticipated the participatory aspects and the potential to reach a wider public, patients and consumers alike. Faced with the massification of societies, Williams deplored the neglect to recognize biochemical and genetic individuality make-ups, individual needs and hence individual suffering as a consequence. Williams' past vision of future was entrenched alike in a context of technocratic population planning and an individual educational, almost humanistic-enlightening effort. Today, in the face of the rapid advances in sequencing technologies, we witness an inflation of a rhetoric centered on first person pronouns and the promise to deepen the knowledge of one's individuality or personality. Private consumer testing and social networks have already taken up slogans like "You make the difference" or "It's all about me". As shown in chapter 4, it is the context of the empowerment of patients and consumers alike that has now become a marketable commodity.

5.4   Conclusion

Chapter 4 and 5 have explored the multi-faceted phenomenon of PM from the perspective of social and historical studies of medicine. Instead of taking sides in the heated debate of whether PM will fulfill its promises in the future, whether it is more "hype" or "hope", the analysis was centered on the expectations, both hopes and fears, past and present. It argues that this specific set of visions, meanings and claims is an essential part of the shaping of PM. The rhetoric of PM emerging since the late 1990s is not a neutral one, but brings into play far-reaching normative claims linked to a historical narrative according to which PM is the logical culmination of scientific progress, deeply entrenched in medical thinking of past time, bridging formerly irreconcilable opposition lines between medicine as an "art" and medicine as a "science". The popular discourse on PM is equally imbued with normative requirements presuming that the way to deepen the understanding of one's individuality and act in the purpose of improving one's health is by joining PM as a digital revolution. The historical narrative is about (information) technological and medical progress and accordingly the myth to push the (last) frontier further in unexplored territory. The appeal of this framing stems from the fact that scientific and technological progress is presented as a joint venture societal movement, and not in a more conventional form by a small, hermetically sealed community of researchers and funding parties.

A closer look at the community of researchers of pharmacogenetics—a core research field of PM—reveals a more differentiated picture. The first journal of pharmacogenetics was launched in 1991 and considered as a logical consequence of the past developments of pharmacogenetics and its technological and scientific advances since the late 1950s. Even after the turn from genetics to genomics, "Personalized Medicine" did not become the overall framing and ultimate goal in this journal. In contrast, the mission statements of the journals founded at the beginning of the twenty-first century not only adopted the rhetorical framing of PM, but also branded pharmacogenomics as a milestone in medical history. Their vision extended to a larger societal context that included not only pharmacology and genetics, but also broad movements of multiple stakeholders within society. In this context, PM could not be reduced to purely technological and scientific innovations, i.e. the emergence of high performative sequencing technologies or the shift from genetics to genomics. Instead, the importance of PM was measured by its capacity to change both medical history and society at large. The invention of PM is therefore a deliberate strategy of rebranding. Drawing on scientific changes and technological innovations within the narrowly defined research area of pharmacogenetics, PM as a concept was invented to enable collaborations and to broaden the research scope beyond questions of technological progress. In fact, pharmacogenetic research strategies labelled as PM are made to appear to be part and parcel of a venerable tradition of past advances, a tradition that seemed to herald a brighter future for (Western) societies.

The increasing importance of pharmacogenomics in research labeled as PM has led us to reconsider the history of pharmacogenetics with a special interest in semantic choices of several spokespersons and alternative framings and concepts before the rise of PM. The question was when, how and why biological individuality has become an epistemological key category for deepening the understanding of biomedical and related health issues. The underlying idea is that turning inter-individual differences into an object of rational research based on the biological science is epistemologically a different approach than to investigate what we all have in common. This point of view is largely disseminated in medical and popular writings. For example, the "Science Journal" recognized in 2007 this difference by claiming that personal genomics constituted the "breakthrough of the year". It is not about "THE genome (as if there were only one!). Instead, it is about your particular genome, or mine, and what it can tell us about our backgrounds and the quality of our futures", declared the editor-in-chief of Science Donald Kennedy (Kennedy 2007). Differences matter, much more than what we all share, this seems to be the appealing message.

If pharmacogenetics and –genomics have contributed to these insights, its history reminds us that the framings of research goals and directions could have taken slightly different forms. Keen to broaden the scope of his findings, Garrod already coined the term of "chemical individuality", while at the same time remaining committed to the normal-pathological divide. He influenced a group of researchers in the late 1950s, albeit only his book "Inborn errors of metabolism", focusing more on abnormalities than on individualities, was widely quoted. Later, Kalow tried to explain human variability by adopting tools and concepts from evolutionary and racial theory. Even though Williams is generally not mentioned together with Garrod, Motulsky or Kalow, he offered a peculiar semantic use, overall framing and underlying concepts of his own research program. His genetotrophic approach—even if not taken up by successive researchers—pretended to recognize and value the importance of "biochemical individualities" for medical research as well as political and societal issues. In Williams' past vision of the future, politicians, urban planners, educators, researchers, and more broadly each citizen, would center their joint efforts on individual needs in order to avoid personal suffering. This overview of the work by researchers carrying out and bringing forward the field of pharmacogenetics, far from being comprehensive, aimed to approach "biological individuality" as a epistemological category in (bio-) medical research. It illustrates that concepts of chemical individuality as defined by Garrod, ways of dealing with human variability as undertaken by Kalow, or even Williams' promise to truly individualize, are not solely bound to technological progress and the emergence of a highly performative sequencing technology, as we would easily believe when approaching PM in a narrower sense of a material intervention. The framings are time-bound as illustrated by the attempts to intermingle the recently rediscovered genetics, body chemistry and the clinic (Garrod 1902, 1923) or pharmacology, (population) genetics and racial and evolutionary theory (Kalow 1962), or even societal problems of overcrowding cities with the possibility to individualize specific, nutritional needs (Williams 1956). An emphasis on the history of survival or oblivion of scientific byways, detours and dead-ends calls attention to the fact that there was no linear technological and scientific development that led to contemporary "Personalized Medicine". Such an analysis sheds light not only on what we gain but also on what we lose by framing specific research projects and traditions as "Personalized Medicine".

Acknowledgments The author would like to thank the the the following people for their helpful comments and suggestions: Daniela Berner, Norbert W. Paul, Till van Rahden, Matthias Speidel and the members of the Department of Social Studies of Medicine at McGill University, especially Thomas Schlich.

References

Ackerknecht EH (1982) Diathesis: the word and the concept in medical history. Bull Hist Med 56(3):317–325

Anonymous (1962) Medicine and genetics. New York Times 13 October:19

Bateson W (1894) Material for the study of variation treated with special regard to discontinuity in the origin of species. Macmillian, London

Bearn AG, Miller ED (1979) Archibald Garrod and the development of the concept of inborn errors of metabolism. Bull Hist Med 53(3):315–328

Bernard C (1878) Leçons sur les phénomènes de la vie communs aux animaux et aux végétaux, vol 1. J.-B. Baillière et fils, Paris

Borup M, Brown N, Konrad K et al (2006) The sociology of expectation in science and technology. Technol Anal Strateg 18(3/4):285–298. doi:10.1080/09537320600777002

Bud R (1993) The uses of life. A history of biotechnology. Cambridge University Press, Cambridge

Cannon WB (1932) The wisdom of the body. W. W. Norton, New York

Dawber TR (1980) The framingham study: the epidemiology of atherosclerotic disease. Harvard University Press, Cambridge

Foster MW, Sharp RR, Mulvihill JJ (2001) Pharmacogenetics, race, and ethnicity: social identities and individualized medical care. Ther Drug Monit 23(3):232–238

Fullwiley D (2007) The molecularization of race: institutionalizing racial difference in pharmacogenetics practice. Sci Cult 16(1):1–30

Garrod AE (1902) The incidence of Alkaptonuria: a study in chemical individuality. Lancet 160(4137):1616–1620. doi:10.1016/S0140-6736(01)41972-6

Garrod AE (1909) Inborn errors of metabolism. Henry Frowde, London

Garrod AE (1923) Inborn errors of metabolism, 2 edn. Oxford University Press, London

Garrod AE (1996) The incidence of alkaptonuria: a study in chemical individuality. Republished. Mol Med 2(3):274–282

Garrod AE (2002) The incidence of alkaptonuria: a study in chemical individuality. Republished. Yale J Biol Med 75(4):221–231

Gaudillère JP, Hess V (eds) (2013) Ways of regulating drugs in the 19th and 20th centuries. Palgrave Macmillian, New York

Gonzalez FJ, Skoda RC, Kimura S et al (1988) Characterization of the common genetic defect in humans deficient in debrisoquine metabolism. Nature 331:442–446. doi:10.1038/331442a0

Grossman I, Avidan N, Singer C et al (2007) Pharmacogenetics of glatiramer acetate therapy for multiple sclerosis reveals drug-response markers. Pharmacogen Genomics 17(8):657–666

Idle JR, Gonzalez FJ (1991) Editorial. Pharmacogenetics 1:1

Jones DS (2013) How personalized medicine became genetic, and racial: Werner Kalow and the formations of pharmacogenetics. J Hist Med Allied Sci 68(1):1–48. doi:10.1093/jhmas/jrr046

Juengst ET, Settersten RA, Fishman JR et al (2012) After the revolution? Ethical and social challenges in "personalized genomic medicine". Per Med 9:429–439

Kalow W (1962) Pharmacogenetics: heredity and the response to drugs. W. B. Saunders, Philadelphia

Kalow W (1965) Contribution of hereditary factors to the response to drugs. Fed Proc 24(6):1259–1265

Kalow W, Tang BK, Kadar D et al (1979) A method for studying drug metabolism in populations: racial differences in amobarbital metabolism. Clin Pharmacol Ther 26(6):766–776

Kalow W (1982) Ethnic differences in drug metabolism. Clin Pharmacokinet 7:373–400

Kalow W (2001a) Historical Aspects of Pharmacogenetics. In: Kalow W, Meyer UA, Tyndale RF (eds) Pharmacogenomics. Marcel Dekker, New York, pp 1–9

Kalow W (2001b) Interethnic differences in drug response. In: Kalow W, Meyer UA, Tyndale RF (eds) Pharmacogenomics. Marcel Dekker, New York, pp 109–134

Kalow W, Motulsky AG (2001) General conclusions and future directions. In: Kalow W, Meyer UA, Tyndale RF (eds) Pharmacogenomics. Marcel Dekker, New York, pp 389–395

Kalow W, Meyer UA, Tyndale RF (eds) (2001) Pharmacogenomics. Marcel Dekker, New York

Kalow W (2006) Pharmacogenetics and pharmacogenomics: origin, status, and the hope for personalized medicine. Pharmacogenomics J 6:162–165. doi:10.1038/sj.tpj.6500361

Kennedy D (2007) Breakthrough of the year. Science 318(5858):1833. doi:10.1126/science.1154158

Knox WE (1958) Sir Archibald Garrod's "inborn errors of metabolism". II. Alkaptonuria. Am J Hum Genet 10(2):95–124

Licinio J (2001) Welcome to the pharmacogenomics journal. Pharmacogenomics 1(1):1–2. doi:10.1038/sj.tpj.6500019

Licinio J, Wong M (eds) (2002) Pharmacogenomics: the search for individualized therapies. Wiley VCH, Weinheim

Lillie EO, Patay P, Diamant J et al (2011) The n-of-1 clinical trial: the ultimate strategy for individualizing medicine? Per Med 8(2):161–173

Morley F (ed) (1958) Essays on individuality. University of Pennsylvania Press, Philadelphia

Motulsky AG (1957) Drug reactions enzymes, and biochemical genetics. J Am Med Assoc 165(7):835–837. doi:10.1001/jama.1957.72980250010016

Motulsky AG (1967) Biochemical genetics in medicine. Acta Paediatr Scand Suppl 172:156–169

Motulsky AG (2002) From pharmacogenetics and ecogenetics to pharmacogenomics. Med Secoli 14(3):683–705

Nebert DW, Zhang G, Vesell ES (2008) From human genetics and genomics to pharmacogenetics and pharmacogenomics: past lessons, future directions. Drug Metab Rev 40(2):187–224

Paul NW, Roses AD (2003) Pharmacogenetics and pharmacogenomics: recent developments, their clinical relevance and some ethical, social, and legal implications. J Mol Med 81(3):135–140

Rothstein W (2003) Public health and the risk factor: a history of an uneven medical revolution. University of Rochester Press, Rochester

Schmeck HM (1962) Heredity Linked to drug effects. New York Times 10 October:62

Schreider E (1966) Typology and Biometrics. Ann N Y Acad Sci 134:789–803

Silber BM (2001) Pharmacogenomics, biomarkers, and the promise of personalized medicine. In: Kalow W, Meyer UA, Tyndale RF (eds) Pharmacogenomics. Marcel Dekker, New York, pp 11–31

Tutton R (2012) Personalizing medicine: Futures present and past. Soc Sci Med 75(10):1721–1728

Vogel F (1959) Moderne Probleme in der Humangenetik. Ergeb Inn Med Kinderheilkund 12:52–125

Williams RJ (1956) Biochemical individuality. The basis for the genetotrophic concept. Wiley, New York

Williams RJ (1958) Individuality and its significance in human life. In: Morley F (ed) Essays on individuality. University of Pennsylvania Press, Philadelphia, pp 125–145

Williams RJ (1967) You are extraordinary. Random House, New York

Part III
Medical Perspectives

Chapter 6
Use of Biomarkers for the Prediction of Treatment Response: Immunoadsorption in Dilated Cardiomyopathy as a Clinical Example

Marcus Dörr, Uwe Völker and Stephan B. Felix

Abstract In this chapter, we outline a clinical example illustrating how integrated analyses of biomarkers might be used for the prediction of treatment response. We report findings from a pilot study that investigated the hemodynamic effects of a novel treatment option, immunoadsorption with subsequent IgG substitution (IA/IgG), in patients with dilated cardiomyopathy. Several previous studies have shown that this treatment leads to a significant improvement of cardiac function and relief of symptoms. Response to this therapy is, however, characterized by a wide inter-individual variability. In a pilot study, we tested the value of clinical, biochemical and molecular parameters for prediction of the response to IA/IgG. This study demonstrated that combined assessment of two markers (negative inotropic activity of antibodies in the blood and gene expression patterns derived from myocardial biopsies) predicts response to this therapy with an extremely high sensitivity and specificity, thereby enabling appropriate selection of patients who most likely benefit from this therapeutic intervention. Further studies will screen for biomarker signatures in blood, which do not depend on endomyocardial biopsies, and will compare responders and non-responders with respect to other molecular markers (e.g. plasma proteome, metabolome and microRNA profiles as well as whole blood transcriptome signatures). In the future, such strategies might facilitate selection of patients with dilated cardiomyopathy who are responders not only to immunoadsorption therapy but also to other heart failure treatments for which a heterogeneous response is observed and thus will help to offer more effective treatments to affected patients.

M. Dörr (✉) · S. B. Felix
Klinik und Poliklinik für Innere Medizin B, Universitätsmedizin Greifswald, F.-Sauerbruch-Straße, 17475 Greifswald, Germany
e-mail: mdoerr@uni-greifswald.de

S. B. Felix
e-mail: felix@uni-greifswald.de

U. Völker
Institut für Genetik und Funktionelle Genomforschung, Universitätsmedizin Greifswald
Friedrich-Ludwig-Jahn-Str. 15a, 17489 Greifswald, Germany
e-mail: voelker@uni-greifswald.de

© Springer International Publishing Switzerland 2015
T. Fischer et al. (eds.), *Individualized Medicine,* Advances in Predictive,
Preventive and Personalised Medicine 7, DOI 10.1007/978-3-319-11719-5_6

Keywords Dilated cardiomyopathy · Immunoadsorption · Gene expression · Negative inotropic activity of antibodies · Prediction of outcome · Biomarker signature · Pilot study

6.1 Dilated Cardiomyopathy—A Common Cause of Heart Failure

Dilated cardiomyopathy (DCM) is a disease of the myocardium characterized by ventricular chamber enlargement and systolic dysfunction (Burkett and Hershberger 2005; Maron et al. 2006). DCM leads to progressive heart failure and a decline in left ventricular contractile function, ventricular and supraventricular arrhythmias, conduction abnormalities, thromboembolism, and is related to an increased risk of sudden or heart failure-related death (Maron et al. 2006). The prognosis for DCM is still poor with limited treatment options, particularly in the end stage of the disease (Stehlik et al. 2012; Miller et al. 2007; Slaughter et al. 2009).

Cardiomyopathies account for one of the three most common causes of heart failure (McMurray et al. 2012), and DCM is the most frequent cause for heart transplantation (Maron et al. 2006). In industrialized Western countries, the prevalence of DCM is approximately 36 patients for every 100,000 population with an incidence of 5–8 new cases per year for every 100,000 population. Recent data even suggest a higher actual prevalence rate of DCM. According to recently published studies, about 30–40 % of patients with congestive heart failure suffer from non-ischemic myocardial heart disease (Zannad et al. 2011; Swedberg et al. 2010; Moss et al. 2009; Tang et al. 2010), among which DCM is the most common entity. Despite advances of drug therapy, long-term prognosis of heart failure has only slightly improved during the last decades and is still poor and is even worse than that of many so-called malignant diseases. Between 1986 and 2003, median survival increased from 1.33 to 2.34 years in men and from 1.32 to 1.79 years in women after the first hospitalization for heart failure (Jhund et al. 2009). Overall, around 40 % of patients admitted to hospital with heart failure die or are readmitted to hospital within 1 year, and approximately 50 % of the patients decease within 4 years (McMurray et al. 2012). The economic burden caused by heart failure is enormous. Heart failure is the cause of 5 % of acute hospital admissions, is present in 10 % of inpatients, and accounts for 2 % of national expenditure on health in Europe, mostly due to the cost of hospital admissions (McMurray et al. 2012). Lifetime costs of medical care after heart failure diagnosis are US $ 110,000/year per individual patient in the US (Dunlay et al. 2011).

6.2 Immunoadsorption as a Therapeutic Option in DCM

The pathogenesis of DCM is heterogeneous. Besides genetic pre-disposition, viral infection and inflammation of the myocardium play a causal role in the disease process of DCM (Maron et al. 2006; Kühl and Schultheiss 2009; Heymans et al. 2009).

Moreover, autoimmune disorders with activation of the cellular and humoral immune system have been implicated in the development of DCM (Heymans et al. 2009; Cihakova and Rose 2008; Kallwellis-Opara et al. 2007). Of particular importance, the presence of cardiac-specific antibodies against various cardiac substructures has been reported in DCM patients (Cihakova and Rose 2008; Jahns et al. 2008). Their pathogenic potential has been proven in animal models by active immunization or by transfer of antibodies against the corresponding epitopes, both leading to the development of a DCM-typic phenotype with dilatation and dysfunction of the left ventricle (Jahns et al. 2004). Interestingly, cardiac antibodies have also been identified as independent predictors of future disease development and progression among healthy relatives of DCM patients (Caforio et al. 2007).

Currently, no specific therapy for treatment of DCM is available. According to current guidelines, patients with DCM should be treated by conventional medical therapy and device-based therapies that are in general recommended also for other entities of heart failure (Hunt et al. 2009; McMurray et al. 2012). The knowledge about the crucial role of cardiac autoantibodies in the development and progression of DCM resulted in a novel treatment approach, namely application of immunoadsorption (IA) therapy by which circulating antibodies can be extracted from the blood. Removal of circulating antibodies by IA has been successfully employed for a number of autoimmune diseases such as Goodpasture's syndrome and lupus erythematodes (Bygren et al. 1985; Palmer et al. 1988). IA is a type of plasmapheresis and can be described as an extracorporal blood purification procedure which uses filter columns to remove circulating autoantibodies (Dörr 2002).

If cardiac antibodies do in fact contribute to cardiac dysfunction in DCM, their removal by IA would be expected to improve the hemodynamics of patients with DCM. In DCM, IA is usually followed by subsequent substitution of standard immunoglobulin-G (IA/IgG) to reduce infection risk following IgG depletion (Roifman et al. 1987), and to block the rebound of antibody production in B cells (Felix and Staudt 2008).

An initial uncontrolled pilot study was performed to characterize the short-term hemodynamic effects of IA in nine patients with DCM and severe heart failure (Dörffel et al. 1997). In this study, extraction of circulating IgGs from the plasma of these patients by anti-IgG columns induced a significant improvement in various cardiac functional parameters (e.g. cardiac index and systemic vascular resistance). Through this pilot study, IA was for the first time identified as an additional therapeutic possibility for hemodynamic stabilization of patients with severe DCM in addition to conventional medical treatment (Dörffel et al. 1997). The findings of this first study were subsequently confirmed by an open randomized controlled pilot study which investigated the hemodynamic influence of IA with anti-IgG columns in patients with DCM and symptomatic heart failure (NYHA class III—IV, left ventricular ejection fraction [LVEF] < 30%), who had received stable oral medication for treatment of heart failure (Felix et al. 2000). In this study, repeated IA/IgG treatment ($n=9$) was accompanied by an improvement of cardiac function as assessed by several invasive and echocardiographic parameters (e.g. cardiac index, stroke volume index, LVEF) compared to a control group without IA ($n=9$) (Felix et al. 2000).

In the meantime, several uncontrolled pilot studies or open controlled studies could reproduce the promising results of the first two using different types of currently available IA columns. Thus, currently available data have demonstrated that IA/IgG resulted in significant increase in cardiac functional and morphological parameters, symptoms' relief, heart failure related clinical biomarkers (Dörffel et al. 1997; Felix et al. 2000; Müller et al. 2000; Felix et al. 2002; Staudt et al. 2002, 2005, 2006, 2010; Doesch et al. 2010; Pokrovsky et al. 2013), increased exercise tolerance (Herda et al. 2010), and improvement of endothelial function (Bulut et al. 2011). Two studies have also shown long-term effects of IA with respect to the improvement of cardiac function, morphology, and the symptom state (Dandel et al. 2012; Müller et al. 2000). In addition, one of these studies could demonstrate that IA may also be associated with reduced rates of heart transplantation and the need for (artificial) ventricular assist devices during a follow-up time of up to 14.7 years (Dandel et al. 2012). However, all these studies were either uncontrolled or open controlled studies which compared the effects of IA to conventional therapy but without any sham intervention in the control group. For the first time, IA as a treatment option in patients with severe heart failure due to DCM is currently evaluated in a large multicenter trial which is based on a double-blind placebo-controlled design. This study will include 200 patients (www.clinicaltrials.gov, NCT00558584).

Experimental studies have shown that the beneficial clinical effects of IA/IgG are accompanied by positive alterations of several pathophysiological processes that are important for the development and progression of DCM. For example, analyses of myocardial tissue have shown a reduction of inflammatory markers (decreased number of CD3 cells, decrease in CD4 and CD8 lymphocytes, reduction of leukocyte common antigen–positive cells, and decreased HLA class II expression) after IA/IgG (Staudt et al. 2001). In addition, a decrease of activated T-cells and an increase of regulatory T-cells has been shown to be associated with hemodynamic improvement after IA/IgG revealing a link between cellular and humoral immunity (Bulut et al. 2010, 2013). Recent data, moreover, indicate that the improvement of hemodynamics may be related to the removal of negative inotropic cardiac antibodies (NIA) (Felix et al. 2002). A further study disclosed that these NIA belong to the IG-3 subclass (Staudt et al. 2002). Another study revealed a novel potential mechanism for the antibody-induced impairment of cardiac function in DCM patients, indicating that the interaction of immunoglobulins obtained from DCM patients with cardiomyocytes implements the binding of DCM-IgG-F(ab')2 to their cardiac antigens, but the Fc part might trigger the negative inotropic effects via a newly detected Fcγ receptor on cardiomyocytes (Staudt et al. 2007). Interestingly, it could also be shown that the degree of hemodynamic improvement after IA/IgG may be related to the genotype of the Fcγ-receptor (Staudt et al. 2010).

6.3   Prediction of Response to IA in DCM by Biomarkers—An Approach to Individualized Medicine

According to currently available data, response to IA/IgG treatment in patients with DCM shows wide inter-individual variability of changes in cardiac function after intervention with responder rates of about 60 % of treated patients (Staudt et al. 2004, 2010; Ameling et al. 2013). In view of the high treatment costs and the invasive character of this therapeutic procedure, determinants that predetermine differential outcome after IA/IgG are therefore of particular interest.

Indeed, several variables have been described to be associated with the beneficial effects after IA/IgG. Thus, patients with shorter disease duration and a more impaired left ventricular function at baseline have been shown to respond with a greater increase in cardiac function as measured by LVEF (Staudt et al. 2010). Moreover, detection of NIA in the plasma of DCM patients, before IA, may predict acute and prolonged hemodynamic improvement during IA/IgG (Staudt et al. 2004; Trimpert et al. 2010). However, these variables did not allow sufficient prediction of the responder state to IA/IgG and thus did not permit the discrimination between responders and non-responders. Therefore, novel diagnostic approaches were needed to allow prediction of therapy response to IA/IgG adequately, thereby enabling the selection of patients that most likely will benefit from this treatment. This treatment strategy appeared ideal for an individualized approach using multiple biomarkers including molecular biomarkers. In recent years, molecular classifiers for the prediction of outcome were primarily developed for cancer patients (Patsialou et al. 2012; van't Veer et al. 2002).

From other myocardial diseases it is already known that molecular markers can contribute to making a diagnosis and to the prediction of therapy response. Thus, predictors of the outcome of patients with suspected myocarditis have been developed using a combination of different clinical parameters and immunohistological signs of myocardial inflammation (Kindermann et al. 2008). Furthermore, transcriptomic approaches have been used for the accurate diagnosis of myocarditis (Heidecker et al. 2011) and the identification of classifiers for individual risk assessment in new-onset heart failure (Heidecker et al. 2008).

In accordance with these approaches we performed a pilot study aiming to test the value of clinical, biochemical and molecular parameters for the prediction of the response to IA/IgG therapy (Ameling et al. 2013). After application of exclusion, criteria, data of 40 DCM patients with left ventricular dysfunction (LVEF < 45 %), and symptoms of chronic heart failure, according to NYHA class II and III, could be included into the analyses. All patients received stable oral medication for at least 3 months before inclusion into the study and throughout the whole study period. From all patients sufficient material from endomyocardial biospies (EMB) obtained at baseline was available for RNA extraction. All DCM patients were treated by IA/IgG using protein-A columns (Immunosorba®, Fresenius Medical Care AG, Bad Homburg, Germany) in one course on 5 consecutive days according to established protocols (Staudt et al. 2006, 2010). After the final session, all patients received 0.5 g/kg polyclonal IgG (Venimmun N, Sandoglobulin®, CSL Behring, Germany) to restore IgG plasma levels (Staudt et al. 2006, 2010). A clinical follow-up examination, including an echocardiographic evaluation of the cardiac function as assessed by LVEF, was performed 6 months after IA/IgG treatment.

Beside the clinical variables the following parameters were available as potential predictors of response to IA/IgG (Ameling et al. 2013):

- Detection of NIA of cardiac autoantibodies as measured by cell shortening in isolated rat cardiomyocytes as described previously (Staudt et al. 2004; Trimpert et al. 2010).
- Transcriptome analyses from EMBs: RNA was isolated and after purification and quality assessment transcriptional profiling of EMBs was performed with GeneChip-Human Genome-HG U133-Plus 2.0-arrays (Affymetrix, Santa Clara, USA) and validated for a subset of genes by qRT-PCR.

At the 6-month follow-up, 24 patients (60%) were classified as responders and 16 as non-responders according to the observed improvement of cardiac function (defined as an increase in LVEF$\geq$20% relative to the baseline value and, in addition, an absolute increase in LVEF$\geq$5%). Responders exhibited an increase in LVEF (from 33 ± 5.7% to 46 ± 6.7%) and a decrease in left ventricular internal diameter (from 67 ± 6.8 to 62 ± 7.4 mm), while these parameters did not change significantly in non-responders during follow-up. Heart failure related symptoms (assessed by NYHA classes) improved in both subgroups, although the improvement was stronger in responders compared to non-responders.

Analyses of the baseline variables that were associated with hemodynamic improvement after IA/IgG and therefore could potentially serve as predictors of the response to this treatment revealed the following variables as independently associated with beneficial hemodynamic effects:

a. a shorter disease duration
b. a lower left ventricular function (LVEF)
c. a larger left ventricular internal diameter in diastole
d. the presence of inflammation in EMBs
e. different myocardial gene expression patterns for various genes related to oxidative phosphorylation, mitochondrial dysfunction, hypertrophy, and ubiquitin proteasome pathway
f. a stronger NIA of antibodies.

With respect to the clinical parameters, analyses revealed that a combination of the four clinical parameters (variables a–d) did not allow a reliable discrimination between responders and non-responders at baseline (Fig. 6.1a). However, gene expression profiling performed at baseline provides a much more detailed snapshot of the molecular state of the hearts of patients at may be of predictive value. Indeed, when differences in myocardial gene expression were considered as predictors of response, a signature of four genes (ras-related nuclear binding protein 1 [RANBP1], regulator of G-protein signaling 10 [RGS10], ubiquitin protein ligase

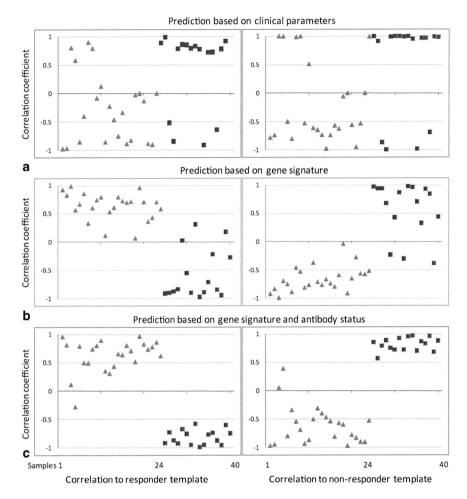

Fig. 6.1 [from (Ameling et al. 2013), reprinted with the permission of Oxford University Press]: Assessment of the value of clinical parameters (**a**), gene signature (**b**), and a combination of gene signature and antibody status (**c**) for classification or responders and non-responders at baseline. Correlation of the individual patients to the responder template is displayed in the left column, and that to the non-responder template in the right column. Green triangles=responders, red squares=non-responders. Validity of classification of patients to responders or non-responders increases with the degree of positive correlation to the corresponding template (*maximum value*) and negative correlation with the other template (*minimum value − 1*)

E3B [UBE3B] and ubiquitin specific peptidase 22 [USP22]) which was determined with two independent algorithms revealed a much better prediction performance (separation of responders and non-responders based on correlation coefficients) than clinical parameters (correlation coefficient cut-off value 0.33 instead of 1) (Fig. 6.1b). The by far best prediction was reached when the four-gene signature and the NIA of autoantibodies were combined (Fig. 6.1c). This approach allowed

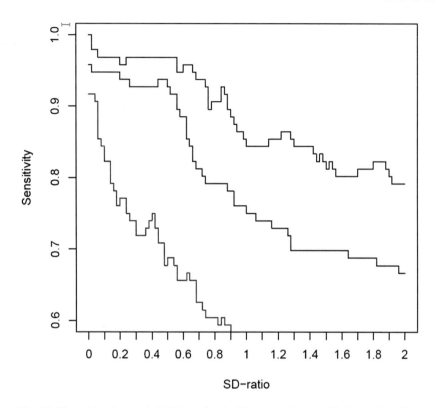

Fig. 6.2 [from (Ameling et al. 2013), reprinted with the permission of Oxford University Press]: Simulation of robustness of prediction of therapy outcome. The robustness of the prediction based on the expression of the four signature genes, NIA of antibodies and their combination was determined by adding a random noise to the parameter values of each sample prior to the classification to simulate the variation of values in the population. Added random noise is displayed on the X-axis as fold-values of the standard deviation (*SD*). Predictions are based on the expression level of four signature genes (*red line*), NIA of antibodies (*blue line*) and a combination of both values (*black line*)

a robust discrimination between responders and non-responders at baseline (sensitivity of 100 % [95 % CI, 85.8 %–100 %]; specificity up to 100 % [95 % CI, 79.4 %–100 %, correlation coefficient cut-off value: −0.28], and was superior to scores derived from clinical parameters, gene expression or NIA of antibodies only.

Evaluation of the robustness of the prediction by an independent test set was not feasible due to a limited number of patients for which EMBs were available, and the invasiveness of this procedure. Therefore, the variation of values in the population was simulated by adding incrementally increasing measurement errors to the data available and by assessing the effect of these errors on classification. These simulations demonstrated that the sensitivity was more error-tolerant when the combined information of molecular signature and NIA of antibodies were exposed to increasing variations in values (Fig. 6.2). The combined score allowed much better assessment of responders than molecular signature or NIA of antibodies alone, which is important in order to assure the appropriate identification of patients that could profit from IA/IgG (Ameling et al. 2013).

The approach that has been chosen in this pilot study represents a promising clinical example for the use of integrated analyses of biomarkers for the prediction of the treatment response for a novel therapeutic option, IA treatment in DCM. Such an approach may help to optimize the application of this treatment with respect to patient selection and improvement of efficacy by treating only patients with high odds of being a therapy responder. Moreover, molecular and biochemical analyses such as those illustrated in this study may also provide new insights in disease pathophysiology and thus constitute important components for the development of therapeutic approaches in individualized medicine. The predictive value of the identified signature should ideally be further validated in independent cohorts. A broad clinical application of this prediction model, however, might be limited due to the fact that it is based on analyses from EMBs, an invasive method, which cannot be used easily in clinical practice. Thus, further development of the approach at best for the use of blood-based molecular signatures is the next necessary step.

6.4   Future Directions/Perspectives for the Prediction of the Response to IA in DCM

Analyses of myocardial gene expression profiles are based on EMBs which are limited to specialized centers. Therefore, for the prediction of the response to intervention in the clinical setting less invasive methods are necessary. In this matter, the application of complementing -omics screens for the thorough characterization of peripheral blood constitute a promising approach. However, these screens are likely to be more complicated because differences between responders and non-responders are likely to be less pronounced in peripheral blood than in heart tissue. Thus, one would probably expect the best chances for signatures integrating information from different layers such as the plasma proteome, miRNAome and metabolome and perhaps also transcriptional patterns of whole blood. Due to the recent advances in technologies such integrated approaches are now feasible. Preferably such screenings should be performed in multicenter settings.

6.5   Conclusions

In this article, we outline a clinical example showing how integrated analyses of biomarkers might be used for the prediction of treatment response. We report findings from a pilot study that investigated the hemodynamic effects of a novel treatment option, immunoadsorption with subsequent IgG substitution, in patients with dilated cardiomyopathy, a common disease with enormous prognostic and public health impact. This study demonstrated that the combined assessment of two bio-markers (negative inotropic activity of antibodies in the blood and gene expression patterns derived from myocardial biopsies) of DCM patients predicts response to this therapy, thereby enabling appropriate selection of patients who most likely ben-efit from this therapeutic intervention. Future studies will need to focus on the blood signatures of responders and non-responders, which are easier to obtain, and thrive to integrate complementing information (e.g. transcriptomic analysis, microRNA, proteome and metabolome profiles). Such strategies will hopefully facilitate selec-tion of DCM patients who are responders not only to immunoadsorption therapy but also to other heart failure treatments for which a heterogeneous response is observed and thus will help to offer more effective treatments to affected patients.

References

Ameling S, Herda LR, Hammer E et al (2013) Myocardial gene expression profiles and cardiode-pressant autoantibodies predict response of patients with dilated cardiomyopathy to immuno-adsorption therapy. Eur Heart J 34(9):666–675

Bulut D, Scheeler M, Wichmann T et al (2010) Effect of protein A immunoadsorption on T cell ac-tivation in patients with inflammatory dilated cardiomyopathy. Clin Res Cardiol 99(10):633–638

Bulut D, Scheeler M, Niedballa LM et al (2011) Effects of immunoadsorption on endothelial function, circulating endothelial progenitor cells and circulating microparticles in patients with inflammatory dilated cardiomyopathy. Clin Res Cardiol 100(7):603–610

Bulut D, Creutzenberg G, Mugge A (2013) The number of regulatory T cells correlates with hemo-dynamic improvement in patients with inflammatory dilated cardiomyopathy after immunoad-sorption therapy. Scand J Immunol 77(1):54–61

Burkett EL, Hershberger RE (2005) Clinical and genetic issues in familial dilated cardiomyopathy. J Am Coll Cardiol 45(7):969–981

Bygren P, Freiburghaus C, Lindholm T et al (1985) Goodpasture's syndrome treated with staphy-lococcal protein A immunoadsorption. Lancet 2(8467):1295–1296

Caforio AL, Mahon NG, Baig MK et al (2007) Prospective familial assessment in dilated car-diomyopathy: cardiac autoantibodies predict disease development in asymptomatic relatives. Circulation 115(1):76–83

Cihakova D, Rose NR (2008) Pathogenesis of myocarditis and dilated cardiomyopathy. Adv Im-munol 99:95–114

Dandel M, Wallukat G, Englert A et al (2012) Long-term benefits of immunoadsorption in beta(1)-adrenoceptor autoantibody-positive transplant candidates with dilated cardiomyopathy. Eur J Heart Fail 14(12):1374–1388

Doesch AO, Mueller S, Konstandin M et al (2010) Effects of protein A immunoadsorption in pa-tients with chronic dilated cardiomyopathy. J Clin Apher 25(6):315–322

Dörffel WV, Felix SB, Wallukat G et al (1997) Short-term hemodynamic effects of immunoad-sorption in dilated cardiomyopathy. Circulation 95(8):1994–1997

Dörr M (2002) Hämodynamische Effekte einer Immunadsorption mit nachfrolgender Immunglob-ulin-G-Substitution bei Patienten mit dilatativer Kardiomyopathie—Ergebnisse einer prospe-ktiven und randomisierten Studie Humboldt-Universität zu Berlin, Berlin

Dunlay SM, Shah ND, Shi Q et al (2011) Lifetime costs of medical care after heart failure diagno-sis. Circ Cardiovasc Qual Outcomes 4(1):68–75

Felix SB, Staudt A (2008) Immunoadsorption as treatment option in dilated cardiomyopathy. Au-toimmunity 41(6):484–489

Felix SB, Staudt A, Dorffel WV et al (2000) Hemodynamic effects of immunoadsorption and subsequent immunoglobulin substitution in dilated cardiomyopathy: three-month results from a randomized study. J Am Coll Cardiol 35(6):1590–1598

Felix SB, Staudt A, Landsberger M et al (2002) Removal of cardiodepressant antibodies in dilated cardiomyopathy by immunoadsorption. J Am Coll Cardiol 39(4):646–652

Heidecker B, Kasper EK, Wittstein IS et al (2008) Transcriptomic biomarkers for individual risk assessment in new-onset heart failure. Circulation 118(3):238–246

Heidecker B, Kittleson MM, Kasper EK et al (2011) Transcriptomic biomarkers for the accurate diagnosis of myocarditis. Circulation 123(11):1174–1184

Herda LR, Trimpert C, Nauke U et al (2010) Effects of immunoadsorption and subsequent immunoglobulin G substitution on cardiopulmonary exercise capacity in patients with dilated cardiomyopathy. Am Heart J 159(5):809–816

Heymans S, Hirsch E, Anker SD et al (2009) Inflammation as a therapeutic target in heart failure? A scientific statement from the translational research committee of the Heart Failure Association of the European Society of Cardiology. Eur J Heart Fail 11(2):119–129

Hunt SA, Abraham WT, Chin MH et al (2009) 2009 Focused update incorporated into the ACC/AHA 2005 Guidelines for the Diagnosis and Management of Heart Failure in Adults. A Report of the American College of Cardiology Foundation/American Heart Association Task Force on Practice Guidelines Developed in Collaboration With the International Society for Heart and Lung Transplantation. J Am Coll Cardiol 53(15):e1–90

Jahns R, Boivin V, Hein L et al (2004) Direct evidence for a beta 1-adrenergic receptor-directed autoimmune attack as a cause of idiopathic dilated cardiomyopathy. J Clin Invest 113(10):1419–1429

Jahns R, Boivin V, Schwarzbach V et al (2008) Pathological autoantibodies in cardiomyopathy. Autoimmunity 41(6):454–461

Jhund PS, Macintyre K, Simpson CR et al (2009) Long-term trends in first hospitalization for heart failure and subsequent survival between 1986 and 2003: a population study of 5.1 million people. Circulation 119(4):515–523

Kallwellis-Opara A, Dorner A, Poller WC et al (2007) Autoimmunological features in inflammatory cardiomyopathy. Clin Res Cardiol 96(7):469–480

Kindermann I, Kindermann M, Kandolf R et al (2008) Predictors of outcome in patients with suspected myocarditis. Circulation 118(6):639–648

Kühl U, Schultheiss HP (2009) Viral myocarditis: diagnosis, aetiology and management. Drugs 69(10):1287–1302

Maron BJ, Towbin JA, Thiene G et al (2006) Contemporary definitions and classification of the cardiomyopathies: an American Heart Association Scientific Statement from the Council on Clinical Cardiology, Heart Failure and Transplantation Committee; Quality of Care and Outcomes Research and Functional Genomics and Translational Biology Interdisciplinary Working Groups; and Council on Epidemiology and Prevention. Circulation 113(14):1807–1816

McMurray JJ, Adamopoulos S, Anker SD et al (2012) ESC Guidelines for the diagnosis and treatment of acute and chronic heart failure 2012: The Task Force for the Diagnosis and Treatment of Acute and Chronic Heart Failure 2012 of the European Society of Cardiology. Developed in collaboration with the Heart Failure Association (HFA) of the ESC. Eur Heart J 33(14):1787–1847

Miller LW, Pagani FD, Russell SD et al (2007) Use of a continuous-flow device in patients awaiting heart transplantation. N Engl J Med 357(9):885–896

Moss AJ, Hall WJ, Cannom DS et al (2009) Cardiac-resynchronization therapy for the prevention of heart-failure events. N Engl J Med 361(14):1329–1338

Müller J, Wallukat G, Dandel M et al (2000) Immunoglobulin adsorption in patients with idiopathic dilated cardiomyopathy. Circulation 101(4):385–391

Palmer A, Gjorstrup P, Severn A et al (1988) Treatment of systemic lupus erythematosus by extracorporeal immunoadsorption. Lancet 2(8605):272

Patsialou A, Wang Y, Lin J et al (2012) Selective gene-expression profiling of migratory tumor cells in vivo predicts clinical outcome in breast cancer patients. Breast Cancer Res 14(5):R139. doi:10.1186/bcr3344

Pokrovsky SN, Ezhov MV, Safarova MS et al (2013) Ig apheresis for the treatment of severe DCM patients. Atheroscler Suppl 14(1):213–218

Roifman CM, Levison H, Gelfand EW (1987) High-dose versus low-dose intravenous immunoglobulin in hypogammaglobulinaemia and chronic lung disease. Lancet 1(8541):1075–1077

Slaughter MS, Rogers JG, Milano CA et al (2009) Advanced heart failure treated with continuous-flow left ventricular assist device. N Engl J Med 361(23):2241–2251

Staudt A, Schaper F, Stangl V et al (2001) Immunohistological changes in dilated cardiomyopathy induced by immunoadsorption therapy and subsequent immunoglobulin substitution. Circulation 103(22):2681–2686

Staudt A, Bohm M, Knebel F et al (2002) Potential role of autoantibodies belonging to the immunoglobulin G-3 subclass in cardiac dysfunction among patients with dilated cardiomyopathy. Circulation 106(19):2448–2453

Staudt A, Staudt Y, Dörr M et al (2004) Potential role of humoral immunity in cardiac dysfunction of patients suffering from dilated cardiomyopathy. J Am Coll Cardiol 44(4):829–836

Staudt A, Dörr M, Staudt Y et al (2005) Role of immunoglobulin G3 subclass in dilated cardiomyopathy: results from protein A immunoadsorption. Am Heart J 150(4):729–736

Staudt A, Hummel A, Ruppert J et al (2006) Immunoadsorption in dilated cardiomyopathy: 6-month results from a randomized study. Am Heart J 152(4) 712:e711–e716

Staudt A, Eichler P, Trimpert C et al (2007) Fc(gamma) receptors IIa on cardiomyocytes and their potential functional relevance in dilated cardiomyopathy. J Am Coll Cardiol 49(16):1684–1692

Staudt A, Herda LR, Trimpert C et al (2010) Fcgamma-receptor IIa polymorphism and the role of immunoadsorption in cardiac dysfunction in patients with dilated cardiomyopathy. Clin Pharmacol Ther 87(4):452–458

Stehlik J, Edwards LB, Kucheryavaya AY et al (2012) The Registry of the International Society for Heart and Lung Transplantation: 29th official adult heart transplant report-2012. J Heart Lung Transplant 31(10):1052–1064

Swedberg K, Komajda M, Bohm M et al (2010) Ivabradine and outcomes in chronic heart failure (SHIFT): a randomised placebo-controlled study. Lancet 376(9744):875–885

Tang AS, Wells GA, Talajic M et al (2010) Cardiac-resynchronization therapy for mild-to-moderate heart failure. N Engl J Med 363(25):2385–2395

Trimpert C, Herda LR, Eckerle LG et al (2010) Immunoadsorption in dilated cardiomyopathy: long-term reduction of cardiodepressant antibodies. Eur J Clin Invest 40(8):685–691

van't Veer LJ, Dai H, van de Vijver MJ et al (2002) Gene expression profiling predicts clinical outcome of breast cancer. Nature 415(6871):530–536

Zannad F, McMurray JJ, Krum H et al (2011) Eplerenone in patients with systolic heart failure and mild symptoms. N Engl J Med 364(1):11–21

Chapter 7
The Role of Pharmacogenomics in Individualized Medicine

Henriette E. Meyer zu Schwabedissen

Abstract Pharmacogenetics aims to use the patients' genetic information in order to treat diseases more efficiently and minimize adverse events. The hereon based concept of Individualized Treatment finds its origin in associations of genetic information with treatment outcome. In this chapter a selection of gene-drug associations are summarized providing details on the underlying mechanisms, the clinical significance, and the current status of clinical implementation. The first example is the association between CYP2D6 and tamoxifen in the treatment of ER-positive breast cancer. A brief summary on the historical development of this research question provides insights in the strengths and difficulties of pharmacogenetics findings. In the context of the summary on the gene-drug association CYP2C19/clopidogrel the difficulties of the clinical implementation process, which have to be encountered, when including genetic testing in the every-day health care are mentioned. However, pharmacogenetics is not only a part of the post-marketing optimization of treatment outcome, but also plays an important role in drug development. Indeed, there are several examples where genetic findings were prerequisite for the following drug development and clinical approval. In this chapter the development of CCR5 antagonists and inhibitors of the bcr-abl tyrosine kinase are summarized. Finally, the novel drug ivacaftors a drug specifically approved for a genetically defined minority of patients with cystic fibrosis, is mentioned in the context of pharmacogenetics.

Keywords Pharmacogenetics · Pharmacodynamics · Pharmacokinetics · Drug development · Individualized Treatment

In a patient cohort it is often observed that a medication with proven efficacy will fail to exert its effect or even will induce adverse events in a subset of patients. Drug efficacy and adverse drug events dose-dependently determine the outcome of drug therapy. Indeed, it is assumed that higher doses will result in higher plasma-levels, which in turn increases the likelihood of experiencing adverse drug reactions. Especially for drugs with a narrow therapeutic window, which is defined by

H. E. Meyer zu Schwabedissen (✉)
Departement Pharmazeutische Wissenschaften, Biopharmazie, Universität Basel,
Kingelbergstr. 50, 4056 Basel, Switzerland
e-mail: h.meyerzuschwabedissen@unibas.ch

© Springer International Publishing Switzerland 2015   93
T. Fischer et al. (eds.), *Individualized Medicine,* Advances in Predictive,
Preventive and Personalised Medicine 7, DOI 10.1007/978-3-319-11719-5_7

the difference between the dose with therapeutic effect and the dose associated with apparent adverse events, inter-individual differences in drug exposure might result in significant changes of the therapeutic outcome. A drug dose within the therapeutic window for the majority of a patient population can be too low or too high for a small number of patients. Those patients have an atypical dose-response curve for this particular drug (Ma and Lu 2011). Atypical dose response curves find their origin in a variety of factors including mechanisms that influence how the organism handles the drug (absorption, distribution and elimination), or changes in the direct response of the organism to the drug. It is one of the aims of indiviudalized treatment concepts to identify the above mentioned response modifying mechanisms, to translate the findings in predictive factors of treatment outcome and to include those in treatment algorithms in order to "individualize" dose regimens or to "individualize" the selection of a specific drug.

The idea of genetics influencing the responsiveness to exogenous compounds (xenobiotics) has definitely not been engendered in the last decades, when pharmacogenetics developed as a central strategy of an individualized therapy concept. Indeed, often cited is a report by Athur Fox, who introduced his accidental findings on the inheritance of sensing the bitter taste of phenylthiocarbamide (PTC) to the public in 1932 (Fox 1932). The genetic variants underlying the lack of perceiving the bitterness of PTC are located in the gene encoding for TAS2R38, one of the receptors expressed in the taste buds of the tongue able to translate binding of the exogenous compound in the sensation of bitter taste (Kim et al. 2004; Drayna 2005).

Assuming PTC would be a drug of clinical use the question "to taste—or—not to taste" would be part of the pharmacodynamics of PTC. Indeed, the pharmacological phase of a drug is divided in pharmacokinetics and pharmacodynamics. In general, pharmacokinetics of a drug describes the change of plasma levels over time, thereby functioning as a monitor of all mechanisms involved in absorption, distribution, metabolism, and elimination. Changes in the mechanisms involved in pharmacokinetics modulate plasma levels and therefore drug exposure. Assuming that the plasma levels reflect the concentration in the biophase, where the drug target is located, changes in plasma levels would certainly reflect changes in biophase drug concentration and thereby in drug response. As mentioned above, pharmacodynamics, which is the second part of the pharmacological phase, associates the mechanism of action (binding to or activation/inhibition of an enzyme, receptor etc.) with drug concentrations (Brunton 2006). Assuming that both the pharmacokinetics and pharmacodynamics of a drug are determined by specific mechanisms, which are under the rule of genetic information, both can be subject to genetically based variation. In conclusion, pharmacogenetics could be defined as a variation of genes involved in pharmacokinetics or pharmacodynamics, which results in changes of the response to a xenobiotic.

The use of patients' genetics in order to treat diseases more efficiently and minimize adverse events is an area within the broader scope of genomic medicine, which promises to contribute to the concept of Individualized Medicine or Individualized Health care. This concept is based on findings associating genetic information with treatment outcome. In the following several examples—but certainly not all—which have contributed to the idea of individualized pharmacotherapy will be summarized in detail.

Drug elimination is governed by a variety of mechanisms including drug transporters, and drug metabolizing enzymes. Most of these mechanisms are substrate specific; accordingly changes in activity due to environmental or genetic factors may result in changes of drug exposure. The sequence variation of genes encoding for proteins involved in drug elimination as a determinant of changes in individual drug exposure has been extensively studied over the last decades, resulting in various examples where specific genes (and their genetic variants) were associated with a specific drug. It is beyond the scope of this section to summarize all examples in this particular field of pharmacogenetics; however reporting details on some of them will help to get a notion of the challenges which are associated with this approach.

It seems noteworthy at this early point of a summary on pharmacogenetics in Individualized Medicine; that efficacy and safety of most of the clinically approved drugs have been empirically proven in large patient cohorts. Accordingly for most of the individuals the drug is safe and will work, assuming a normal distribution of a dose-response curve in a population the major contribution of pharmacogenetics to an individualized drug therapy concept is to provide explanations for extremes, or in other words the less likely event in regard of a patient cohort.

One of the enzymes involved in drug metabolism is CYP2D6, this member of the protein family of cytochrome P450 enzymes was one of the first CYP enzymes where inheritance of changes in activity has been studied due to observations of a multimodal pattern in response to certain drugs such as the anti-hypertensive debrisoquine (Mahgoub et al. 1977), and the anti-arrhythmic drug sparteine (Eichelbaum et al. 1979). CYP2D6 metabolizes about 25 % of the drugs on the market and its polymorphisms are assumed to significantly affect the metabolism of about 50 % of these drugs (Ingelman-Sundberg et al. 2007). Importantly, activity of CYP2D6 is categorized in different phenotypes including the ultra-rapid metabolizer, which has above-average activity, the extensive metabolizer with normal activity, the intermediate metabolizer which are individuals exhibiting lower CYP2D6 activity than average, and finally poor metabolizer with no CYP2D6 activity. Importantly, those phenotypes derive from different genetic variations, especially a poor metabolizer can harbor a variety of different polymorphisms and/or mutations all resulting in the observation of the poor CYP2D6 activity phenotype (Owen et al. 2009).

One of the drugs metabolized by CYP2D6 is the estrogen receptor (ER) antagonist tamoxifen, which is used in the treatment of ER positive breast cancer. This drug is extensively metabolized predominantly by the CYP system resulting in several primary and secondary metabolites, some of which have enhanced anti-estrogenic activity in breast cancer cells compared to tamoxifen itself (Desta et al. 2004; Lim et al. 2005). The latter provides reason for considering tamoxifen as a prodrug, which has to be metabolized to exhibit its full pharmacological activity. Especially hydroxylation of N-desmethyltamoxifen, the metabolite of CYP3A4/CYP3A5-mediated N-demethylation of tamoxifen, to endoxifen has been shown to be mediated by CYP2D6 (Stearns et al. 2003). Concluding from a mechanistic point of view, reduced activity of CYP2D6 would be associated with decreased endogenous formation of the active metabolite endoxifen resulting in reduced tamoxifen efficacy in individuals harboring low function alleles of CYP2D6. In accordance with this mechanistic consideration there are findings of retrospective studies showing that genetic variants resulting in the phenotype of poor CYP2D6 metabolism are associated with a reduced outcome of postmenopausal women treated with the ER-antagonist (Goetz et al. 2005; Bonanni et al. 2006; Borges et al. 2006; Lim et al. 2007; Schroth et al. 2007; Schroth et al. 2009; Kiyotani et al. 2010). Even if several studies observed an association of the poor metabolizer genotype with the tamoxifen treatment outcome of postmenopausal women, there were several others unable to support the idea of a CYP2D6 genotype based treatment decision (Nowell et al. 2005; Wegman et al. 2005; Wegman et al. 2007). Finally, secondary analyses from the Arimidex, Tamoxifen Alone or in Combination (ATAC) trial (Rae et al. 2012) and the Breast International Group (BIG) 1–98 study (Regan et al. 2012) in which an association between CYP2D6 genotype and the outcome of women was not observed in large study populations, resulted in an expert debate on the clinical significance of the findings. In response to the evolving debate the international tamoxifen pharmacogenomics consortium (ITPC) formed to address the questions which had been raised during the discussions on "hypes and hopes" (Klein et al. 2013). The debate which was potentially biased on the one hand by the fact of the data "fulfilling" the expectation based on the current understanding of tamoxifen metabolism and on the other hand by the need for a real—statistically significant—proof in order to implement a change in clinical practice. However, the recent meta-analysis of ITPC revealed some issues in the procedure of study design, especially the standardization of inclusion criteria (only postmenopausal women, adjuvant tamoxifen monotherapy with 20 mg/day intended for a five year treatment, a follow up every year etc.) at the different sites and the method of detecting poor metabolizer (germline genotype vs. tumor genotype; and importantly comprehensive genotyping of all potential polymorphisms resulting in the poor metabolizer phenotype) were reported to contribute to the heterogeneity of results (Province et al. 2014). The meta-analysis of the different pharmacogenetics studies revealed a meta-analytical hazard ratio of 1.25 (95 %CI 1.06–1.47) and 1.27 (CI 1.01- 1.61) for invasive disease free survival and breast cancer free survival, respectively. However, the statistically significant impact of CYP2D6 on the outcome of tamoxifen treatment in postmenopausal women is only seen in a few of the included studies, resulting in the conclusion that only prospective studies will help to shed light on the question whether genotype-guided selection of hormonal therapy improves clinical outcome of postmenopausal women with ER-positive breast cancer (Province et al. 2014). The need for prospective studies in order to provide proof for the amelioration of treatment outcome is a prerequisite for the clinical implementation of genetic testing in treatment guidelines, which are issued within the paradigm of evidence-based medicine by the professional organizations of the respective clinical discipline.

However, the current guidelines for adjuvant endocrine therapy in women with ER-positive breast cancer published in 2010 by the American Society of Clinical

Oncology recommend the treatment with an alternative class of drugs namely the aromatase inhibitors for a maximum duration of five years in postmenopausal women as the first-line treatment. There is no recommendation for CYP2D6 genotyping in this guideline (Burstein et al. 2010). Similar recommendations have been published by the German Society of Gynecology and Obstetrics (Deutsche Gesellschaft für Gynäkologie und Geburtshilfe, DGGG), which can be viewed at http://www.awmf.org (AWMF-Number: 032—045OL).

Even if the question on the implementation of a CYP2D6-genotyping as a basis for therapeutic decisions has not been readily solved from a scientific point of view, there is clearly an important contribution to individualized therapy concepts of studies driven by the association of CYP2D6 and tamoxifen. One of the consequences is the awareness for the contribution of CYP2D6 to tamoxifen bioactivation and the notion of a potential impact of changes in activity on tamoxifen efficacy. Those can not only derive from genetic differences, but also from the inhibition of the enzyme by environmental factors. Accordingly, co-administration of another substrate of CYP2D6 can significantly alter endoxifen formation, as shown for selective serotonin reuptake inhibitors such as paroxetine and fluoxetine (Stearns et al. 2003; Jin et al. 2005). Ten to twenty-five percent of women diagnosed with breast cancer experience symptoms of mild to major depressive disorders in the first year after diagnosis (Fann et al. 2008), and are therefore treated with antidepressants. Some of the antidepressants currently used in clinical practice are potent (strong) inhibitors of CYP2D6. Accordingly, findings were showing an increase in the rate of recurrence of breast cancer in women taking tamoxifen and strong CYP2D6 inhibitors (Chubak et al. 2008; Kelly et al. 2010). Even if other studies were not able to find a similar impact on the outcome of women with ER-positive breast cancer (Cronin-Fenton et al. 2010; Lash et al. 2010) it has been recommended to avoid concomitant treatment of tamoxifen with potent CYP2D6 inhibitors (Burstein et al. 2010). This recommendation was issued in 2010, however a recent survey on revealed only a slight reduction of concomitant prescriptions of tamoxifen and CYP2D6 inhibitors declining from 6.75%, observed in January 2006, to 4.87% in May 2012. Importantly, even if the percentage of women prescribed tamoxifen and paroxetine significantly declined (2.72–1.36%), there is still a significant percentage of women taking the combination of tamoxifen with other strong inhibitors as observed in this Belgium survey (Dieudonne et al. 2014). Similar results have been observed in the Netherlands (Binkhorst et al. 2013), suggesting that in order to implement changes of treatment habits such as the avoidance of a specific drug combination needs more than the awareness for a specific mechanism.

Another example of a CYP2D6 substrate, where genotype-based pharmacotherapy has been recommended due to genetic variation, is the opioid analgesic codeine, which is also used as an antitussive. The O-demethylation of codeine into morphine is catalyzed by CYP2D6. Even if this catalytic step represents a minor pathway and accounts for only 5–10% of codeine clearance in the majority of the population (extensive metabolizers with average enzyme activity), this conversion is assumed to be essential for the opioid activity of codeine (Thorn et al. 2009). Importantly, the metabolite morphine is assumed to play a pivotal role in the development of adverse events observed in association with codeine treatment. Increased activity of CYP2D6 as observed in ultra-rapid metabolizers can result in significantly enhanced codeine to morphine conversion (Gasche et al. 2004). Based on several case reports where severe toxicity (resulting in respiratory or circulatory depression, respiratory arrest, or cardiac arrest) of codeine was observed in ultra-rapid metabolizers, the Clinical Pharmacogenetics Implementation Consortium (CPIC) recommend that the use of codeine should be avoided in ultra-rapid metabolizers (Crews et al. 2014). Whether genotyping is sufficient for the identification of all patients at risk is currently unknown, as it has been previously reported that a combination of genotyping and phenotyping is necessary to identify all individuals with increased morphine formation (Lotsch et al. 2009). It remains unclear, whether the recommendation of the consortium will be implemented in clinical practice, particularly as there are several reports uncovering the challenges during the process of implementing pharmacokinetic based genetic testing in clinical practice. Indeed, several sites are currently testing the implementation of the CPIC guidelines on clopidogrel and CYP2C19 genotyping. Some of them have recently reported their experiences (Shuldiner et al. 2014; Weitzel et al. 2014), these will be described in more detail below. The gene-drug association of CYP2C19 and clopidogrel has been highly discussed in the last years. Clopidogrel is an antiplatelet agent which is often used in cardiovascular medicine. Clopidogrel is a prodrug, which has to be hepatically metabolized to form its pharmacologically active metabolite (R-130964). Up to 90 % of the absorbed prodrug are converted to an inactive metabolite (SR26334) catalyzed by the carboxyesterase-1, and only the remaining part can be bioactivated. This bioactivation is a two-step process, where CYP2C19 appears to have the most prominent role. The active metabolite exerts its effect by the irreversible binding to the P2Y12-receptor of thrombocytes, thereby inhibiting the ADP-induced thrombocyte activation (Siller-Matula et al. 2013). Despite the observation of dual-antiplatelet therapy being highly effective in the prevention of cardiovascular events after acute coronary syndrome (ACS) and percutaneous cardiovascular intervention (PCI), 25 % of treated individuals exhibit reduced clopidogrel response, summarized in the phenotype of high-on-treatment-platelet-reactivity (HTPR). It has been reported that this phenotype is associated with adverse ischemic events, particularly with short-term thrombotic events like the acute and sub-acute stent thrombosis (recently reviewed by (Siller-Matula et al. 2013)). In accordance with the above mentioned role of CYP2C19 in the bioactivation of clopidogrel naturally occurring genetic variants resulting in the reduced activity of this particular enzyme, namely the frequently occurring CYP2C19*2 and the less frequent CYP2C19*3 variant, have been repeatedly reported to be associated with reduced response, at least in high-risk patients where clopidogrel efficacy was measured by a reduction of adverse cardiovascular events (summarized by Shahin and Johnson 2013). Furthermore the CYP2C19*17, a genetic variant assumed to result in enhanced enzyme activity, has been associated with augmented clopidogrel activation and enhanced risk of bleeding, a symptom of higher efficacy of an antiplatelet agent (Harmsze et al. 2012). Even if CYP2C19 is the strongest individual predictive factor for HPTR, the CYP2C19 loss of function (LOF) carrier status explains only 5–12 % of the variability (Trenk and Hochholzer 2014). This fact and the lack of prospective randomized studies are subject to controversial discussions on the need for the implementation of genetic testing prior to starting a therapy with clopidogrel (Pare et al. 2011; Sibbing et al. 2011). In 2010 the FDA published a boxed warning, indicating that clopidogrel effectiveness depends on CYP2C19 and that "tests are available to identify a patient's CYP2C19 genotype; these tests can be used as an aid in determining therapeutic strategy. Consider alternative treatment or treatment strategies in patients identified as CYP2C19 poor metabolizers" (FDA-3-12-2010). This boxed warning might be considered as premature at that time point due to the lack of large randomized prospective studies as summarized in the reaction of the American College of Cardiology (ACC) on this alert (Holmes et al. 2010). In 2011 CYP2C19 genotyping has been included as class IIb classification in the ACC guidelines suggesting "a selective, limited approach to platelet genotype assessment and platelet function testing" (phenotyping). In this manuscript the writing committee emphasizes that the "recommendations for the use of genotype testing and platelet function testing seek to strike a balance between not imposing an undue burden on clinicians, insurers, and society to implement these strategies in patients" and the benefit for the patients (Wright et al. 2011). However, in the same year the CPIC issued recommendations for the use of CYP2C19 genetic information to guide clopidogrel therapy in patients with ACS or PCI if genetic information on CYP2C19 is available. It is suggested to personalize treatment by increasing clopidogrel dosage or using alternative antiplatelet agents such as prasugrel or ticagrelor in patients harboring LOF alleles (Scott et al. 2011). In 2012 the results of a proof of concept study (RAPID GENE) were reported. In this study patients undergoing PCI for ACS or stable angina were randomly assigned to CYP2C19*2 allele screening at randomization or conventional treatment with subsequent CYP2C19*2 genotyping. In the rapid genotyping group patients with the LOF allele were treated with prasugrel (10 mg/d), while non-carriers and patients in the conventional treatment arm received clopidogrel 75 mg daily. In order to monitor clopidogrel efficacy the HTPR phenotype was assessed one week after treatment start. This phenotyping revealed that none of the CYP2C19*2 carriers in the rapid genotyping group treated with prasugrel had this risk factor of adverse outcome, while 30 % of the CYP2C19*2 patients in the conventional treatment group exhibited HTPR. However, 18 % of non-carriers of CYP2C19*2 alleles in both groups had HTPR after one week of clopidogrel treatment (Roberts et al. 2012). Based on the current guidelines the option to perform genetic testing is up to the individual clinician. However, Trenk and Hochholzer, recently concluded that the strategy of one-size-fits all by using clopidogrel treatment alternatives such as prasugrel and ticagrelor might most likely be favored above a genotype based treatment in the guidelines of ACCF/AHA and ESC for the acute coronary syndrome (Trenk and Hochholzer 2014). With growing evidence it has been repeatedly mentioned by experts in the field that a gradual shift to a more individualized antiplatelet therapy approach of routine care might be expected (Xie et al. 2013; Stimpfle et al.

2014). This gradual shift is hampered by the lack of prospective genotype-directed randomized clinical trials, which validate the advantage of using genetic based dosing over standard of care, potentially also resulting in the paucity of clear recommendations for pharmacogenetic testing by professional associations. Several groups have recently reported issues which have to be encountered when including genetic testing in the every-day health care. According to the authors the challenges are of a broad nature, including financial, infrastructural, and organizational problems (documentation systems, logistics of genotyping; infrastructure for genotyping reimbursement), the lack of decision support for genomic medicine which are accompanied by ethical and medico-legal concerns (Pulley et al. 2012; Shuldiner et al. 2014; Weitzel et al. 2014), which certainly have to be addressed prior to the enforcement of regulations.

In addition to drug metabolizing enzymes, there is increasing evidence that drug transporters are part of the network of genes determining drug elimination and drug exposure. Involved in the transmembrane transport of drugs are efflux and uptake transporters, both classes are harboring a huge number of different proteins. One uptake transporter that has been extensively studied with focus on pharmacogenetics in the last decade is the hepatic uptake transporter OATP1B1 (SLCO1B1). This particular transporter was first cloned and functionally characterized in 1999 (Abe et al. 1999) and transports a variety of compounds in clinical use (Meyer zu Schwabedissen and Kim 2009). In 2001 Tirona et al. identified several genetic variants, which influence activity of the transporter (Tirona et al. 2001). One of the variants, namely the 521T>C (rs4149056), was observed to frequently occur in different populations, thereby setting the stage for a variety of studies reporting the influence of this particular genetic variant on the hepatic clearance of substrate drugs (summarized in Kalliokoski and Niemi 2009; Gong and Kim 2013). In accordance with the impact of a reduced function of the transporter on hepatic clearance, resulting in increased drug exposure was a report in 2009 where simvastatin induced myopathy was shown to be linked to the above mentioned genetic variant of OATP1B1 (Link et al. 2008). Myopathy is a known adverse event of the treatment with statins, which is assumed to be dose-dependent. Statin induced muscle toxicity has generally been found at low frequency in clinical trials and large observational studies with definite myopathy (defined as CK levels $>10 \times ULN$ with muscle symptoms) in about 0.01–0.3 % of the patients, and rhabdomyolysis in approximately 0.003–0.01 % of the patients. However, symptoms of mild myopathy (CK elevation 5–10 and muscle symptoms) are reported more frequently with 5–20 % depending on the study population and the definition of myopathy (Stewart 2013). The role of OATP1B1 as a determinant of statin toxicity was further supported, when the association of genetic variants of the hepatic uptake transporter and the adverse event (determined by CK-elevation) was replicated in a cohort recruited using a medical record system (Carr et al. 2013). Similarly Ferrari et al. reported that patients harboring the above mentioned genetic variant had an odds ratio for statin-induced elevated serum CK levels of 8.86 (Ferrari et al. 2014). Even if Canestraro et al. recently mentioned that the body of evidence surrounding the association of statin related myopathy with SLCO1B1 is strong enough to propose dosing algorithms for clinical use (Canestaro et al. 2014), there are doubts about the clinical significance of this association (Stewart 2013). In 2012 the CPIC drafted an OATP1B1 genotype based personalization of simvastatin dosing (Wilke et al. 2012). Whether the genotype-based treatment decision will find its way into the clinical setting remains to be seen. However, it seems noteworthy that the pharmacological target is located intracellularly in hepatocytes. Assuming that OATP1B1 is involved in hepatic uptake it seems likely that changes in transport activity might result in changes of statin efficacy. Even if there are studies supporting this notion, this topic is far beyond the scope of this summary on pharmacogenetics.

In order to provide a comprehensive summary on the role of pharmacogenetics in Individualized Medicine, it seems inevitable to mention that there are various examples, where genetic findings were prerequisite for the progress in drug development. Indeed, in some cases genetic findings have significantly contributed to the understanding of disease development and/or progression and thereby served as the rationale in subsequent drug development. One example is the development of human immune deficiency virus (HIV)-entry inhibitors. Even if the surface protein CD4 was identified early as the primary receptor facilitating the entry of the virus into immune cells, it soon became evident that co-receptors namely the G-protein-coupled 7-transmembrane chemokine receptors CXCR4 (CXC-chemokine receptor 4) and CCR5 (CC-chemokine receptor 5) are also significantly contributing to the cellular entry, a prerequisite for virus replication (Deng et al. 1996; Feng et al. 1996). Supported by the finding that a naturally occurring 32bp-deletion in the coding region of CCR5, which results in a frame shift and thereby in a modification of the extracellular loop of the receptor (Samson et al. 1996), results in a natural strong but incomplete resistance to HIV-1 transmission in homozygote carriers, CCR5 was considered as a potential target in HIV treatment (Blanpain et al. 2002). The idea of this co-receptor being a safe drug target was further supported by the fact that the CCR5 deletion variant occurs in 1 % of Caucasian populations, without obvious influence on human health. Nowadays the understanding of the contribution of co-receptors to HIV entry and thereby cellular tropism of the virus has tremendously increased. Indeed, the CCR5 is the co-receptor which is used by a specific HIV-1 subtype—the so-called R5-strain, while the CXCR4 co-receptor defines cell tropism of another viral subtype, namely the X4-strain. It is assumed that the R5-virus is preferentially transmitted and is therefore mostly present in early stages of the disease. However X4-viruses emerge in 40–60 % of HIV-positive individuals. HIV entry inhibitors have been developed and clinically tested. Miraviroc was the first approved compound in this class (Este and Telenti 2007; Maeda et al. 2012). In patients considered for treatment with an CCR5 antagonist the presence of mixed or X4-virus strains has to be excluded by additional laboratory tests in order to ensure efficacy of CCR5 inhibitors (Lin and Kuritzkes 2009). Accordingly the use of CCR5 antagonist can be considered as a "Individualized Treatment", where detection of a molecular marker is predictive for treatment outcome. As previously stated by Bob Carlson an individualized drug could be defined as a medical treatment, where the desired response can be predetermined by a molecular diagnostic test (Carlson 2008). To sum up, the HIV-resistance which had been observed in a certain genetically defined subpopulation was the basis of the further drug development of CCR5 antagonists. Even if this is not a classic example of pharmacogenetics, it shows that pharmacogenetics significantly contributes to the understanding of disease development and progression, and therefore plays an important role in future developments of pharmacology.

Even if the implementation of genetic testing of PK-associated genotypes in clinical practice is still under debate, genotype-based personalization is already used in other fields of medicine, where the concept has more or less quietly entered the health care system. Genetic variation resulting in changes of pharmacokinetics and thereby modulating drug efficacy and toxicity is by far not the only focus of individualized drug therapy. Especially in oncology a variety of "targeted" or "individualized treatments strategies" are readily used. The so-called "targeted drugs" exhibit substantial benefits in small, molecularly defined, pharmacologically relevant subsets of patients. In accordance with the need to identify patients of the molecularly defined subset treatment with a "targeted" drug is associated with a molecular diagnostic, including the detection of genetic variations. The "targeted" drugs represent approximately one in five original new molecules approved by the FDA since 2010 (Pacanowski et al. 2014).

An early example of a targeted drug concept where genetic findings in a tumor entity contributed to drug development and treatment selection, was the identification of a balanced reciprocal translocation between chromosome 9 and 22 in chronic myeloic leukemia (CML). This translocation results in a short chromosome also known as the Philadelphia chromosome [t(9;22)(q34;q11)], and was first described by Rudkin et al. in 1964 (Rudkin et al. 1964). This genetic aberration is nowadays considered as a hallmark of CML development and progression. Without going into all molecular details, the translocation results in the chimeric *BCR-ABL1* oncogene gene which encodes for a fusion protein and results in a constitutively active Abelson kinase (ABL kinase). The constitutive activity overrides the tightly regulated homeostatic molecular circuits that normally govern ABL kinase activity and thereby the growth and differentiation of hematopoietic progenitors (Quintas-Cardama and Cortes 2009). In conclusion, the genetic aberration triggers the development and progression of leukemia, thereby providing the rationale for developing inhibitors of the bcr-abl tyrosine kinase. In 2001 STI571 (also known as imatinib mesylate or Glivec) an inhibitor of the fusion protein was introduced into clinics, which significantly improved the clinical outcome of CML patients harboring this specific genetic variation in their tumor cells (Mauro and Druker 2001a; Mauro and Druker 2001b). The combination of a genetic testing as the basis for the use of imatinib makes this particular molecule an "Individualized Drug" (Carlson 2008). Even if the bcr-abl targeting therapy has significantly improved drug response and thereby clinical outcome 20–50% of the patients develop resistance to imatinib (Hughes et al. 2006; Lau and Seiter 2014). Several mechanisms are assumed to contribute to the development of this drug resistance including changes in cellular uptake mediated by a drug transporters (Crossman et al. 2005), activation of alternative pathways (Li and Li 2007), clonal evolution (Cortes et al. 2003), and persistence of imatinib insensitive stem cells (Corbin et al. 2011). In addition to these drug target independent mechanisms, there are a variety of genetic variants of the bcr-abl tyrosine kinase (Jabbour et al. 2013) of which some significantly modulate binding of imatinib. The understanding of how the genetic variants impact imatinib binding was a prerequisite for the development of new tyrosine kinase inhibitors, as recently summarized in an excellent overview by Brian Druker (Druker 2008). The mechanistic studies were basis for the development of second generation tyrosine kinase inhibitors such as desatinib, nilotinib, or bosutinib, which found their way into the treatment guidelines as first or second-line therapy (Baccarani et al. 2013). Even if the development of the second generation TKIs was partly driven by genetic variants, the currently used TKIs, especially desatinib and nilotinib, are not able to override the resistance conferred by all mutations, new drugs and treatment regimens are under investigation to reduce the incidence of treatment failure (Druker 2008; Jabbour et al. 2013). It remains to be speculated that some of them will be given after molecular diagnostics of the factor contributing to resistance.

Another "Individualized Drug" developed due to genetic findings in tumor biology is trastuzumab. The development of this antibody was based on the finding that the amplification of the gene encoding for HER2 results in overexpression and ligand independent activation of the encoded tyrosine kinase growth factor receptor. In accordance with the ligand independent and enhanced activity of this modulator of cell differentiation and proliferation, HER2 overexpression was identified as a prognostic factor independently predicting time to relapse and survival among women with breast cancer (Slamon et al. 1987). In 1998 trastuzumab (Herceptin®) was approved for treatment of metastatic HER2-positive breast cancer. This monoclonal antibody, which binds to an extracellular domain of the receptor inhibiting its activity was the first HER2-targeted treatment of cancer (McKeage and Perry 2002). With the introduction of the HER2 antibody in breast cancer therapy, detection of HER2 overexpression by a combination of methods for detecting the protein (immunohistochemistry) and the DNA (like fluorescence in situ hybridization) became fundamental in the process of planning the individualized therapeutic strategy for women with breast cancer (Tafe and Tsongalis 2012; Perez et al. 2014; Reynolds et al. 2014). Additional drugs targeting the amplified HER2 tyrosine kinase are in development or have been already approved for metastatic breast cancer including pertuzumab, lapatinib, neratinib, and afatinib (Li and Li 2013). Amplification and increased expression of the proto-oncogene HER2 has been reported in several other tumor entities (Bofin et al. 2004), in some of those the addition of HER2 inhibitors to conventional treatment regimens is currently investigated or has been shown to improve the clinical outcome (Bang et al. 2010; Yang et al. 2014). In addition, the concept of using HER2 overexpression as a strategy for the antibody-mediated delivery of small molecules has recently resulted in the approval of trastuzumab-emtansine (kadcyla®). Trastuzumab-emtansine is an antibody-drug conjugate (ADC), where the cytotoxic compound is conjugated to the HER2-targeting antibody, which is released from the antibody after HER2 binding thereby enhancing tumor targeting of the compound (LoRusso et al. 2011). The treatment with this novel compound is also based on the detection of HER2 amplification/overexpression in the tumors.

One of the younger examples of a "targeted" or "individualized" drug is vemurafenib. This compound is a potent inhibitor of mutated BRAF and approved for the treatment of melanomas. BRAF is one of the Raf kinases, which are an intracelullar part of the mitogen-activated protein (MAP) kinase signal-transduction pathway that transmits mitogenic signals from activated cell surface growth factor receptors to the nucleus under normal physiologic conditions. The identification of genetic variants in this kinase resulting in signal-independent activation was the basis for developing vemurafenib, which efficiently inhibits mutated BRAF (Vultur et al. 2011). It is of note that about 40–60 % of cutaneous melanoma carry mutations in BRAF that lead to the constitutive activation of downstream signaling through the MAPK pathway (Davies et al. 2002). Patients carrying the BRAF V600E variant significantly benefit from a treatment with this compound (Chapman et al. 2011). Another recently approved drug where the subset of patients benefiting from therapeutic treatment is identified by genetic testing is ivacaftor. This CFTR potentiator is used in the treatment of cystic fibrosis (CF). CF is a genetic disease that is caused by loss-of-function mutations of the ABC-Transporter Cystic Fibrosis transmembrane conductance regulator (CFTR), an anion channel that is critical for epithelial ion and fluid transport (Riordan 2008). About 2,000 different mutations have been identified to result in the disease of CF (Sosnay et al. 2013), ivacaftor is licensed for clinical use in patients with one particular genetic mutation namely the G551D (rs75527207) gating mutation of CFTR which is prevalent in about 5 % of the patients (Bobadilla et al. 2002; Clancy et al. 2014). Referencing to a very recent commentary by Ian M. Balfour-Lynn entitled "Personalized medicine in cystic fibrosis is unaffordable" ivacaftor seems a good example to introduce an aspect of individualized drugs, which is on the level of economics (Balfour-Lynn 2014). In the case of CF, where genotyping is often performed during prenatal diagnostics the reimbursement of the genotyping is not reason for the debate, however, the cost for a presumable life-long treatment is assumed to be a major obstacle to this drug, which undoubted is expected to enhance the quality of life of several patients (Whiting et al. 2014). Designing drugs for a small number of patients (so-called orphan drugs) which have to be tested for efficacy and safety during drug development will inevitably result in increasing treatment costs, a topic that is currently discussed in association with the approval of ivacaftor (O'Sullivan et al. 2013).

In addition to genetic variants, which can be directly linked to proteins involved in pharmacokinetics or -dynamics, there are examples of modifying genetic variants, which indirectly affect drug responses (wanted or unwanted). One example is the association between a certain MHC haplotype and the occurrence of adverse drug events with abacavir. This drug is a nucleoside with anti-HIV activity. Treatment with this drug is associated with severe hypersensitivity reactions, which occur in 5–8 % of the treated patients within the first six weeks of treatment (Hetherington et al. 2001). In 2002 Mallal et al. first reported an association between a specific MHC class II haplotypes and the occurrence of the abacavir-induced hypersensitivity reaction. Especially the HLA-B*5701 variant, which was detected in 14 of 18 patients with confirmed abacavir hypersensitivity, was introduced as a predictive genetic factor for this safety issue. According to the data by Mallal et al.

carriers of this genetic variant were more than 100 times more likely to experience the adverse event compared to individuals lacking this particular polymorphism (Mallal et al. 2002). Even if the observed sensitivity of 72 % was not reproduced in similar studies by Hetherington et al. and Hughes et al. these retrospective studies reported that the presence of the HLA-B*5701 polymorphism was the most significant predictor of abacavir hypersensitivity (Hetherington et al. 2002; Hughes et al. 2004). In 2006 the first prospective study was published reporting a significant decrease of hypersensitivity reactions from 8 to 2 % by high-resolution HLA class I and class II typing and avoiding treatment in HLA-B*5701 carriers (Rauch et al. 2006). In 2012, Illing et al. reported a rationale for the observed phenotype as structural analyses showed that abacavir non-covalently binds to HLA-B*5701, where it alters the binding of other peptides triggering the immonological event assumed to be associated with the hypersensitivity reaction (Illing et al. 2012). HLA typing is now globally recommended prior to abacavir treatment, thereby providing an example on how genetic testing can be used to reduce the risk of side effects (Daly 2014).

Polymorphisms in the MHC class II region have been repeatedly reported to be associated with drug induced adverse events. Indeed, the HLA-B*5701 polymorphism has also been linked with flucloxacillin induced liver injury (ILI) (Daly et al. 2009). However, as stated by Philips & Mallal, the low prevalence of flucloxacillin induced liver injury, with an estimated 8.5 cases in 100,000 treated patients and the positive predictive value of HLA-B*5701 for flucloxacillin ILI being only 0.12 %, would translate in almost 14,000 patients that would need to be tested for HLA-B*5701 and excluded from flucloxacillin to prevent a single case of flucloxacillin ILI (Phillips and Mallal 2013). It seems noteworthy at this point, that flucloxacillin is an antibiotic, which is often given only for a short period of time. However, an analysis like the above mentioned calculation of the number need to genotype (treat) in order to prevent one case, names one of the major challenges of pharmacogenetics in prediction of rare adverse events. Assuming that a particular drug for most of the patients in a population is considered efficacious and/or safe, then for most drugs adverse drug events have to be considered as a rare event. In addition, adverse drug events are often multifactorial and the contribution of one factor might be minor. With all these assumptions in mind, prevention of adverse drug events is always the work with minorities. Another example in this context is the finding that the risk of venous thromboembolic events is significantly increased in women carrying the coagulation factor V Leiden variant and taking oral contraceptives (Sass and Neufeld 2002). Although this association is well accepted, no genotyping for factor V is currently performed before prescribing oral contraceptives. Stingl Kirchheiner and Brockmöller recently mentioned that it had been "argued that the benefits gained by reducing the incidence of thrombosis would not outweigh the extensive burden of genotyping millions of women taking oral contraceptives"(Stingl Kirchheiner and Brockmöller 2011).

It is the aim of the concept of an individualized drug therapy to identify predictive factors for the individual outcome, which includes the therapeutic intended effect, unwanted side effects or even toxicity of the drug, and to include the predictive factors in a therapeutic decision tree. It has been noted previously by Turner and Pirmohamed, that pharmacogenetic candidate gene studies might be substantially more successful in identifying replicable common variants of appreciable effect size compared with candidate gene investigations in disease genetics. The authors explain this by a greater understanding of pharmacological pathways compared with that of disease processes (Turner and Pirmohamed 2014). Considering the above summarized examples of drug-gene associations, which have been presented in the context of our current understanding of the underlying mechanisms, this statement might be supported. However, there are several obstacles for the direct translation of the mechanistically supported findings in clinical practice, which have been also mentioned in the context of the examples above. A major concern of translating the findings in individualized health care is often the lack of prospective randomized clinical trials. Especially in the case of drug-pharmacokinetic gene associations, this fact is a major concern, which might be in part explained by the effect size, the frequency of the genetic variations and the number of patients with a certain disease (Stingl Kirchheiner and Brockmoller 2011). However, translation of genetic findings into clinical care not only depends on the sensitivity, predictive value and effect size, but also on the clinical scenario which includes the relevance of early detection for the patient, availability of alternative treatments and the accessibility of genotyping (Manolio 2013). It seems noteworthy in the context of a summary on pharmacogenetics in Individualized Medicine that the fact of implementation of an additional laboratory test (such as the HIV tropism test) in the treatment decision for HIV has been previously rated as one of the disadvantages of a HIV drug (Maeda et al. 2012). Included in this deliberation by Maeda et al. was certainly the fact of the current geographic distribution of HIV infections, with most of the infections occurring in middle and low income countries (http://www.unaids. org/), where financial considerations are assumed to certainly influence treatment decisions. However, economics are not only concerns in low income countries. The current debate on the costs for life long treatment with an individualized drug like ivacaftor certainly reflects a similar concern in other countries (O'Sullivan et al. 2013). Efforts are now being made to study the process of implementing genotyping in clinical practice. Often mentioned is the lack of expert guidance for clinicians in every day care. A comprehensive resource that curates knowledge about the impact of genetic variation on drug response is the Pharmacogenetics Knowledge Base (https://www.pharmgkb.org), in addition the Clinical Pharmacogenetics Implementation Consortium provides guidelines for the use of genetic data in drug therapy. Those guidelines are written by experts and are published in peer reviewed journals, in order to meet this particular need.

Taken together pharmacogenetics plays an important role in individualized concepts, and is in part pioneering in the process considering the obstacles and advantages of the underlying strategies.

References

Abe T, Kakyo M, Tokui T et al (1999) Identification of a novel gene family encoding human liver-specific organic anion transporter LST-1. J Biol Chem 274(24):17159–17163

Baccarani M, Deininger MW, Rosti G et al (2013) European LeukemiaNet recommendations for the management of chronic myeloid leukemia: 2013. Blood 122(6):872–884

Balfour-Lynn IM (2014) Personalised medicine in cystic fibrosis is unaffordable. Paediatr Respir Rev 15(Suppl 1):2–5

Bang YJ, Van Cutsem E, Feyereislova A et al (2010) Trastuzumab in combination with chemotherapy versus chemotherapy alone for treatment of HER2-positive advanced gastric or gastro-oesophageal junction cancer (ToGA): a phase 3, open-label, randomised controlled trial. The Lancet 376(9742):687–697

Binkhorst L, Mathijssen RH, van Herk-Sukel MP et al (2013) Unjustified prescribing of CYP2D6 inhibiting SSRIs in women treated with tamoxifen. Breast Cancer Res Treat 139(3):923–929

Blanpain C, Libert F, Vassart G et al (2002) CCR5 and HIV infection. Receptors Channels 8(1):19–31

Bobadilla JL, Macek M Jr, Fine JP et al (2002) Cystic fibrosis: a worldwide analysis of CFTR mutations-correlation with incidence data and application to screening. Hum Mutat 19(6):575–606

Bofin AM, Ytterhus B, Martin C et al (2004) Detection and quantitation of HER-2 gene amplification and protein expression in breast carcinoma. Am J Clin Pathol 122(1):110–119

Bonanni B, Macis D, Maisonneuve P et al (2006) Polymorphism in the CYP2D6 tamoxifen-metabolizing gene influences clinical effect but not hot flashes: data from the Italian Tamoxifen Trial. J Clin Oncol 24(22):3708-3709 (author reply 3709)

Borges S, Desta Z, Li L et al (2006) Quantitative effect of CYP2D6 genotype and inhibitors on tamoxifen metabolism: implication for optimization of breast cancer treatment. Clin Pharmacol Ther 80(1):61–74

Brunton LL (ed) (2006) Goodman and Gilman's the pharmacological basis of therapeutics. McGraw-Hill, New York

Burstein HJ, Prestrud AA, Seidenfeld J et al (2010) American Society of Clinical Oncology clinical practice guideline: update on adjuvant endocrine therapy for women with hormone receptor-positive breast cancer. J Clin Oncol 28(23):3784–3796

Canestaro WJ, Austin MA, Thummel KE (2014) Genetic factors affecting statin concentrations and subsequent myopathy: a HuGENet systematic review. Genet Med. doi:10.1038/gim.2014.41

Carlson B (2008) What the devil is personalized medicine? Biotechnol Healthc 5(1):17–19

Carr DF, O'Meara H, Jorgensen AL et al (2013) SLCO1B1 genetic variant associated with statin-induced myopathy: a proof-of-concept study using the clinical practice research datalink. Clin Pharmacol Ther 94(6):695–701

Chapman PB, Hauschild A, Robert C et al (2011) Improved survival with vemurafenib in melanoma with BRAF V600E mutation. N Engl J Med 364(26):2507–2516

Chubak J, Buist DS, Boudreau DM et al (2008) Breast cancer recurrence risk in relation to antidepressant use after diagnosis. Breast Cancer Res Treat 112(1):123–132

Clancy JP, Johnson SG, Yee SW et al (2014) Clinical pharmacogenetics implementation consortium (CPIC) guidelines for ivacaftor therapy in the context of CFTR genotype. Clin Pharmacol Ther 95(6):592–597

Corbin AS, Agarwal A, Loriaux M et al (2011) Human chronic myeloid leukemia stem cells are insensitive to imatinib despite inhibition of BCR-ABL activity. J Clin Invest 121(1):396–409

Cortes JE, Talpaz M, Giles F et al (2003) Prognostic significance of cytogenetic clonal evolution in patients with chronic myelogenous leukemia on imatinib mesylate therapy. Blood 101(10):3794–3800

Crews KR, Gaedigk A, Dunnenberger HM et al (2014) Clinical pharmacogenetics implementation consortium guidelines for cytochrome P450 2D6 genotype and codeine therapy: 2014 update. Clin Pharmacol Ther 95(4):376–382

Cronin-Fenton D, Lash TL, Sorensen HT (2010) Selective serotonin reuptake inhibitors and adjuvant tamoxifen therapy: risk of breast cancer recurrence and mortality. Future Onco 6(6):877–880

Crossman LC, Druker BJ, Deininger MW et al (2005) hOCT 1 and resistance to imatinib. Blood 106(3):1133-1134 (author reply 1134)

Daly AK (2014) Human leukocyte antigen (HLA) pharmacogenomic tests: potential and pitfalls. Curr Drug Metab 15(2):196–201

Daly AK, Donaldson PT, Bhatnagar P et al (2009) HLA-B*5701 genotype is a major determinant of drug-induced liver injury due to flucloxacillin. Nat Genet 41(7):816–819

Davies H, Bignell GR, Cox C et al (2002) Mutations of the BRAF gene in human cancer. Nature 417(6892):949–954

Deng H, Liu R, Ellmeier W et al (1996) Identification of a major co-receptor for primary isolates of HIV-1. Nature 381(6584):661–666

Desta Z, Ward BA, Soukhova NV et al (2004) Comprehensive evaluation of tamoxifen sequential biotransformation by the human cytochrome P450 system in vitro: prominent roles for CYP3A and CYP2D6. J Pharmacol Exp Ther 310(3):1062–1075

Dieudonne AS, De Nys K, Casteels M et al (2014) How often did Belgian physicians co-prescribe tamoxifen with strong CYP2D6 inhibitors over the last 6 years? Acta Clin Belg 69(1):47–52

Drayna D (2005) Human taste genetics. Annu Rev Genomics Hum Genet 6:217–235

Druker BJ (2008) Translation of the Philadelphia chromosome into therapy for CML. Blood 112(13):4808–4817

Eichelbaum M, Spannbrucker N, Steincke B et al (1979) Defective N-oxidation of sparteine in man: a new pharmacogenetic defect. Eur J Clin Pharmacol 16(3):183–187

Este JA, Telenti A (2007) HIV entry inhibitors. The Lancet 370(9581):81–88

Fann JR, Thomas-Rich AM, Katon WJ et al (2008) Major depression after breast cancer: a review of epidemiology and treatment. Gen Hosp Psychiatry 30(2):112–126

Feng Y, Broder CC, Kennedy PE et al (1996) HIV-1 entry cofactor: functional cDNA cloning of a seven-transmembrane, G protein-coupled receptor. Science 272(5263):872–877

Ferrari M, Guasti L, Maresca A et al (2014) Association between statin-induced creatine kinase elevation and genetic polymorphisms in SLCO1B1, ABCB1 and ABCG2. Eur J Clin Pharmacol 70(5):539–547

Fox AL (1932) The relationship between chemical constitution and taste. Proc Natl Acad Sci U S A 18(1):115–120

Gasche Y, Daali Y, Fathi M et al (2004) Codeine intoxication associated with ultrarapid CYP2D6 metabolism. N Engl J Med 351(27):2827–2831

Goetz MP, Rae JM, Suman VJ et al (2005) Pharmacogenetics of tamoxifen biotransformation is associated with clinical outcomes of efficacy and hot flashes. J Clin Oncol 23(36):9312–9318

Gong IY, Kim RB (2013) Impact of genetic variation in OATP transporters to drug disposition and response. Drug Metab Pharmacokinet 28(1):4–18

Harmsze AM, van Werkum JW, Hackeng CM et al (2012) The influence of CYP2C19*2 and *17 on on-treatment platelet reactivity and bleeding events in patients undergoing elective coronary stenting. Pharmacogenet Genomics 22(3):169–175

Hetherington S, McGuirk S, Powell G et al (2001) Hypersensitivity reactions during therapy with the nucleoside reverse transcriptase inhibitor abacavir. Clin Ther 23(10):1603–1614

Hetherington S, Hughes AR, Mosteller M et al (2002) Genetic variations in HLA-B region and hypersensitivity reactions to abacavir. The Lancet 359(9312):1121–1122

Holmes DR Jr, Dehmer GJ, Kaul S et al (2010) ACCF/AHA clopidogrel clinical alert: approaches to the FDA "boxed warning": a report of the American College of Cardiology Foundation Task Force on clinical expert consensus documents and the American Heart Association endorsed by the Society for Cardiovascular Angiography and Interventions and the Society of Thoracic Surgeons. J Am Coll Cardiol 56(4):321–341

Hughes DA, Vilar FJ, Ward CC et al (2004) Cost-effectiveness analysis of HLA B*5701 genotyping in preventing abacavir hypersensitivity. Pharmacogenetics 14(6):335–342

Hughes T, Deininger M, Hochhaus A et al (2006) Monitoring CML patients responding to treatment with tyrosine kinase inhibitors: review and recommendations for harmonizing current methodology for detecting BCR-ABL transcripts and kinase domain mutations and for expressing results. Blood 108(1):28–37

Illing PT, Vivian JP, Dudek NL et al (2012) Immune self-reactivity triggered by drug-modified HLA-peptide repertoire. Nature 486(7404):554–558

Ingelman-Sundberg M, Sim SC, Gomez A et al (2007) Influence of cytochrome P450 polymorphisms on drug therapies: pharmacogenetic, pharmacoepigenetic and clinical aspects. Pharmacol Ther 116(3):496–526

Jabbour EJ, Cortes JE, Kantarjian HM (2013) Resistance to tyrosine kinase inhibition therapy for chronic myelogenous leukemia: a clinical perspective and emerging treatment options. Clin Lymphoma Myeloma Leuk 13(5):515–529

Jin Y, Desta Z, Stearns V et al (2005) CYP2D6 genotype, antidepressant use, and tamoxifen metabolism during adjuvant breast cancer treatment. J Natl Cancer Inst 97(1):30–39

Kalliokoski A, Niemi M (2009) Impact of OATP transporters on pharmacokinetics. Br J Pharmacol 158(3):693–705

Kelly CM, Juurlink DN, Gomes T et al (2010) Selective serotonin reuptake inhibitors and breast cancer mortality in women receiving tamoxifen: a population based cohort study. Bmj 340:c693

Kim UK, Breslin PA, Reed D et al (2004) Genetics of human taste perception. J Dent Res 83(6):448–453

Kiyotani K, Mushiroda T, Imamura CK et al (2010) Significant effect of polymorphisms in CYP2D6 and ABCC2 on clinical outcomes of adjuvant tamoxifen therapy for breast cancer patients. J Clin Oncol 28(8):1287–1293

Klein DJ, Thorn CF, Desta Z et al (2013) PharmGKB summary: tamoxifen pathway, pharmacokinetics. Pharmacogenet Genomics 23(11):643–647

Lash TL, Cronin-Fenton D, Ahern TP et al (2010) Breast cancer recurrence risk related to concurrent use of SSRI antidepressants and tamoxifen. Acta Oncol 49(3):305–312

Lau A, Seiter K (2014) Second-line therapy for patients with chronic myeloid leukemia resistant to first-line imatinib. Clin Lymphoma Myeloma Leuk 14(3):186–196

Li S, Li D (2007) Stem cell and kinase activity-independent pathway in resistance of leukaemia to BCR-ABL kinase inhibitors. J Cell Mol Med 11(6):1251–1262

Li SG, Li L (2013) Targeted therapy in HER2-positive breast cancer. Biomed Rep 1(4):499–505

Lim YC, Desta Z, Flockhart DA et al (2005) Endoxifen (4-hydroxy-N-desmethyl-tamoxifen) has anti-estrogenic effects in breast cancer cells with potency similar to 4-hydroxy-tamoxifen. Cancer Chemother Pharmacol 55(5):471–478

Lim HS, Ju Lee H, Seok Lee K et al (2007) Clinical implications of CYP2D6 genotypes predictive of tamoxifen pharmacokinetics in metastatic breast cancer. J Clin Oncol 25(25):3837–3845

Lin NH, Kuritzkes DR (2009) Tropism testing in the clinical management of HIV-1 infection. Curr Opin HIV AIDS 4(6):481–487

Link E, Parish S, Armitage J et al (2008) SLCO1B1 variants and statin-induced myopathy—a genomewide study. N Engl J Med 359(8):789–799

LoRusso PM, Weiss D, Guardino E et al (2011) Trastuzumab emtansine: a unique antibody-drug conjugate in development for human epidermal growth factor receptor 2-positive cancer. Clin Cancer Res 17(20):6437–6447

Lotsch J, Rohrbacher M, Schmidt H et al (2009) Can extremely low or high morphine formation from codeine be predicted prior to therapy initiation? Pain 144(1-2):119–124

Ma Q, Lu AY (2011) Pharmacogenetics, pharmacogenomics, and individualized medicine. Pharmacol Rev 63(2):437–459

Maeda K, Das D, Nakata H et al (2012) CCR5 inhibitors: emergence, success, and challenges. Expert Opin Emerg Drugs 17(2):135–145

Mahgoub A, Idle JR, Dring LG et al (1977) Polymorphic hydroxylation of Debrisoquine in man. Lancet 2(8038):584–586

Mallal S, Nolan D, Witt C et al (2002) Association between presence of HLA-B*5701, HLA-DR7, and HLA-DQ3 and hypersensitivity to HIV-1 reverse-transcriptase inhibitor abacavir. Lancet 359(9308):727–732

Manolio TA (2013) Bringing genome-wide association findings into clinical use. Nat Rev Genet 14(8):549–558

Mauro MJ, Druker BJ (2001a) STI571: targeting BCR-ABL as therapy for CML. Oncologist 6(3):233–238

Mauro MJ, Druker BJ (2001b) STI571: a gene product-targeted therapy for leukemia. Curr Oncol Rep 3(3):223–227

McKeage K, Perry CM (2002) Trastuzumab: a review of its use in the treatment of metastatic breast cancer overexpressing HER2. Drugs 62(1):209–243

Meyer zu Schwabedissen HE, Kim RB (2009) Hepatic OATP1B transporters and nuclear receptors PXR and CAR: interplay, regulation of drug disposition genes, and single nucleotide polymorphisms. Mol Pharm 6(6):1644–1661

Nowell SA, Ahn J, Rae JM et al (2005) Association of genetic variation in tamoxifen-metabolizing enzymes with overall survival and recurrence of disease in breast cancer patients. Breast Cancer Res Treat 91(3):249–258

O'Sullivan BP, Orenstein DM, Milla CE (2013) Pricing for orphan drugs: will the market bear what society cannot? Jama 310(13):1343–1344

Owen RP, Sangkuhl K, Klein TE et al (2009) Cytochrome P450 2D6. Pharmacogenet Genomics 19(7):559–562

Pacanowski MA, Leptak C, Zineh I (2014) Next-generation medicines: past regulatory experience and considerations for the future. Clin Pharmacol Ther 95(3):247–249

Pare G, Eikelboom JW, Sibbing D et al (2011) Testing should not be done in all patients treated with clopidogrel who are undergoing percutaneous coronary intervention. Circ Cardiovasc Interv 4(5):514-521. (discussion 521)

Perez EA, Cortes J, Gonzalez-Angulo AM et al (2014) HER2 testing: current status and future directions. Cancer Treat Rev 40(2):276–284

Phillips EJ, Mallal SA (2013) HLA-B*5701 and flucloxacillin associated drug-induced liver disease. Aids 27(3):491–492

Province MA, Goetz MP, Brauch H et al (2014) CYP2D6 genotype and adjuvant tamoxifen: meta-analysis of heterogeneous study populations. Clin Pharmacol Ther 95(2):216–227

Pulley JM, Denny JC, Peterson JF et al (2012) Operational implementation of prospective genotyping for personalized medicine: the design of the Vanderbilt PREDICT project. Clin Pharmacol Ther 92(1):87–95

Quintas-Cardama A, Cortes J (2009) Molecular biology of bcr-abl1-positive chronic myeloid leukemia. Blood 113(8):1619–1630

Rae JM, Drury S, Hayes DF et al (2012) CYP2D6 and UGT2B7 genotype and risk of recurrence in tamoxifen-treated breast cancer patients. J Natl Cancer Inst 104(6):452–460

Rauch A, Nolan D, Martin A et al (2006) Prospective genetic screening decreases the incidence of abacavir hypersensitivity reactions in the Western Australian HIV cohort study. Clin Infect Dis 43(1):99–102

Regan MM, Leyland-Jones B, Bouzyk M et al (2012) CYP2D6 genotype and tamoxifen response in postmenopausal women with endocrine-responsive breast cancer: the breast international group 1-98 trial. J Natl Cancer Inst 104(6):441–451

Reynolds K, Sarangi S, Bardia A et al (2014) Precision medicine and personalized breast cancer: combination pertuzumab therapy. Pharmgenomics Pers Med 7:95–105

Riordan JR (2008) CFTR function and prospects for therapy. Annu Rev Biochem 77:701–726

Roberts JD, Wells GA, Le May MR et al (2012) Point-of-care genetic testing for personalisation of antiplatelet treatment (RAPID GENE): a prospective, randomised, proof-of-concept trial. The Lancet 379(9827):1705–1711

Rudkin CT, Hungerford DA, Nowell PC (1964) DNA contents of chromosome Ph1 and chromosome 21 in human chronic granulocytic leukemia. Science 144(3623):1229–1231

Samson M, Libert F, Doranz BJ et al (1996) Resistance to HIV-1 infection in caucasian individuals bearing mutant alleles of the CCR-5 chemokine receptor gene. Nature 382(6593):722–725

Sass AE, Neufeld EJ (2002) Risk factors for thromboembolism in teens: when should I test? Curr
Opin Pediatr 14(4):370–378

Schroth W, Antoniadou L, Fritz P et al (2007) Breast cancer treatment outcome with adjuvant
tamoxifen relative to patient CYP2D6 and CYP2C19 genotypes. J Clin Oncol 25(33):5187–
5193

Schroth W, Goetz MP, Hamann U et al (2009) Association between CYP2D6 polymorphisms
and outcomes among women with early stage breast cancer treated with tamoxifen. Jama
302(13):1429–1436

Scott SA, Sangkuhl K, Gardner EE et al (2011) Clinical Pharmacogenetics Implementation
Consortium guidelines for cytochrome P450-2C19 (CYP2C19) genotype and clopidogrel therapy. Clin Pharmacol Ther 90(2):328–332

Shahin MH, Johnson JA (2013) Clopidogrel and warfarin pharmacogenetic tests: what is the
evidence for use in clinical practice? Curr Opin Cardiol 28(3):305–314

Shuldiner AR, Palmer K, Pakyz RE et al (2014) Implementation of pharmacogenetics: the
university of Maryland personalized anti-platelet pharmacogenetics program. Am J Med Genet
C Semin Med Genet 166C(1):76–84

Sibbing D, Bernlochner I, Kastrati A et al (2011) Current evidence for genetic testing in clopidogrel-
treated patients undergoing coronary stenting. Circ Cardiovasc Interv 4(5):505-513 (discussion
513)

Siller-Matula JM, Trenk D, Schror K et al (2013) Response variability to P2Y12 receptor inhibitors:
expectations and reality. JACC Cardiovasc Interv 6(11):1111–1128

Slamon DJ, Clark GM, Wong SG et al (1987) Human breast cancer: correlation of relapse and
survival with amplification of the HER-2/neu oncogene. Science 235(4785):177–182

Sosnay PR, Siklosi KR, Van Goor F et al (2013) Defining the disease liability of variants in the
cystic fibrosis transmembrane conductance regulator gene. Nat Genet 45(10):1160–1167

Stearns V, Johnson MD, Rae JM et al (2003) Active tamoxifen metabolite plasma concentrations
after coadministration of tamoxifen and the selective serotonin reuptake inhibitor paroxetine. J
Natl Cancer Inst 95(23):1758–1764

Stewart A (2013) SLCO1B1 Polymorphisms and Statin-Induced Myopathy. PLoS Curr 5.
doi:10.1371/currents.eogt.d21e7f0c58463571bb0d9d3a19b82203

Stimpfle F, Karathanos A, Droppa M et al (2014) Impact of point-of-care testing for CYP2C19 on
platelet inhibition in patients with acute coronary syndrome and early dual antiplatelet therapy
in the emergency setting. Thromb Res 134(1):105–110

Stingl Kirchheiner JC, Brockmöller J (2011) Why, when, and how should pharmacogenetics be
applied in clinical studies?: current and future approaches to study designs. Clin Pharmacol
Ther 89(2):198–209

Tafe LJ, Tsongalis GJ (2012) The human epidermal growth factor receptor 2 (HER2). Clin Chem
Lab Med 50(1):23–30

Thorn CF, Klein TE, Altman RB (2009) Codeine and morphine pathway. Pharmacogenet Genomics
19(7):556–558

Tirona RG, Leake BF, Merino G et al (2001) Polymorphisms in OATP-C: identification of multiple
allelic variants associated with altered transport activity among European- and African-
Americans. J Biol Chem 276(38):35669–35675

Trenk D, Hochholzer W (2014) Genetics of platelet inhibitor treatment. Br J Clin Pharmacol
77(4):642–653

Turner RM, Pirmohamed M (2014) Cardiovascular pharmacogenomics: expectations and practical
benefits. Clin Pharmacol Ther 95(3):281–293

Vultur A, Villanueva J, Herlyn M (2011) Targeting BRAF in advanced melanoma: a first step
toward manageable disease. Clin Cancer Res 17(7):1658–1663

Wegman P, Vainikka L, Stal O et al (2005) Genotype of metabolic enzymes and the benefit of
tamoxifen in postmenopausal breast cancer patients. Breast Cancer Res 7(3):R284–R290

Wegman P, Elingarami S, Carstensen J et al (2007) Genetic variants of CYP3A5, CYP2D6, SULT1A1, UGT2B15 and tamoxifen response in postmenopausal patients with breast cancer. Breast Cancer Res 9(1):R7

Weitzel KW, Elsey AR, Langaee TY et al (2014) Clinical pharmacogenetics implementation: approaches, successes, and challenges. Am J Med Genet C Semin Med Genet 166C(1):56–67

Whiting P, Al M, Burgers L et al (2014) Ivacaftor for the treatment of patients with cystic fibrosis and the G551D mutation: a systematic review and cost-effectiveness analysis. Health Technol Assess 18(18):1–106

Wilke RA, Ramsey LB, Johnson SG et al (2012) The clinical pharmacogenomics implementation consortium: CPIC guideline for SLCO1B1 and simvastatin-induced myopathy. Clin Pharmacol Ther 92(1):112–117

Wright RS, Anderson JL, Adams CD et al (2011) 2011 ACCF/AHA focused update of the guidelines for the management of patients with unstable angina/Non-ST-elevation myocardial infarction (updating the 2007 guideline): a report of the American College of Cardiology Foundation/ American Heart Association Task Force on practice guidelines. Circulation 123(18):2022–2060

Xie X, Ma YT, Yang YN et al (2013) Personalized antiplatelet therapy according to CYP2C19 genotype after percutaneous coronary intervention: a randomized control trial. Int J Cardiol 168(4):3736–3740

Yang W, Raufi A, Klempner SJ (2014) Targeted therapy for gastric cancer: molecular pathways and ongoing investigations. Biochim Biophys Acta. doi: 10.1016/j.bbcan.2014.05.003

Part IV
Concept-Based Ethical Questions

Chapter 8
On a Philosophy of Individualized Medicine: Conceptual and Ethical Questions

Konrad Ott and Tobias Fischer

Abstract This chapter deals with the conceptual layers of Individualized Medicine (IM). These layers are analyzed as follows: First, the method of critical reconstruction is outlined. Second, the concept of IM is analyzed according to its constitutive momenta (stratification, diagnosis, prediction, prevention, risk). Third, IM is related to four different medical approaches (lifeworld, conventional medicine, alternative ways of healing, genomics/genetics). Fourth, the discussion of its ethical implications focusses on three crucial topics: "cura sui" (Foucault), solidarity and informed consent. Finally we want the composed IM framing to be determined as a "Fröhliche Wissenschaft" (Nietzsche). The chapter adopts not a classical bioethical perspective, but tries to show how IM refers to philosophical and even anthropological questions. This rather uncommon perspective however might lead to contribute to an improved, i.e. more reflective and critical self-understanding of medicine.

Keywords Individualized Medicine · Conceptional analysis · Philosophy of medicine · Ethics · Anthropology · GANI_MED · Stratification · Prediction · Prevention

8.1 Introduction

The task of the philosophical sub-project of "GANI_MED" is to produce a discourse-rationale and a conceptual reconstruction of the approach of IM, i.e. its framing, concept and model, and its epistemic, ethical, and healthcare-political implications, as well as their potential consequences. We aim to find out what processes are occurring with IM and what kind of macro-innovation IM could be, within the

K. Ott (✉)
Philosophisches Seminar, Christian-Albrechts-Universität zu Kiel,
Leibnizstr. 6, 24118 Kiel, Germany
e-mail: ott@philsem.uni-kiel.de

T. Fischer
Department für Ethik, Theorie und Geschichte der Lebenswissenschaften,
Universitätsmedizin Greifswald, Walther-Rathenau-Str. 48, 17475 Greifswald, Germany
e-mail: tobias.fischer@uni-greifswald.de

© Springer International Publishing Switzerland 2015
T. Fischer et al. (eds.), *Individualized Medicine,* Advances in Predictive,
Preventive and Personalised Medicine 7, DOI 10.1007/978-3-319-11719-5_8

realm of the health care sector (Flessa and Marschall 2012). Such a reconstruction involves the analysis of a fast changing field. It is not clear what is involved in IM and what IM could have to offer, in spite of all of the efforts which have been made in defining IM (cf. Langanke et al. chap. 2) and concise short formulas such as "-omics knowledge plus prevention equals IM." We therefore try to offer a rating evaluation which takes the current academic discussion into account. Our sub-project relies on the assumption that IM is still in early development (a statu nascendi), but could be of momentous medical-philosophical, health-economical and "bio"-ethical relevance in the future.

It is still, within both the scientific and social debate, highly controversial, whether the concept of "Individualized Medicine" is just a hype or rather a long-term and profound shift within the field of medicine, that is to speak an alteration, which affects the fundaments of medicine (Bartens 2011a; Grill and Hackenbroch 2011; Juengst et al. 2012; Hempel 2009). Evaluations extend over a wide range, from "old wine in new skins" (Schleidgen and Marckmann 2013) to a "great transformation in medicine" (Snyderman and Langheier 2006). It is unclear, even within the disciplines, what role IM should play (Nature Biotechnology 2012). The concept of IM lies somewhere between the rejection of the "one size fits all" approach (Mancinelli et al. 2000) and the healthcare-political question of what kind of medicine a continuously aging society actually needs (Vollman 2013).

At first sight, the pattern of hype and disillusion seems to reoccur again and again. A lot of research and development has been undertaken under the label of "IM", which would probably have been undertaken in different disciplines if it wasn't for IM—simply under different labels. The term "IM" is also an attractive catch-phrase in successful research proposals. And the term "IM" is of interest for pharmaceutical companies regarding the development of new products and business models (Ott and Fischer 2012).

Furthermore, various stakeholders define IM in their own interest. Therefore a sober analysis of IM is necessary. It can be shown that almost all the separate conceptual aspects of IM are to be found in traditional concepts and scopes of medicine. IM is therefore not new. Gadebusch and Michl (2010) have shown several precursors of Individualized Medicine in medical history, which are to be located within the broader scientific context of the individual. But it is not a disadvantage to IM if, in the historical scientific perspective, it had had precursors, which simply were not able to establish themselves at their own time. Prior concepts of medicine had an awareness of the tension between nomothetic and idiographic disciplines (Windelband 1907), which could be characteristic of IM, and reflected this in a medical-philosophical way (most recently Buzzoni 2003): "It is right to assert that medicine—like all sciences of man—cannot do without the singularity of the patient in a sense utterly unknown to the natural sciences" (Buzzoni 2003, p. 6). However, it does not speak against an IM framing of medicine, if in addition to the positive and new knowledge of -omics in research; it can update other topics within medical philosophy, which reach back into ancient philosophy. IM could become the reason and motivation for health professionals to address medical philosophy beyond bioethics, especially because IM asks and deals with fundamental questions.

The real value of something is often underestimated once it has been falsely overestimated. This could be the case with IM, with its real value thus being trivialized. It could be that IM will not trigger new questions with regard to its content but IM could at the same time work as a new theoretical "framing". Such a "framing" neither simply exists nor can it be deducted from empirical research results. Rather, it has to be spelled out conceptually. For that matter, a scientific-sociological observation of the discourse is only of limited help. Certainly (with Foucault 1973; Knorr-Cetina 1984; etc.), statements on IM can be collected and assigned to networks of agents and groups of stakeholders. However, this approach cannot reach beyond a collection and evaluation of opinions, and statements on IM. We are not interested in such a phraseology.

Of course, such a "framing" is not a new paradigm. We do not fall into line with the inflationary manner of speaking of paradigms and their transformations, which escalated referring to Thomas Kuhn's *Structure of Scientific Revolution* (1962). As a matter of principle, it is questionable whether participants of current scientific controversies can anticipate whether their research will result in a paradigm shift. Kuhn argues that this kind of knowledge is reserved to the scientific historian only. It is therefore dishonest for scientists to attest themselves to a paradigm shift.

What we would like to adopt from Kuhn is the often overseen concept of "achievements". This concept claims that a paradigm shift does not only cause epistemic disruptions, but also leads to theoretical achievements which, in the later development of the science, cannot any longer be arbitrarily reversed. This contradicts the relativistic interpretation of Kuhn's concept of a paradigm shift and links Kuhn's theory with the theories of scientific progress. Not only paradigms, but nascent "framings" can entail achievements as well. What could therefore be a possible achievement of an "IM framing"? Maybe a new way of thinking about medicine as Gadebusch and Michl (2010) suggest, following on from Ludwik Fleck? Or rather a possible macro-innovation, which could even unfold its potential beyond the health care system as Flessa and Marschall (2012) suggest by arguing that through IM "Based on -omics sciences a multi-cause-multi-effect paradigm becomes action-guiding, i.e. the complete network of genetics, metabolism, habitat and behavior is considered in the overall picture as a stochastic risk system" (translation by the authors).

IM does not happen; rather, it is interpreted, regulated, and practiced by physicians and health care politicians. Thereby, in its abstract universality, all IM agents bear a fundamental formal responsibility. This responsibility is to be separated from the question of which personal and joint responsibility each person is assigned to regarding his or her own health. This higher level responsibility can be regarded as "scientific-ethical" responsibility, which means the medical responsibility of asserting a (joint) responsibility for a person's own health. Metaphorically, IM could be understood as a kind of melody in the field of medical dance. This melody is not played to us by others; rather we have to learn how to play it ourselves. In other words, a "framing" is not to be discovered, but rather "composed".

Such a composition is put together by a discourse-rational redemption of hypotheses, which count as contributions to the IM discourse which the "framing" is eventually formed of.

Ever since its beginnings philosophy stretches the distinction of using a word in common speech, of using the definition of a scientific term in individual sciences, and the concept formation, which counted as philosophy's domain for a very long time. Concepts are meanings, which cannot be defined arbitrarily or "be set up" with a strategic intention, but are to be rather formed and reflected within the "Arbeit am Begriff" (Hegel 1970).

Common speech, however, is the starting language and final meta-language of all concept formations. In many simple cases (such as "chair" or "window"), the use of a word as in common speech is sufficient as the basis for concept formation. Wittgenstein, who is known to be the primary representative of a conventionalist understanding of a concept that is aligned with common speech, however, never claimed that the meaning of all concepts is equal with their use in common and everyday speech. Mostly, the decisive paragraph of his *Philosophical Investigations* is reflected deceptively. He says: "For a large class of cases of the employment of the word 'meaning'—though not for all—this word can be explained in this way: the meaning of a word is its use in the language" (Wittgenstein 2009, § 43).

However, this differentiation causes problems for concepts such as "sustainability", "justice/fairness", and "Individualized Medicine", which tend to overstrain a conventionalism of concepts. How do we know, which word falls into Wittgenstein's big class and which does not? If, in conferences, clinical practice, and a working committee of experts (i.e. how the word is used in medical speech), we listen to the way IM is used as a concept, we might get closer to an understanding of what IM is and entails, but this approach can never replace the conceptual work that is needed. We think that words which, because of their action-guiding character and their mission statement function are always in danger of being strategically "engaged", can prima facie not be considered to belong to this big class. Hence, IM does not belong to his big class of cases, for which the use of a word is sufficient for the determination of its meaning. However, that is a feature IM shares with all philosophical and ethical concepts.

In many areas of the sciences, as well as in medicine, a "stipulatory" definition is common practice: Terms are comprehensible abbreviations, which are determined and whose correct use can be acquired e.g. in classes on medical terminology. For philosophy and the systematic social sciences, however, such a stipulatory approach to concept formation is not convincing, as it gives too much room to a strategic arbitrariness, as for example, definitions of IM, which correspond to the research priorities of an individual university hospital. This would be equivalent to the following "logic": IM is usually known as something positive. Here, we do x. We define IM in a way that x is among it. Hence it follows: We do something positive. Philosophy must never associate with such "logic". The strategic definition of terms and concepts for personal aims and interests (linguistic politics) is irreconcilable with the practice of philosophy.

The meaning of concept is hence linked to the "reasons", we can adduce for a specific definition and/or a specific use of a word within a science, in politics, or in everyday language (Ros 1990). Such reasons are to be given to others. The label-like headline "IM" becomes a concept of Individualized Medicine because we assure each other of the good reasons that are indicative of using IM with certain meanings in our language. A potential IM framing can only be composed of these meanings.

8.1.1 Proposition

We claim that the "framing" called "IM" could update the idea of health for our time. If medicine is understood as a discipline, which is bound to the human practice of healing and hence, as Max Weber (1968, p. 38) puts it, is "an Werte gekettet" ("chained to values", translated by the authors), it stands in the tradition, which goes beyond the "classic" traditions like the Chinese, Indian, Egyptian, and Greek medical concepts and even beyond shamanism and other archaic attempts of healing. Nosologic taxonomies as well as their underlying concepts of "corporeality" transform over time and it is far from certain which taxonomy and which according concept of "corporeality" matches IM framing best. Supposedly, it could be a concept doubled in itself, which grasps an objectifiable body and a feeling corporeality as complementary perspectives. According to this concept, IM does not blank out the perspective of the person, who receives medical help (and who, traditionally, is called "patient"). As no one only is "healthy" or "ill", but one feels "healthy" or "sick" too.

Concepts of health are expressed in mytho-poetic images like the "Fountain of Youth". The architectural sketches of a "constructed art of healing", which are part of the history of hospitals (Buß 2012; Wagenaar and Mens 2010), presuppose ideas of how human beings can find their way from sickness and crisis to recovery and a renewed, maybe even "enhanced" health, assisted by a healing expert. In this respect, the history of the concept of health, which is often concealed under nosologies, is part of a general medical history.

Insofar as the concept of health is pragmatically implied in the practice of medicine as an art of healing ever since its beginnings, IM is nothing new in this fundamental regard. As much as IM originates in and corresponds to the "-omics" research strategies of contemporary bioethics, it shares a conceptual research design, which is *sub specie* to be reconstructed within the concept of health.

Ideas are neither subjective individual ideas, nor are they time-transcending entities. But they, not unlike the concepts of freedom, justice and sustainability, are rather related to human affairs and their organization and regulation. Ideas are never ethically indifferent, but always relate to something good or to a distinction between good and bad that lies within this idea of health vs. disease, freedom vs. slavery, sustainability vs. overexploitation. Ideas are to be understood in such categorical oppositions. Therefore, IM has an intrinsic relation to the scientific doctrine of diseases (nosology) as well as to medical-philosophical deliberations on health.

In this inescapable opposition of sickness and health, IM could relocate medicine on the side of health. In this regard, IM would be possible (and desirable) as a new "framing" aiming for a possible science of health and its related phenomena (healing, recovery, cicatrization, and strengthening). Speaking of IM as "health care" insinuates this as a general orientation.

By the emphasis on health, IM moves within close vicinity of medical traditions like salutogenesis (Antonovsky 1987) and philosophical approaches like Gadamer's (Gadamer 2010), which untie medicine from its fixation on diseases and which postulate the complex phenomenon of health as the central topic of medicine. We will reflect on this later in the chapter. According to our thesis, an IM framing does not necessarily correspond to the tendency to generate new diseases and all kinds of disorders. The epistemically as well as ethically relevant debate on the augmentation of diseases and disorders as well as the tightening scopes of the healthy are relevant for such an IM framing. However, it is not a foregone conclusion whether an IM framing will be forever spinning the net of diseases and disorders into a tight web. Maybe, this is just a problem while IM is relatively young, which will disappear once IM has matured or ripened.

In a philosophy that refers to Hegel's dialectic approach, ideas are to be thought of as real ideas, i.e. to be thought of as to be engaged in affectively. The idea in its actual immediacy is the healthy body itself. In its immediacy, the idea is beautiful and good, as in Greek: *kalagathos*. Diseases and nosologies referring to those diseases represent the idea in its stage of negativity. They determine the ways in which the idea can be wrecked by the rigor of nature and society (e.g. during a war). Concepts or framings such as IM are theoretical momenta, in which the idea is realized in human practice, which can become social realities (e.g. in every day clinical practice and in health politics). By consequently understanding IM as the reality of health, we can identify the following central themes in their philosophical and ethical perspective: the concept has inner tensions, ambiguities, and contradictions. In the dialectic dynamics, contradictions are not seen as fallacies, rather the matter itself is contradictory in the sense of intrinsic dynamics, which is to be followed compositionally.

Thereby, the method of this chapter is outlined as the reconstruction of the proposition, the concept of IM in its conceptual momenta (stratification, diagnosis, prediction, prevention, risk), the mediating attribution to medical approaches (lifeworld, conventional medicine, alternative ways of healing, genomics/genetics), the discussion of its ethical implications as the conjunction between maturity and "*cura sui*" (Foucault) and finally, the idea of health, which has been analyzed with the conceptual moments and hence gone through a transformation. In the end, we want the composed IM framing to be determined as a "Fröhliche Wissenschaft" ("gay science") of *great health* (Nietzsche 1974, 1999a, Aph. 382).

8.1.2 Idea and Utopia

In Ernst Bloch's sketch of medical ideals, which belong to the utopian inventory of human kind (as does the above mentioned Fountain of Youth), he mentions the utopia of Sweven's "*Limanora, The Island of Progress*" (Bloch 1985, p. 530). On this island, the inhabitants have overcome the idea of medicine having primarily a healing focus. The medical care of *Limanora* has developed beyond "the crude state of mere cure of disease" (Sweven 1903, quoted in Bloch 1985). Medicine as practiced on *Limanora*, contains active enhancement, eugenics, and above all preventive measures.

Concepts of IM seem to resemble this utopia. IM is hence the conceptual realization of an old medical utopia, which preventively avoids diseases before they outbreak and need to be fought. Diseases are supposed to be deterred to outbreak. That is what the term "prevention" refers to, which is part of the conceptual inventory of IM. When diseases are averted, the state of health is maintained. No person ever becomes a patient. Of course, the concept of prevention is not a new one either. Efforts of preventing diseases (including prevention of accidents) and health prevention measures are numerous and well established. Nobody needs to be taught via IM that brushing your teeth prevents cavity and that regular exercise prevents cardio-vascular diseases. Nobody is seriously denying the effect of prevention. Nevertheless, as Gadamer claims, medicine that pursues a curative approach (especially in its pharmaceutical and surgical efficiency) might also have weakened the sensibility for the importance of prevention:

> Wir müssen uns leider eingestehen, daß dem Fortschritt der Wissenschaft der Rückschritt der allgemeinen Gesundheitspflege und der Prävention auf dem Fuß gefolgt ist. "We unfortunately have to admit that the progress of science is followed by the regression of general health care and prevention" (Gadamer 2010, p. 136, translated by authors).

Although the necessity of preventive measures was and is officially emphasized, an unofficial thought pattern among the population has become established, in which curative measures are substitutes for preventive measures. Within the paradigm of a powerful conventional medicine, prevention seems to be redundant; IM could (and should) be able to correct this false impression of medical omnipotence.

Certainly, no one takes the view that a preventive IM could replace the traditional curative medicine. In case of an outbreak of diseases despite preventive measures, they should still be treated with specific medications that work with the maximum effectiveness and palatability. With an eye to curative medicine, IM is guided by the aim to develop as many effective and well-tolerated medications as possible, i.e. medications that individual patients react upon well with as few side-effects as possible. These, however, are ordinary aims of pharmaceutical research and are not specific for IM. Such a pharmaceutical research might have positive therapeutic effects as well as ambivalent economic implications for insurance systems, but they are conceptually less relevant than the concept of a preventive IM. In this regard, research is part of the improvement of medication's effectiveness and palatability, but, no matter how worthwhile, it is not part of the conceptual core of IM.

However, utopias like the one of prevention, mostly are dialectic; so in every utopia, there is a dystopian moment. During the twentieth century, dystopias have outstripped utopias. In "*1984*", the all-inclusive, preventive and curative society turns into a prison under full surveillance for every one; in "*Brave New World*", individual freedom must subordinate to societal stability; and independent thinking is eliminated in "*Fahrenheit 451*". A society that tries to prevent criminal acts by all means could not be a liberal society. This dialectic also appears in the concept of IM (which critics of IM have no doubt about): The nightmares of permanent surveillance and control, the "soft" pressure to "compliance" for all kinds of tests and check-ups, the subtle erosion of the principle of "informed consent", "Bio-power/-politics", "Bio-capitalism", weakened solidarity etc. are provoked by several authors (Lemke 2008; Rajan 2006). An IM framing would delegate more decisive power and responsibility to the individual, but would also reflexively combine this with a claim for more self-control (Ott and Fischer 2012; Langanke et al. 2013a, b).

That is why a critical discourse on IM was quickly established, which considers the issues of anxiety, threat, and at least the risks that could be linked to the IM concept (Ott and Fischer 2012). However, these morally motivated anxieties are not supposed to be misunderstood as immediate truths about IM. Moral concerns and anxieties towards IM have to acknowledge their own potential insularity. If IM carries components which would be a moral evil today, it does not mean future patient collectives will still consider this aspect as a moral evil.

The apparently dramatic dialectic of a utopia of prevention and the dystopia of control by "biopower", could in fact relax within the reality of IM, if the level of clinical routine is achieved: In the end, things are never as bad as they seem and IM could turn out to be much more "normal" that it was hoped or suspected to be. In composing an IM framing, we could avoid the mistake to hypostatize one single conceptual momentum like the one of prevention. Sober-minded and pragmatic positions, which might be less spectacular, but possibly be medical-philosophically and ethically more sustainable, could replace dramatic oppositions and scenarios, with which some people try to give themselves importance in the media. Our proposition includes a rather normalized viewpoint of IM: The originality of IM does not have to be something spectacular; maybe the truths of IM are not more spectacular than "an apple a day keeps the doctor away".

As mentioned above, conceptual and theoretical transformations will not simply be derived from new data. Clinical studies and their results, as well as the formation of hypotheses, theoretical concepts, problems in the clinical practice, and philosophical deliberation are in a dynamical reciprocal relation to each other. Hence, we turn against a positively abridged comprehension of science, which considers itself superior in the view of all the biomedical research including its richness of data and an ethical reflection. Rather, we see a different relation: The more medical research proceeds towards an IM, the more urgent becomes the recollection of medicine to its philosophical and ethical fundaments. This conclusion was conceded in many debates among the colleagues within the GANI_MED project. Thus, the realization of IM as a science of health would blow in a light breeze of philosophy into the more prosaic everyday realities of clinics and public health departments.

8.2 A Presuppositional Analysis of IM

8.2.1 A Guiding Principle of IM (in Greifswald)

The guiding principle of the University Medicine Greifswald is "Individualized Medicine". This sounds promising and is even supposed to. But, what does a "guiding principle" actually mean here? What does "guiding" mean? And what does "principle" mean? "Guiding" is to be understood as "guiding action and giving orientation" (cf. Stegmaier 2008). Guiding principles have manifoldly been studied in moral sociology. Even in technical sociology guiding principles as the "paperless office", the "car-adapted city", and the "artificial intelligence" etc. were studied. However, we also know about historically powerful but "evil" guiding principles like the one called "*Lebensraums im Osten*" or the one of the "purity of arian blood". This implies that guiding principles can shape and modify human performances and attitudes and ultimately realities without vouching for an unmistakable moral certitude. Rather, they have to be embedded in broader conceptions of morality and decency. Along these lines, it can be said that as a new guiding principle, IM needs to be fitted into such comprehensive medical-ethical conceptions. This is the ethical presupposition of the concept of a guiding principle.

Guiding principles like IM pragmatically imply expectations. Expectations of IM could be articulated from different roles and interests, whereas initially, we accept all interests as such. Interests are abundant: Patients are interested in individual care, in understanding and considering their individual situation, and they are interested in better medical care without their insurance payments rising. Researchers are interested in new insights in etiology, the association of nosological entities, and statistic acquisition of courses of diseases within larger populations up to their terminal points, but are also interested in their own careers as scientists and their own academic reputation. Physicians are interested in predicative diagnostics and specified therapeutic agents and supposedly above all in a higher compliance of patients. Insurance companies are interested in more precise risk-collectives ("risk-orientated scales"). Biomedical-pharmaceutical industries are interested in new attractive products and business models regarding successful generic drugs. Health care politicians are interested in reducing health care-political conflicts regarding their various economic interests. Because of all these interests and their heterogeneous expectations, IM is health care-politically over-determinate, and hence, all the hype about it is not a coincidence. Possibly, currently existing expectations and those, which can be justified and answered will differ widely. A reconstruction of an IM concept must therefore not simply confirm existing expectations, but will have to rectify some of them. However, the revision of some false and exaggerated expectations does not have to end in the mere negation of expectations, but can also "form" expectations. In this sense, forming a concept will eventually also help to form an IM-specific "health concept".

We doubt that IM will profoundly change the way we think about the classic questions asked in medical ethics, mostly concerning the beginning and the ending periods of human life. For a long time, medical ethics was orientated to rules and morals rather than to values and culture (e.g. Düwell 2008). It was all about rights, duties, contingent and absolute permissions, protection of privacy, "informed consent" etc. The literature on that is abundant. However, questions on eugenic indication for abortion, debates on the concept of contradiction in case of organ donation or assisted suicide are not part of the ethics of IM. For an IM framing, the principles developed by Beauchamp and Childress (2008), for example, do not need a fundamental revision. Mutatis mutandis, the same applies to questions about the fair allocation of scarce medical treatments. IM does not tell anything about how much a treatment is allowed to cost, which can for example delay the growth of lung tumors for several months without completely defeating it. Even the idea of decision-making autonomy of patients ("informed consent") is not questioned fundamentally. Nevertheless, IM could have the power to relocate medical ethics more strongly back into the realm of eudemonistic ethics; hence give rise to old questions of a good and succeeding life, how to handle the individual corporeality and its mortality. Those, who have primarily dealt with normative questions of medical ethics (Ott 2008), also had this focus in mind for IM and hence, overlooked these eudemonistic points of IM. Questions about the good life are related to the concept of health. We do not want to deny that people with chronic diseases and handicaps can lead a subjectively satisfying and meaningful life, which is in that regard a "good" life. Health however might be defined as neither a necessary nor a sufficient condition of a good life. However, following our intuitions, health is more than just a contingent accessory of any concept of a good life, namely some kind of "equipment". Diseases and disabilities are constraints of an individual's scope of action and mostly are experienced as such. If health manifests itself phenomenally as a lust for life and a thirst for action, it is part of the fundamental experiences of a succeeding life. In that matter, the hypotheses that the concept called IM could update the contemporary idea of health in a genuinely dialectic way, entails a primarily eudemonistic significance (cf. sect. 8.4).

8.2.2 Methodology

Now we have to explain the methods we use to argue for our proposition. Essentially, our methodology is part of a reconstructive theory of science, as it is practiced by Hucklenbroich in the realm of medical theory (Hucklenbroich 2010, 2013). The ideas, terminology, concepts, models etc. that are used in the IM discourse are analyzed in two directions. First, regarding their presuppositions; secondly regarding their likely consequences. The analysis of presuppositions relies on the scheme: *X presupposes a, b, c ...n*. We call this kind of analysis "p-analysis". The analysis of consequences relies on the scheme: X will (probably) have the following consequences: $c_1, c_2, \ldots c_n$. We call this kind of analysis "c-analysis". The analysis of presuppositions is methodologically stricter, because it does not rely on probabilities, but on conceptual relations. Saying a ball is red implies that the ball has a color. The "c-analysis" by contrast, cannot do without probabilistic considerations, which require life experience, insight into human nature, and strength of judgment. Moreover, c-analyses constantly touch the fine lines between prediction and what is desirable. Hence, it is necessary to distinguish carefully between developments, which are considered to be likely/unlikely and developments, which are considered to be desired/undesired. The ethical concept of c-analyses allows it to point out to IM's scopes, which are very much appreciated from today's medical ethical perspective. Of course, everything can turn out differently. Unacceptable, however, are cloudy confusions of predictions and hopes.

Certainly, a prognostic c-analysis always makes general assumptions on how modern human beings predominantly behave when faced with innovations and impositions of IM. We can assume that future patients will take for granted what seems like huge innovation to us today: There is a good case to assume that human beings get used to innovations and come to value the positive aspects of ambivalent developments more strongly. It is insinuated that rebels against biopower are and will stay a small minority. Supposedly, the "right to know", for example, which in an IM framing might remain in abstract form, but which will lose some of its normative significance, if for the affected persons, the advantages of an improved knowledge about their individual predisposition for diseases will predominate. These suppositions are realistic assumptions about how human beings react upon what is presented to them as medical progress. These assumptions can be confirmed or refuted by quantitative and qualitative social research (cf. Erdmann chap. 12).

As far as the desirable consequences of IM are concerned, the idea of a responsible patient, as predominant in the post-paternalistic paradigm of medical ethics, cannot be denied through IM anymore. The shape of freedom endures in an IM framing. The reasons speaking against paternalism do not become obsolete by IM. IM hence does not lead back to an old paternalism. Rather, in an IM framing, the ideas of responsibility and health could fuse to new constellations and could suggest an interpretation of patient-doctor-relation that is orientated towards the concept of an unconstrained consultation (also shared decision making). The stochastic-probabilistic dimension of IM is in contrast to clear-cut therapy recommendations. It is also thinkable that in the wake of IM, confident patient collectives arise, whose members consider themselves as experts of their own diseases and dispositions and who critically engage in debates with IM experts. Such patient collectives could have the power to transform the principle of the "informed consent", which in IM is mostly geared to the individual patient, to new forms of health-related collective action. We would absolutely welcome such consequences.

8.2.3   Reconstruction

After these preliminary considerations, starting from the standard setting TAB-study (Hüsing et al. 2008), we want to attempt a first approach to a reconstructive concept-forming of IM. For the authors of the TAB-study, "individualization means" a

subdivision (which goes beyond the status quo) of patient populations into clinically relevant subgroups. The process is known as stratification. There is the constant assumption
that the more target-orientated diagnosis, risk specification and interventions there are, the
more criteria there is, or the more specific the criteria is, which can be used for grouping.
For this subdivision, new and specific biomarkers are used in Individualized Medicine,
which emerge from the genomic and post-genomic research (Hüsing et al. 2008, p. 9, translated by the authors).

The TAB-study distinguishes between a number so-called "concepts of individualization", which can be understood as strategies of theming. They are concepts,
which reappear in the context of IM. Thus, from the presupposing-analytic perspective it is asked, which terminological investments have been adopted into the
theming of IM. The more precise terms are: (1) stratification (classification), (2)
biomarker, (3) risk specifications, (4) (predictive) diagnostics, (5) findings, (6)
intervention, (7) responsibility, autonomy and maturity.

The idea of health is always included in these terms, which specify the IM framing. These terms now need to be specified.

(1) Stratification
IM groups single patients into groups with mutual features. Stratification turns the
individual into a "token of a type". It happens based on typing schemata. Group-
forming stratification does not happen arbitrarily—"everyone who wears a tie when
coming to a doctor's consultation"—but rather with reference to certain diseases
and their causes. A nosology is still presumed. IM hence, aims for stratifications,
which are geared towards specific goals. The group-formation would be "individualizing" in a strong sense if in the end "ideally" every group would consist of one
individual. In reality, this is impossible, "so that a more accurate choice of term
would be 'stratified medicine'" (Hüsing et al. 2008, p. 10, translated by the authors).
This definition seems accurate to us and is well established by now. Thus, the term
"Individualized Medicine" is objectively to be understood as a stratified medicine,
which classifies individuals into certain groups (types) (cf. Kohane 2009). Thanks
to the term "stratification", there is no necessity to reconstruct the philosophical
concept of individuality any closer. There are rather old medical concepts on stratificatorial typification, which are interesting as an antecedent of IM (cf. Langanke
et al. chap. 2).

These concepts proceeded from affinities between the constitution and the con-
figuration of health and sickness. According to that, IM would be "a stratificatorial
science of health". What is it that makes and keeps an individual, who belongs to
a certain type, healthy? In this sense, in the p-analysis, the idea emerges that there
might be more than just one health as Nietzsche argued (cf. Danzer et al. 2002).
Accordingly, it would be necessary to develop a type-related hygienic in addition
to the existing nosology. While conventional medicine determines health by the
general assigning of certain threshold values, those threshold values would be seen
critically in IM (cf. Gadamer 2010, p. 138). This would also apply to nosological
concepts like the "metabolic syndrome", which, first of all, is constituted by such
critical values.

Stratified groups do not necessarily have to be patients in the traditional sense. Particular groups consist of people, who "have" a certain disposition for the outbreak of a disease. Often, these people are "healthy sick patients" with pre-clinical results. These people feel healthy and are objectively free of complaints, but are not totally fit, but rather threatened by certain diseases, which are "on their way". The results show a future disease or disorder. Health is not a state, which is threatened by exogenous diseases (like infections), but in which many endogenous, among them a lot of genetically co-conditioned, tendencies work against health, and whose danger is underrated by the subjective appearance of health. By that, the experienced health gains a moment of false dawn.

IM, in its consequence, can hence lead to a loss of security and to everyone asking if and how healthy they "still" are. From the subjective perspective, this can lead to the loss of trust in the actual stability of health, which could have further negative consequences. Could it be an unpleasant consequence of IM that it undermines the basic trust, which is so necessary for health? Having basic trust in one's own body is not a mental state, which relates to the declaration of good health, but is rather a momentum of healthy life itself. It is not about watching one's "body" meticulously with one's own mind and then generalizing the basic trust from the positive results, but rather the basic trust is located precedent to the distinction between "mind" and "body". If this basic trust is a momentum of healthy life itself and if, under different circumstances, it were possible for IM to undermine this basic trust, IM could in certain cases actually be harmful. Hence, IM would need to be practiced in ways, which reinforce the basic trust in one's body, which carries us through life.

Insofar, the c-analysis can make plausible, why IM could support the spread of hypochondria. The disposition to hypochondria, which itself is considered to be a pathological disorder, could become acute in many people. Medical-ethical considerations of IM would have to take these consequences into account. It would furthermore be unclear how to evaluate the pervasive techniques of technological self-monitoring of health-parameters (cf. "www.quantifiedself.com") namely either as an expression of rational self-management or as an expression of a disorder induced or reinforced by IM. What might look like the ideal compliance from a physician's perspective might appear as the high-tech version of hypochondria from a more critical-reflective point of view.

With IM, the concept of the "patient" changes as well. In the common sense, patients are persons, who suffer from manifest diseases, who receive a treatment and are certified sick, hence are not allowed to work anymore. Law certainly needs a binary code of "sick" and "healthy", in order to regulate legal consequences of sickness. On the legislative level it is sufficient for a patient to be certified unfit for work by a physician. But we leave this juridical perspective out of consideration here. Populations of IM, however, do not only consist of patients in the "classic" or juridical sense of the word, but are composed of people without any acute clinical symptoms. How, then, does the concept of IM change the concept of the patient? In simple terms, stratification turns an individual into a "candidate" for a future disease: "X is a typical c-candidate" would be a physician's statement, which blends in well with IM. This status of being a candidate gets expanded by "studies on correlations". Many of these studies are in fact subsumed under IM. The idea that diseases do not occur in an isolated way, but actually interrelate is nothing new and can be found in many medical theories. This thought however, can now be stated more precisely by clinical studies too. The actual aim of such research on correlations is to be seen as a causal understanding of "science" and to recognize the deeper etiology of correlations. Complex relative risks arise from a patient's perspective.

The conceptual difference between candidate and patient can be captured in a distinction between patients in the wide sense, i.e. candidates, and patients in the strict sense. By that, the term patient becomes ambiguous. We terminologically suggest to introduce "recruit" as the generic term and to distinguish between sick "patients" and "candidates". Everyone, who comes into contact with IM, either because he/she participates in epidemiological studies or studies on correlations, or by clinical routine results etc. are in a certain sense recruited. After the abolition of the compulsory military service, it is terminologically acceptable to introduce the term "recruit" into the civil-medical context. Practically IM aims for candidates not to become real patients (or at least as late as possible). That way, the idea of prevention is settled, which might be the guiding principle of IM (see below). Part of the dimension of IM is: Recruits are hired and stratified into groups of candidates in order to reduce the number of patients and their treatments.

The stratification of recruits can only take place under (epidemiological) research or medical (physician supervised) perspectives. This practically makes a big difference, which also entails fundamental ethical problems for research subjects. Epidemiologically seen, the typification of a candidate was true and prognostically correct, if disease C outbreaks in X at the time of t, and X then becomes a patient. From the physician's perspective, this outbreak is to be avoided. From the medical perspective the prediction is of practical use, while it makes epistemic sense from an epidemiological perspective. Both can run afoul of each other. That is how the problem results (among them incidental findings) emerges, which occur in the context of epidemiological study, but which would diminish the study's validity if they cause a research subject to be redefined as a patient, who then needs to be taken out of the study. Is the idea of prevention taken seriously in IM, many potential patients would have to be taken out of clinical studies due to the idea of prevention and rather be assigned to a curative therapy, which helps to keep them healthy. Basically, IM would be conceptually irreconcilable with the aims of epidemiology. The interest of IM might thus have been generated in epidemiological contexts ("context of discovery"), but the epistemic aims of epidemiology would be problematic in the medical-ethical "context of justification" of IM, because individual health risks mostly have to be ignored in epidemiological research (cf. Erdmann chap. 12).

The participants of today's epidemiological studies sacrifice themselves for the sake of a future IM, while they might believe they are extensively medically cared for. This is a medical-ethical problem which is however not specific to IM.

(2) Biomarkers
Biomarkers are "objective measurement parameters" which are used to evaluate regular and pathological processes as well as reactions upon therapeutic and preventive

interventions. Biomarkers are supposed to give information, which are useful for the process of stratification. The concept of biomarkers therefore seems clear at first sight. However, the search for biomarkers is more difficult than initially thought, if it is not about etiological questions, but—according to the idea of IM—about "valid" biomarkers, which can be used clinically, either for prevention or prediction (Poste 2011). There are many things which can be understood or defined as a biomarker. Everything can be a biomarker from a gene sequence to the age at which a woman falls pregnant. The biomarker "male", for example, represents a life expectancy reduced by several years. The term "biomarker" therefore appears to have a very wide use. That however, would imply a very low significance of the term. The wide use of the term constitutes a new area of research and it is very likely that a lot of publications on biomarkers (and SNP's will flood the academic market in the coming years). Research on biomarkers might lead to a more precise determination of relative risks and to a better etiological understanding of diseases and their correlations. But what are physicians expected to do with all these biomarker studies? The correlation between biomarkers and such studies for the sake of stratifications probably leads to more individuals being counted into several groups and types. But into how many groups can an individual as a potential candidate possibly fall?

Maybe within the IM framing we will see, after a critical evaluation of many studies on biomarkers and after a reflection of criteria of validity, reason that the essential, if still rough, biomarkers have long been found and known in everyday life: age, overweight, malnutrition, consummation of drugs, lack of exercise, medical family history etc. Ironically speaking, in relation to these "rough" biomarkers, the "subtle" biomarkers, which most research is focused on, are practically less significant. The IM framing could hence lead to a dialectics of research and healing, which can hardly be reconstructed by IM researchers: IM searches for subtle biomarkers, while as an art of healing, it can and should confirm the significance of the rough biomarkers. These discussions could lead to a future IM, which grasps itself as the "Fröhliche Wissenschaft", and which considers the research on biomarkers as a huge data graveyard.

(3) Risks

IM identifies the individual risk profiles of recruits. A risk is related to damage; it can be understood as a combination of probability of occurrence (PO) and the extent of damage (ED). PO and ED can be influenced by IM (can be reduced, delayed, slowed down, alleviated etc.); of course, ED can always only be evaluated from the candidates' point of view. Health related risks are among general risks in life. If the term risk is seen in this sense, then it is like the compensation claims side of a venture. In ventures we take certain risks to achieve something. If a lot of risks occur on life's path, it can be asked what the venture actually is. The venture is lived life itself. All health related risks however, which might be determined and evaluated take place in the realm of the venture called life. The IM framing will however have to deal with specific and relative risks, but it must not wipe away the realm of lived life. If it did, IM would no longer see the wood for the trees.

These objectives are very often found in IM contexts:

Through genotyping and multi-parameter diagnostics individual risk profiles should soon be made with clinical recognizable disease symptoms, as well as statements of probability about the future health development of an individual (Hüsing et al. 2008, p. 11, translated by the authors).

A precondition for a treatment as early as possible is the profound knowledge of all factors relevant to health. The probabilistic statements refer to a "health-wise" development. The adjective "health-wise" implies certain ideas of health. It becomes obvious again how conceptually underdetermined an IM framing would be without an explicit concept of health. In order to identify these relevant factors, manifold "biomarker-based" data like environmental factors (e.g. nutrition, contact with environmental pollutants, pathogenic agents), lifestyle and socioeconomic status, genetic conditions, physical and mental constitution as well as prior medical treatments is necessary. We see that by expressions like "relevant to health", an epistemic impulse is generated, to generate information on the individual and his/ her specific risks. This impulse does not need to elaborate the invested concepts of health and relevance in order to use techniques of generating knowledge. The positivity of many findings and risk profiles, however, will always be linked to the idea of health.

To build an individual risk profile, an extensive and "thorough" knowledge of the person is presupposed, which ranges from the genome to her lifestyle and family history. This knowledge needs to be generated and saved, which evokes the medical-ethical questions of the pledge of secrecy and juridical questions of the protection of personal rights and data (cf. Bahls et al. chap. 11).

For the evaluation of risks, the distinction between "absolute" and "relative" risk is essential in the IM framing. The increase of a relative risk is miscalculated by most patients (cf. Erdmann chap. 12). A risk increase of 25 % does not equal a risk of 25 %, but refers to the population. If the absolute risk is 10 %, an increase of the relative risk "of up to 25 %" means the individual risk is about 12.5 %. Many of the relative risks can be ignored. Part of the ethics of IM is a thorough obligation to inform candidates and patients about relative risks. In consultations between candidates, patients and physicians, the sharing of knowledge about the relative importance of many risks is to be aimed for (cf. Erdmann chap. 12).

In the case that a relative risk of mortality is reduced, within a mortality rate of 100 %, another risk necessarily increases. At the same time, the medical effort to reduce the risk of death caused by certain diseases, can lead to many people living longer and healthier. Insofar, the efforts of lowering the deaths caused by x, leads to an increase in the percentage share of those, who die of y as well as a prolongation of life in good or acceptable quality. IM ultimately confirms that everybody has to die from something, as well as the fact that there is a life before death in return.

Very often, IM research is done in studies of correlations. In the correlations of diseases, probabilities are linked in a risk perspective. If a person P's risk to suffer from a disease D-1 at the time of t is x % higher than for the whole population and if D is correlated to disease D-2 with y %, the increased relative risk of P related to D-2 can be calculated. The recruit hence becomes a multiple candidate. But how are persons supposed to deal with such statistical information of multiple relative risks?

The mental effort to constantly have an eye on one's own statistical risk profile seems to be high (or too high) in everyday life. One could say, it would be an imposition and a structural overload, to have to integrate one's own risk profile into one's lifestyle. This can only be achieved selectively. For that, one has to compare relative risks with each other and be able to evaluate its ability to influence one's health. These strategies can be quite conventional: enough sleep, exercise, moderate consumption of alcohol and tobacco etc.—hence things we all know and nobody would deny. Beyond that, certain vulnerabilities could be emphasized and for people to be accordingly warned of them.

With the help of IM, individuals can know better, what they primarily should be on their guards against. IM could provide a kind of identification of somatic vulnerabilities (weak spots). Nobody can permanently have an eye on all their risks, but should pay attention to one's weak spots, of which everyone has.

In this way however, IM does not move away far from other "alternative" approaches of healing.

Conceptually considered, IM could nevertheless also give the all clear for many risks, i.e. it could inform people about their relative risk to suffer from certain diseases being lower than for the average population. On the other hand, pragmatically viewed, these all-clears could also encourage carelessness and recklessness. Physicians with their traditional habitus do not tend to give all-clears and refuse to lull patients into a false sense of security. They are reluctant to tell a patient that he/she is not the classical type for diabetes or that smoking in moderation (well, okay, three cigarettes a day aren't all too bad) is okay. Statements of this kind are considered to be medically negligent or even irresponsible. The physician's ethos contains duties to warn. But does this kind of ethically justified habitus not require a modification within and because of IM? Should IM only warn about higher, but not inform about lower health risks? This is a question for medical ethics. Thus when from a physician's perspective is it recommended to say: "Regarding x, you do not need to worry."

In any case, there is a wide field of the ethics of IM concerning risk opening up here: Is there an actual threat of a widespread disease mongering a whole population? When are typifications and diagnosis supposed to take place? Starting with a person's birth? In Germany, newborns are examined following a strict regimen over their first years of life (U1, U2 etc.). IM could tie in with that practice. New IM testing could be obligatory or optional. Should stem cells be gathered as a precaution? Could obligatory mass- or population wide-screenings be considered? Who is allowed to offer the testing (maybe online?) and what kind of professional qualifications would somebody have to have in order to be allowed to offer such services? How continuous should such a monitoring become? Which tests are considered to be useful? How valid are those tests regarding the "false positives" and "false negatives"? Does someone count as irresponsible, if he/she flinches from certain practices of diagnosis? What if one does not have a preference for this kind of "monitoring-mill"? How could the so much hoped strengthening of patients' autonomy in this area be realized? Should there be a prior consultation about what one is getting into with IM? Does IM change the attitude towards the well-known phenomenon of non-compliance? Does IM shift the way, in which physicians and health insurance policies tend to react upon non-compliance?

We do not claim to be able to give answers to all these questions. But, in sect. 8.4, we would like to present a normative argument of why relative risks, which can arise from all these possible practices of diagnosis, do not give good reason to give up on a principle of an insurance system, which is orientated on mutual solidarity for the sake of a system of self-responsibility. The more we calculate and standardize relative risks, the less sense it makes to enumerate relevant factors or hold against relevant factors to each other.

(4) Predicative Diagnosis and Prevention
Usually, a diagnosis is defined as the detection of something going on. In the classic case, the physician diagnoses a disease in a patient. Within the context of IM, the concept of diagnosis is relocated within the realm of predication and probability, so that it is correctly called "predicative-probabilistic diagnosis". Diagnosed is hence a possibility (risk) for a future disease, which can be specified with statistically generated numeric values. In principle, this is not new either. In many recommendations, physicians presuppose threatening future diseases: "But O dear, O dear, O deary; when the end comes sad and dreary …!" These expressions are meant predicatively. New is simply the form of technology, which can achieve a higher level of accuracy of predications. There is no doubt, however, that any prognosis becomes less accurate the further it reaches into the future. Which disease will dominate in old age, remains uncertain. Long-term predications in the area of relative risks, which rely on subtle markers (see above) are subject to this logic of predication. In consequence, it would be epistemically unjustified to generate a prognostic security for the recruits, which actually does not exist. Physicians should rather more often talk about existing insecurities instead of risks, to which a value of probability cannot be applied.

It is a mistake to assume to increase the security of a prognosis by increasing the number of predicative parameters. The expression "multi-parameter diagnostics" presupposed a number of parameters, which are numerically open. Principally, an individual can be preventively-predicatively examined and diagnosed accordingly based on a numerically indefinite great number of parameters. This opens the path into a kind of infinity which has to be pragmatically limited. Insofar, a multi-parameter diagnostic procedure needs to be limited as to what is medically reasonable. We think that a reduction of complexity is necessary. IM, in its epistemic dimension, sharpens the senses for the complexity of the incidence of diseases as well as health, but it needs to reduce this complexity in regard to its medical dimension in order to increase the recruits' willingness to participate in these predicatively orientated practices of diagnosis based on biomarkers.

As reported in the TAB study, the willingness to participate in a widespread predicative diagnostics is generally questionable among the population:

> The hoped for individual and collective health effect will only be achieved through Individualized Medicine when citizens are willing to take part in tests not only for the investigation of their individual risk of disease, but also when the test results will be used, in a medical and health-political perspective, 'meaningful' and appropriately medically related way (Hüsing et al. 2008, p. 21, translated by the authors).

These two forms of willingness hence need to be awoken and nurtured, thereby health-political questions arise about which positive and negative incentives seem to be acceptable. Here, we are only mentioning the question of what a "meaningful" and "appropriate" action could be. It is not trivial that the parenthesis in the quotation above bestows a privilege on the medical and health-political perspective. By that, the appropriate health-related action is handed over to the sovereignty of experts. As a consequence however, a paternalistic momentum sneaks into the IM framing. The candidates are supposed to "implement" what seems appropriate sub specie in the light of tests and diagnostics from a medical and health-political perspective. This seems questionable to us.

It sounds trivial to say that actions are supposed to be "meaningful and appropriate". But appropriate in what regard? The experts' answer could be: appropriate to a certain life plan, which can but does not have to contain the goal to stay healthy as long as possible and to live as long as possible. Even if the highest objective of a majority would be to live as long as possible it doesn't follow that, on an individual basis, an appropriate behavior toward health affecting risks must follow this objective. To respect the liberty of other individual persons means not to project the majority's objectives upon them. But it is not up to physicians and health-politicians to decide the appropriateness of such health-related behavior in regard to various life plans. Obviously, many people risk their health for goals that mean a lot to them. However, it can also be beneficial of health, to achieve these goals in life: Those who successfully pursue an academic career, are likely to be rewarded with the high expectation of the life of a public official, which balances out the stress of earlier years. Those, who risk their health by engaging in competitive sports and are spared from severe injuries, stay fit until old age. Within an IM framing, wouldn't it be worth thinking about health as something that can be reproduced, maybe even stabilized by the endurance of certain strains? This idea seems to be obnoxious. It sounds like Nietzsche's infamous quote: "What does not kill him makes him stronger," (Nietzsche 1968, 1999b, Maxims and Arrows, Aph. 8) which belongs to the context of Nietzsche's idea of "great health". This does not equal the ideology of "praised be what firms". Rather, it is about a performative-dynamic element of health.

This secular perspective can be complemented with a biblical one, it can be understood as the meaning of and goal in life to be satisfied with and by lived life. "Being full of life" does not mean "getting as old as possible". This will be discussed further in sect. 8.4.

(5) Findings, Dispositions, and Prevention
Important for IM is the unsuspicious talk of "findings". IM could have the effect of reducing the appearance of recruits "without finding". A "finding" is something suspicious, which must not slip medicine's attention and which needs to be clarified. IM could become a concept of medicine, whose practices of findings could be guided by the ideal that no suspicious finding is ever missed. Without going into further detail of medical-theoretical discourses on biostatic methods, it can be noted that a finding is something which does not correspond with prior expectations of normality. A finding is something that lies outside of a normal range, which is

defined by boundary limits, or in close vicinity of this threshold. Findings hence are *eo ipso* suspicious and need to be clarified by the means of laboratory diagnostics. It is linked to the impulse to clarify it as a precaution. It is considered to be a certain pre-suspicion, which justifies or even requires further examinations. By that a criminalistics element enters the IM framing. Therefore, IM critics indeed point out a true aspect, when they make the analogy between IM and secret service undercover operations (Lemke 2008, p. 171; cf. Ott and Fischer 2012, p. 196).

IM hunts for evidences for risks and prototypes of manifest diseases in stratified cohorts of recruits. IM produces incidental findings as a byproduct of studies of correlations and needs to develop regulations of how to deal with these incidental findings. These kinds of regulations are to be institutionalized in an ethically satisfying manner (Rudnik-Schöneborn et al. 2014).

Findings depend on standardizations. They are not simply discovered by medicine, but rather get epistemically constituted. They are determinations, which are certified by clinical studies. In most cases, it is about statistically significant correlations between the transgression of boundary values and the corresponding relative risk to suffer from the disease. The connection of multi-parameter-diagnostics including such setting of boundary values inevitably leads to an increase of findings. Practicing physicians must not ignore such standardizations, which originate from clinical research, even if they think critically about them. Since the physician's experience and power of judgment became devaluated by "*evidence-based medicine*", now critical physicians can only invoke to something they cannot officially rely on. We cannot evaluate whether that situation might change if all studies, which were not able to identify significant correlations, were published. But we want to emphasize that the current practice of publication could lead to systematical distortions.

Findings indicate dispositions for increased relative risks. Inevitably, terms of dispositions (like "virtual" diseases, tendency, latency, susceptibility, association etc.) are adopted into a predicative IM diagnostics. Terms of dispositions like "explosive", "fusible", "potent", "mortal" etc. are "tricky" and their use is full of pitfalls. How is health conceptually related to various susceptibilities? Are virtual diseases already real diseases in *statu nascendi*? Is a virtual disease already there in some way? Are there useful analogies to incubation periods of infections before they outbreak? Are values outside of the normal range per se pathological or only pathogen? Health seems to be surrounded by pathogen dispositions. Further research could be conducted on the epistemological problem of these terms of dispositions within an IM framing in a subsequent project.

Given the various dispositions, the principle of prevention inevitably comes into play. Prevention and prophylaxis of diseases has always been right and good, but thanks to IM it moves right into the core area of medicine; the principle to prevent diseases from breaking out in a candidate instead of curing diseases. Health is supposed to be maintained preventively instead of being restored curatively. Staying healthy is better than recovering from a disease. This fundamental "betterness relation" seems to be contained in the concept of health itself. New again, is not the principle as such, but rather the role that is conceptually adjudged to individual prevention in the framework of IM—which, thanks to molecular genetic knowledge and imaging methods, is done with good reason. Consequently, this leads to ethical questions about which measures were permissible, if one aims to nurture or require this preventive tendency. In the IM framing, the idea of health is connected to old and new practices and techniques of prevention. There is a broad range of such practices and techniques, which serve and support the maintenance of health. In the context of such practices, we have deferred to the idea of "*cura sui*", which can be found in the later work of Foucault (Foucault 1985, 1986, 1990; Ott and Fischer 2012). We go into further detail on this matter in 8.4.

The shift from a curative to a preventive medicine means that not only sick people come to see physicians, but that physicians examine healthy people as well and that recruits have to want such examinations: "In order to be able to help the patients (or potential patients) better, we have to search the healthy for certain pre-dispositions" (Golubnitschaja 2009, translated by the authors). For that, healthy people, who sub specie are supposed to be stratified, need to show a willingness to participate, which can be generated with various different incentives (personal interest, moral pleas, financial incentives or threatening with a penalty). It is interesting in the quote cited above that beyond dispositions, there also seem to be "pre-dispositions".

The long known fact that a lot of people, especially men, rather tend to see a doctor later than sooner, applies to preventive IM with convincing force. For IM, it can never be "too early". Many curative treatments will only become necessary, because a patient has ignored the options of early diagnosis (Wikler 1987; Árnason 2012). Many treatments will become apparent as such, which could have been avoided preventively. Omitted prevention becomes the cause for a disease. Omissions, however, are actions, hence are to be evaluated differently than infections, which only happen to someone. Hence, new attributions of moral responsibility become acceptable (Langanke and Fischer 2012). However, the ethical attitude of leniency could equally remain valid in the IM framing. IM does not have to be unrelenting towards those who were "sloppy" with their individual prevention. Lifeguards also rescue reckless swimmers and mountain climbers. If, in an IM framing, prevention is considered the first best solution and curative treatments are considered the second best solution for the sake of health, which is ethically unquestionable, in the moralizing perspective, it does not follow that people who need medical help are always considered negligent regarding their neglected prevention.

Strangely enough, it seems that recently, the conceptual idea of prevention has become less important for IM. It has become peculiarly quiet concerning the idea of prevention, most probably because research has failed so far to present "useful" biomarkers for this approach (Fischer et al. 2015).

(6) Intervention

The concept of intervention includes therapies in the traditional sense, but also forms of prevention, tests, monitoring, and early diagnosis. In the IM framing, the therapeutic impulse of medicine is shifted and displaced forward to forms of prevention, tests and monitoring and "check-ups". Interventions take place earlier in order to prevent more intensive interventions later. These early interventions could indeed be less burdening for the patients than later therapeutic interventions (medications, surgery etc.). Samples are taken obligatorily and in a standardized way the moment a patient is hospitalized. The concept of intervention has a broader potential implementation than the concept of therapy. The broader potential implementation of intervention is at the same time, necessarily less specific. To clarify a finding is a form of intervention, too.

In which way are medical rules of indication transformed by this broad concept of intervention? Rules of indication are a central issue in medical theory. One says in the medical practice that it is indicated (or shown) to do x from a medical perspective. This indication marks the shift to a therapeutic practice. Something being indicated means it is inappropriate not to be done. These speech acts are recommendations of prudence in the sense of Kant, which patients are allowed to expect to be given in accordance to the Hippocratic Oath, i.e. in the patient's best interest. IM does not alter this. But how can physicians be sure their manifold offers of intervention really are in the particular patient's best interest? Does a broader offer of interventions increase the security of recommendations or not? Would it be a deficit for a physician to admit that given all the uncertainties regarding relative risks, dispositions etc., he/she cannot be sure which intervention would be indicated in this very specific case? Could it not even be said that in a concept of IM a good physician is considered to be one who admits corresponding uncertainties towards his or her patient? Is it not even thinkable that IM physicians are motivated to dissociate themselves from the image of the demigod in white, who always knows what needs to be done? Most of all, these more relaxed and observant attitudes also gain the status of an intervention. Keeping an observing eye on a patient does not mean doing nothing, but rather a "wait and see". If thanks to IM a concept of intervention which can be implemented is suggested, it would be mistaken to intentionally fill it with the sense of a therapeutic "actionism". Even a cure, a vacation, fasting, dancing and singing, biophilia oriented activities like gardening, etc. can be recommendable interventions for disease prevention.

Could IM not also have the power to change general attitudes towards interventions in another more unexpected direction? A new concept of disease, which takes into account the probabilistic nature of the disposition, has to free the physician from the "mandatory appeal" to intervention. In that sense, it could be argued that the probabilistic character of IM actually strengthens the option of "wait and see". It is not necessary to interfere immediately, but rather time is needed for the observation as to how a disposition develops. The impulse to intervene immediately could be replaced with calmness. Other things being equal, it would also be necessary to accept an individual who decides to neglect a preventive option of intervention and willingly accepts the risk of a later therapy. From the candidate's perspective it does not have to be irrational to reject interventional offers, even if that increases the risk of more intense interventions later. This is comparable to the relation between risks of diagnostic procedures and the risk of possible surgery later. It leads us back to the question of which kind of risk behavior is rational to one's own corporeality.

(7) Responsibility, Autonomy and Maturity
The questions of medical responsibility and the scope of shared responsibility of the individual for his or her own health are traditional questions of medical ethics. A

patient's lacking compliance with therapies, to which they should ideally agree to, is a lack of responsibility often lamented on by physicians. IM does not fundamentally change any of these issues of shared responsibility and compliance. Even without IM, issues of ethics of allocation are part of the uncomfortable health-political debates. Especially since IM did not pose any new questions in reality. The discourse on justice and solidarity in a health care system has to be led independently of IM. At most, IM could lead to bringing the ideal of "responsibility" into realization (Maio 2012; Kollek and Lemke 2008; Aurenque 2012; Langanke and Fischer 2012; Wallach and Meyer-Abich 2005; Gefenas et al. 2011).

The idea of a mature patient is neither refuted nor devalued by IM. It is implemented in IM framing onto the individual in the different roles of recruit, candidate and patient. The principle of informed consent must also be extended to that effect. The principle will at any rate not be suspended with, it could and should instead come to have a more central role *a fortiori* in IM contexts (cf. sect. 8.4.2).

8.3   IM in the Context of Medical Approaches

8.3.1   Introduction

For our research approach, it is necessary to look at the changes initiated by IM regarding the understanding of health and sickness in a broader philosophical framework. It is crucial to take the pluralism of theories in medicine into account, if IM is to be widely comprehended and evaluated (Wiesing 2004; Franke 2012). IM is a framing or concept in relation to other medical concepts. This might sound trivial; less trivial, however, is the exact determination of these relations like extension, complement, replacement, influence, superimposition, dominance, exclusion, contrariety. IM seems to lie on a more fundamental level than *evidence-based medicine*, but does not stand in contradiction to it. The attempts of stratification ironically have the effect that due to smaller cohorts the "evidence base is much more complex" (Flessa and Marschall 2012, p. 55, translated by the authors).

The logics of relations (classification) is an important field of medical philosophy, since relational-logical mistakes can have effects for medical-ethical and health-political debates. The most common mistake is using an incorrect exclusiveness, which is not hard to recognize. This is easily shown with the problem of the relation between disease and health. Here, dichotomic, polar, and orthogonal relations are a good example. IM should not come into contradiction with conventional and traditional medicine. Rather, as we have shown, IM is nothing particularly new, but rather a new constellation of well-known elements within a new framing, whose components we have outlined earlier. After the analysis of presuppositions, we now want to focus on attributions. These attributions are made according to the idea of health. We did show that and why this idea is essential for IM, but we have not yet determined it in any further detail, which is to be done in the following section.

8.3.2 Health

In the nineteenth century, science was turning away from ontological and towards more analytical concepts of diseases. This process seems to be promoted by the progresses of biomedical research and IM. As a consequence, concepts we use to extrapolate the field of diseases and health(s), become questionable. Even the field itself seems to transform regarding its expansiveness. We discover diseases not simply as something existing, but we are permanently making definitions, of what is supposed to be (not) pathological or count as a disease (or not). The controversy surrounding nosological terms follows automatically from its growing constructive character. For example, this controversy is apparent in the debate among physicians whether the "metabolic syndrome" is a new and real disease or rather a "construct". The borderlines between health, disorder and disease become fuzzy also in the realm of mental states, as the example of Asperger syndrome may indicate.

If so, IM enhances this general constructivist character of nosological concepts.

The concept of health and disease as well as related concepts (disorder, disability, age, and disposition) are considered so-called "theoretical terms" in scientific medicine; i.e. terms, whose meaning is only accessible through a comprehensive theory of medicine or broad concepts like IM, but which is also constituted by the cultural ideas of values. Hence homosexuality is no longer regarded as a disease. There are a lot of cases which were differently pathologized, which can be shown very well based on the doctrine of mental disorders and diseases. Of course, the IM framing must not amount to a new dissenting pathology. Nobody would deny that on this abstract level. However, we need to examine all the new diseases and disorders sub specie the idea of health very carefully because they might lead to a dissenting pathology and socially induced behavioral problems. The discourse on ADHS and Ritalin is a model example of such a development.

Insofar, we do not need a new nosology for IM, but rather a critical attitude towards nosologies, i.e. we need critical nosological discourses within IM! IM, which can meet this demand, alters and shifts borderlines of health and disease, weakening these borders which previously seemed to be fixed. The conceptually induced transformation of terms, which partly takes place behind the backs of those involved, which is mixed with strategic speech, stipulatory definitions and rhetorical advertisements, cannot be allowed to happen but needs to be answered to.

Traditionally, disease is defined as a displacement of physical or mental homeostasis, which negatively influences—subjectively or inter-subjectively perceivable—a human being's performance and well-being. The question of how the concept of disease in IM is related to other concepts of disease is discussed in Põder and Assel (chapter 9). For our discussion, we want to draw on the well-known definition:

> A disease is an illness, it is serious enough to be incapitating and therefore is (i) undesirable for its bearer; (ii) a title to special treatment; and (iii) a valid excuse for normally criticizable behavior (Boorse 2004, p. 84).

This definition meets the distinction between "disease" and "illness". Such distinctions can be helpful to replace the (abstract) binary scheme of "healthy-sick" by a more complex concept, as done by Galen in ancient times. Bernard Gert's concept of *malady* might be taken into account as well (Gert et al. 1981). Hence, it is true: "IM has the potential to mutate ideas of health and disease" (Flessa and Marschall 2012, p. 64, translated by the authors).

This might be articulated in a strange way, as there is no "mutation" happening, but rather a loosening of established concepts and fixed ideas of models and the attempt to make altered ideas plausible in an IM framing.

There is no universally accepted definition of health. According to the famous WHO definition, health is "a state of complete physical, mental and social well-being and not merely the absence of disease or infirmity" (WHO 1946); it is hence a lot more than the mere absence of illness. The absence of illness is a necessary, but not a sufficient requisite. The WHO definition has been widely criticized, because it defines health as some ideal state, which no one ever reaches. Anschütz (1987, p. 100) refers to the WHO definition as "philistine" and "naïve". For more critical opinions see Wehkamp (2012). The WHO definition can by all means only be an ideal of absolute health, which can be understood as one pole of a continuum, at whose other end, there is the state of severe chronic disease, which leads to the "basin of attraction" of death. In the middle of this bi-polar continuum, other various different "orthogonal" (Franke 2012) configurations of moments of health and sickness exist. Among them are those which are discussed under the paradigm of preclinical findings. The appropriate representation of these orthogonal relations within one continuum requires a complex model of causations and mechanisms of action, as they are implicitly presumed in the research of correlations. At least, the categories of correlation and of positive and negative "feedbacks" need to permitted. In this sense, Flessa and Marschall (2012) demand a multi-cause-multi-effect-model.

The modeling of multiple causations and mechanisms of action brings up the question of what actually counts as a reason for something. It is common knowledge that Aristotle's complex field of causation has been reduced to the efficient causality ("causa efficiens") within a concept of the modern natural sciences. Possibly, an IM framing would reflect this reduction critically. Beside the various causes for diseases, it needs to be asked what the reasons for health(s) actually are. So, what could be reasons for health? At this point, it is not absurd to ask about "top-down" causalities, i.e. whether mental attitudes can cause physiological states, as this is claimed in psychosomatic approaches. Of course, believing in God, does not help against kidney stones, but the understanding of a lot of diseases and disorders, which refuse to be captured within a "one-cause-one-effect" model (Hyland 2011), could possibly be promoted by "top-down" causalities.

8.3.3   Four Approaches to Medicine

We would now like to interrelate the concept of health and an IM framing (relating to the two first paragraphs of this chapter) with various approaches of medicine and the art of healing. We distinguish four approaches: (1) the lifeworld, (2) the traditional-conventional, (3) the human-ecological and "alternative", and (4) the molecular-genetic approach. To our assessment, it would be premature, to match IM with a molecular-genetic approach or to claim a new, unverified approach for Individualized Medicine. We rather claim that IM cannot clearly be attributed to the molecular-genetic approach.

(1) The Lifeworld/Cultural Approach

The linguistic game of health on the one hand, disease and sickness on the other, and everything somewhere in the middle of these two (minor ailments, disorders, healings, recovery, recrudescence etc.) reaches into the natural history of human beings, but is far more than just a relic of earlier times. Medicine as an art of healing is incapable of substituting this linguistic game, which is deeply rooted anthropologically (e.g. via pain); it is hence bound to the fundamental values, which are implicit in the "grammatical" difference between "health" on the one hand and "sickness", "illness", and "disease" on the other hand. This axiological linguistic game is more fundamental than nosological systems of classifications and cultural schemata of corporeality, which are prone to historic transformations.

The concept of a lifeworld entails deep cultural beliefs and basic patterns of interpretation (Habermas 1981). Strictly speaking, there are many cultures, but only one lifeworld. Being corporally constituted and existing in time as rational beings, human beings have always shown interest in their own health and that of others. Hence, the concepts of health and disease (from childhood diseases to the defects of old age) refer to anthropologically fundamental experience in the lifeworld, which are experienced and processed in the dialectics of being and having a body (Plessner 1975).

For our birthdays, we wish each other above all good health, as this is necessary, though not sufficient for all activities, which we want to pursue. If we hear about someone being seriously ill, we initially feel obliged to feel sorry for him or her. The outbreak of a disease is usually commiserated, not welcomed. "Healthy" is something that's usually better than "sick" and "ill"; there is a fundamental "betterness-relation" in this lifeworld experience (which is only denied in some variations of Christianity).

Young, healthy, well turned out human beings with their erotic charisma personify health. One would say, they are "bursting with health and power". When aging, the perception of health is shifted to "sprightliness". Sick people, who are restricted in practical matters, cannot simply be consoled by telling them they have a lot of time to relax, meditate, contemplate, and pray. (We would consider such talk as inappropriate). Rather, we express our hope that the sick person might quickly recover completely.

Health is in the lifeworld perspective a valuable asset, but an asset that is to be categorically set apart from property and possessions. Health is a grounding for a lived life. Therefore health and disease are asymmetrical in a lifeworld perspective: disease is noticeable and invasive, health is discreetly hidden –it manifests itself by escaping our attention: "But what exactly is health, this secret something, which we all are familiar with and do not know exactly, just because it is so wonderful to be healthy?" (Gadamer 2010, p. 141, translated by authors).

Something wonderful is nothing that contradicts the laws of nature, instead something unlikely. In this sense it is wonderful to stay healthy for decades. With health it is about the continuance and therefore we wish each other to *stay* healthy. But how do we understand health as a continuance?

In the phenomenological perspective, health is something that automatically shadows and resonates in the manifold corporal and physical performances. At the same time, it is not undetermined. Even though undisclosed very often, health discloses itself in the feeling of well-being, joy of life, and drive etc. If, by contrast we say, we do not feel well, we take a perspective into account, which is not the one of scientific observation. "Feeling" does not equal "emotionality", but rather an immediate self-awareness (following Herder and Schleiermacher).

A feeling here expresses on the corporal-phenomenological level: "to feel oneself as healthy". We do feel healthy, when our organs basically work silently or at most make some pleasant sounds, like a pleasant fatigue after an exhaustion; that is also why Canguilhem defined health as life with silent organs (Canguilhem 2002), which is certainly a little bit abridged. From the phenomenological perspective, we understand health as feeling well. After longer consideration, the current well-being transforms into stable health, which can be called robustness, strength of resistance or resilience and which medically speaken, is related to efforts of the working immune system. IM framing wonders which kind of lifestyles are conducive to a stable health. The question does not only bear upon what is to avoid, but also upon stimuli, challenges, and thrills etc.

Unfortunately, locations of health are not locations of medicine. The architectonical of a constructed art of healing (Buß 2012; Wagenaar 2010) is founded in the motivation to rehabilitate the locations of medicine as places of recovery, which can be experienced without anxieties. If IM is research of health, then it needs to leave the clinics. IM should be aware that the *societal loci* of health are outside of clinics.

Health and diseases are model examples of the "structure of care of the being" as Heidegger argues in *Sein und Zeit* (Heidegger 1977). Caring is existential. To care about one's own health as well as about the health of children, wives and husbands, friends and family is a well-known phenomenon. In our lifeworld we care about more than just our own health. To care about one's health in an anthropic way presupposes the precariousness of being human. Here, precariousness means an exposure to evils, which can be alleviated, but not completely removed. At any time, something can happen to you. Especially health seems to be precarious, even if one—correctly or incorrectly—ascribes oneself with a robust constitution. We all share this corporal constitution with each other, even though everyone is left alone in and with their body after all. Even the intuition that worries might be inappropriate and exaggerated, is no stranger to us. We often tend to say that others should not worry so much. That way, in the experienced lifeworld, human beings warn each other of an exaggerated fear of diseases, which is hypochondria. In order to find the right amount of worrying and caring for one's own and others' health a power of judgment and worldly wisdom are needed. Health is the ordinariness and precariousness, which we feel as well-being and whose possible loss makes us worry. There are justified worries, but also a deep existential fear, to lose one's health permanently. Fear however is depressing and hypochondriac phobias are pathological. Care and caution are considered to be reasonable when dealing with objects in one's own possession; they also apply to the good way to deal with one's corporeality. In an Aristotelian sense, care is in the moral degree of care, whereas recklessness and fear or phobias are the opposite vices.

Within our lifeworld, we know that health needs to be looked after and nurtured and that it can, in return, be risked, that one is to some degree responsible for one's own health and that it should not be misused. In this essential sense, every person carries a certain amount of responsibility for his or her own health. In that regard, discourses on health related personal responsibility can indeed draw onto lifeworld knowledge (cf. sect. 8.4.2).

(2) The Traditional/Conventional Approach

Today, in its core, medicine's directive is therapeutic. This applies to many patients' attitude to health as well: people go to see a doctor when they are sick. The physician considers every patient as an individual patient. In a way, the cautious care for the right medication for this very patient was and is a form of IM; whether on the basis of "trial and error" or in accordance to "evidence-based probabilities": Does the patient tolerate a certain medication or not? Does specific therapy show effectiveness or not? Certainly, in today's conventional medicine, prevention, precaution and early diagnosis are accepted goals, and hence nothing new.

Conventional medicine, simply understood as medicine *lege artis*, is actually better than its reputation; its results are generally impressive and very often monumental (e.g. in intensive care or neonatology). Even those who support alternative ways of healing concede that for the majority of clear and specific diseases, a conventional therapy is called for. Conventional medicine takes the approach, according to which separate body parts (units) are infested by certain diseases (Hyland 2011). Causes of diseases are often designed as pathogens, which is true for all infectious diseases. This led to a culture of medical specialists for separate organs in conventional medicine. This culture of specialists offers a lot of advantages in regard to the treatment of many specific diseases (e.g. ophthalmology, urology, dermatology etc.). As a research discipline, medicine is analytical, as the concept of "science" generally is. In its analysis, it decomposes and then inevitably reaches to a molecular-biological level.

The ethos of conventional medicine is refreshingly pragmatic: The physician is not interested in treating symptoms, but wants to treat causes; he or she wants the patient to be healed and healthy after the treatment. A good clinician joyfully sees a patient leave the clinic healthy, but he or she does not make a fuss about it; it is just his or her job. Even if conventional medicine needs to surrender in its fight for health, in palliative care, it can still offer various treatments for alleviation. The idea of prevention is no stranger to conventional medicine; rather, it accepts and represents it as well. However, from the economical point of view, it mostly consists of the treatment of diseases than of prevention. Hence in principle, conventional medicine can hardly neglect an IM framing; however it needs more attractive incentives to foster prevention and precaution. Conventional medicine leads to a better provision of health care services for the whole population (at least in industrialized countries), to many successful cures, to an increasing life expectancy, to a higher quality of life in old age. There are also drawbacks to these successes, namely a number of very old people, who suffer from a multitude of discomforts, disorders, disabilities, and diseases. The stages of life of very old people are consequences of conventional medicine, independent of IM. Hence, IM with its predictive-preventive directive, cannot solve these problems, but can only slow them down for the individual. The best prophylaxis against dementia can lead to people starting to suffer from dementia after they are 90 years old. But that does not solve the problem of senile dementia. In reference to an aging society we cannot expect IM to perform miracles. The issue of how to finance pensions and health care in an aging society cannot be solved by IM, but rather has to be dealt with on a political level.

From the medical point of view, it can be asked, in what regard IM could be superior to conventional medicine. "Better is the enemy of good"—but what are the improvements of IM compared to conventional medicine? IM approaches often seem to be simply melioristic in regard to prevention, early diagnosis, side effects and "compliance". IM appears as an unspectacular reformism of what is common practice anyway and not as a revolution in and of medicine.

(3) Alternative Approaches

The heterogeneous concepts of the so-called alternative medicine, in which the old medical ethical problem of mountebankery certainly plays a role, criticize the mechanistic or "Cartesian" abstractions of traditional medicine and of general socio-cultural assumptions about what makes who sick and what keeps who healthy. The corporeality model of a complex mechanism and the "modular approach" to disease are rejected (Hyland 2011). The human body is understood as an intrinsic system, which, under the right conditions, has the power of self-healing. *Medicus curat, natur sanat.*

The approach, which is going back to Rudolf Virchow and which matches specific diseases with organismic units, is not at all wrong, but rather leads to a good therapeutic result in three-quarters of all treatments. The remaining diseases and symptoms however, could be treatable with different forms of healing, hence "by the art or 'other part' of medicine, not the science of modern medicine" (Hyland 2011, p. 8). From the perspective of conventional medicine, the field of a "hyphen-discipline" like "psycho-neuro-immunology" means, from the perspective of alternative forms of healing, the field of a "holistic" psycho-somatic medicine, which has been made accessible by medical "outsiders" like Samuel Hahnemann, Victor von Weizsäcker, Aaron Antonovsky and others. The competition between conventional and alternative medicine should not be argued out on the abstract-confrontational level of anthropologies and ideologies, but rather, as the new expression of "complementary medicine" suggests, on the level of shared labor-like cooperation, in order to find the best health solutions for recruits, candidates, and patients. Conventional and alternative medicine contradict each other as much as clinic and sanatorium do. Correctly understood, the IM framing could solve the dogmatic confrontations and mediates between both approaches. Especially approaches of alternative medicine emphasize the importance of the prevention of diseases. Combinations of IM, salutogenesis and "*cura sui*" hence do not appear as conceptually contradictory (cf. sect. 8.4).

Maybe secretly, IM is much closer to alternative healing concepts than many of its protagonists would officially admit. The presupposition of studies of correlation could correspond to the premises of "holistic" approaches. It might similarly apply to a multi-cause-multi-effect-model as well. We thus suspect a secret conceptual vicinity of IM and alternative medicine, which cannot be admitted by either side. IM, understood as the "Fröhliche Wissenschaft", however, does not need to be afraid of or shamefacedly try to hide this vicinity.

In the context of prevention guided strategies of lifestyle, as well as on the level of publications on alternative healing methods, the broad range of alternative healing approaches have gained great importance to many people. Mostly, these strategies turn against "stressors", which, broadly speaking, are linked to lifestyles in career-focused achieving societies (how to deal with time, lifestyle of high mobility, sleep and nutritional behavior, consummation of alcohol and tobacco etc.). They also turn against a high-tech medicine and represent medical ideals of the "prudent general practitioner", who has built up a long-term relation of trust to his or her patients. This medical ideal is very well combinable with the expectations that many link to IM.

Alternative healing approaches put a big emphasis on the phenomena of recovery, convalescence, relaxation, refreshment and exercise: *Salutogenesis* has become a winged word for that. The underlying idea seems not only to be getting off the sickbed "well-re-patched", but even refreshed. Alternative medicine draws upon (often gender-specific and psycho-somatic) expectations and role requirements, which are lined to careers and social constraints. It draws attention to conflicts between supposed constraints in modern work life and the personal times and rhythms of a human body. It allows the body a certain right to veto against all the expectations and requirements which can harm health (cf. Schäfer 1993). The dialectics of stimulant and stress are conceptualized as one of "eustress" and "distress".

Alternative healing approaches set big, yet not unlimited trust in the body's self-healing capacities. Environment, metabolism, time management etc. have a big impact on this capacity. What a successful medication or surgery is to conventional medicine, the recreative cure is to alternative medicine. While conventional medicine focuses on body units, alternative medicine relates to the corporeal organism and its environment. As far as health promoting environments are concerned, there are cross connections between alternative medicine, human-ecology, and environmental ethics. Experimental studies on the so-called biophily-hypothesis, for example, have shown that healing processes are more successful in greener atmospheres than in sterile settings. The theory of "nature's power of healing" goes back to the times of Goethe. Healing baths, which are decorated with the label "state-approved", were already places of regenerative convalescence back then. Later, sanatoriums entered the stage with their gardens and parks as well as little pavilions and hermitages (Buß 2012). The life reform movement adopted several themes of alternative healing approaches and of what later became known as "human ecology". The major failure of this romantic medicine (Schelling, Carus) is no reason for IM to discreetly ignore the inventory of possible traditions (Wiesing 1995).

(4) Genomic-Genetic Approaches

IM is mostly constituted in the context of molecular-genetic research strategies in the life sciences. IM relies on the knowledge of -omics, which is gathered on the molecular and genomic level. There is a terminological distinction between "genomics" and "genetics". While genomics examines the universality of the human genome, (as in comparison to animal genomes), genetics determines the genetic features of single individuals. At first IM is both genomics and genetics. At this point a real paradigm is often spoken of. This genomic "paradigm" of medicine is, similar to IM, related to high expectations. "The major impact of the genome project will be a slow but steady conceptual evolution—a change in the way we think about disease and normal physiology" (Watson and Cook-Deegan 1990). This slow but steady conceptual evolution could be consolidated by IM. As a matter of fact, IM and genomics form a research-strategic alliance. Some believe basically that IM is nothing but the PR side of the genomic paradigm.

The influence of genetic factors on health and diseases has been discussed over 100 years. In abstract universality, there is indeed such an impact. Nobody would claim genes have no influence on incidences of diseases; monogenetic congenital disorders are absolutely irrefutable. Problematic however is it to make this abstract validity a fundament for a paradigm, i.e. to conclude the basic causality of many or even all congenital disorders from just one single factor of influence. From the perspective of an extreme bio-genetic paradigm, which assumes a hereditary co-conditionality of almost all diseases, something like this is claimed: Virtually all disease becomes genetic disease. A genetic diagnosis involves individualization in the description of specific dispositions to and risks of certain diseases. Thus it could become an obligation, to know one's genes and to behave both genetically correctly and in awareness of your health. "Recognize yourself" would become genetically realized.

In the genetic paradigm, causes for diseases would be shifted to its specific code or lack of code. Diseases would be internally inherent and would only secondarily, if at all, be caused externally (triggered). The so-called "primary cause" would hence be found in a complex environment with all its risk factors. The "ultimate cause", however, would be found in the genetic foundation of every individual's somatic existence. Consequently the question arises of whether IM is or should be defined through a genetic understanding of disease. We do see the hazard that IM gets laid down in a procrustean bed of genetic reductionism.

8.3.4   Conclusion

Lifeworld, conventional medicine, alternative healing approaches and genomics/ genetics all compete for the interpretational sovereignty of the linguistic game of health and sickness, and with that eventually for an idea of health itself. The idea of health is an "essentially contested concept", which is in this multipolar tension and which does not merge completely into one of these approaches. Insofar, the idea is subversive. So we basically claim that IM makes all the other approaches move. Hence, it does not come as a surprise that all approaches try to win over IM to their side, i.e. try to conceptually monopolize IM for their cause. But IM itself reacts upon these competing approaches. Research-strategically, IM might be assigned to a rather molecular-genetic approach. However, there seems to be a peculiar vicinity to alternative approaches of healing. Hence, nobody can claim to have an indisputable monopoly on the sovereignty of the concept of health. Right now, IM might be a "wild card", a catalyst and a key for a synthetic (maybe "human-ecological") understanding of health, which contains all the different momenta of truth. Maybe, IM could and should take up a mediating role between the various outlined approaches. IM would then be meditating between the polarizations and its oppositions, by loosening them up. At first, an IM framing is a loosening of existing concepts and approaches in medicine. That is why it is with things staying the same neither possible nor desirable to turn IM into a rigid concept, let alone a doctrine. IM requires creative thinking. In this sense IM is a "Fröhliche Wissenschaft" for lateral thinkers!

8.4   The Ethical Dimension of IM: Informed Consent, "Cura Sui", and Solidarity

8.4.1   Introduction: Medicine and Health

Medical Ethics does not lie outside social developments. From a medical historical perspective, people's health has improved tremendously compared to earlier times. Today, we can hardly imagine the misery in pre-modern pauperism and modern proletarianism. Even in the period of industrialization, there were serious health risks, which were unequally distributed and which motivated physicians such as Rudolph Virchow to consider himself as the advocate for the poor, whose job it was to criticize conditions of housing and work that ruined people's health (tuberculosis, cancer caused by anilin vapors; scurvy in seafaring etc.). However, death also threated people in the upper classes. The President of Germany Friedrich Ebert died of an appendicitis, still in 1923, the environmentalist Hugo Conwentz died of a furuncle, the successful merchant dies of a rotten tooth in Thomas Mann's novel "Buddenbrocks". From the perspective of the poor, death often appeared as the big leveler, from whom nobody can escape, no matter how rich they are. In Marx, there is a significant note, in which a proletarian fights against the working conditions, which significantly shorten his life expectancy, while he had been able to reach an age of 50 (Marx 1962). Even Foucault acquired something positive out of the life-prolonging efforts of modern biopower (Ott and Fischer 2012).

At the beginning of the twenty-first century, health related inequalities still exist, among them inequalities between the poor and the rich, male and female, and between north and south on a global perspective. The inequalities, which existed between Western and Eastern Germany in 1990 are almost balanced out today. Whereas a connection between wealth and life expectancy, which could also be an aspect of IM correlation studies, can still not be denied (Bartens 2011b; Fuhr 2013; Griggs 2009): Chronic diseases can cause poverty and chronic poverty can promote diseases and disorders. All these inequalities are not to be trivialized, but it is fair to say that they are inequalities that have been seen throughout history and that causes for different morbidity and mortality are manifold (education, nutrition, relationships, environment etc.).

However, there is no obvious reason, why IM should be insensible and ignorant towards sanitary differences and their complex of causation. As argued in the paragraph above, IM can conceptually freely be combined with ideas of traditional medicine and human ecology, as long as the IM framing is not scientistically subordinated to a genetic approach. A scientist subordination would exist, if IM uncritically and without reflection deflects to one medical approach, which is closest to "science".

Overall, middle class people in northern countries live in a rich, aging society with low birth rates and with a medicine, whose achievements can only be sincerely admired. We are on the cusp of a wish-fulfilling medicine, cosmetic medicine and enhancement. Attractive combinations of IM and enhancement will probably be developed. Trans-humanism is likely to have discovered IM already.

Many seem to be obsessed with the idea of keeping up their youthful appearance, especially their shape. Biologically, this is the preservation of neotony. Consequently, all these efforts lead to cosmetic surgery. From a culturally critical perspective, these tendencies are easily criticized, e.g. by saying it is fundamentally perverse that an aging society is so guided by fictive images of youthful beauty (such as in advertising). These critics are justified, but they misjudge the fact that most of us do not stoically accept all the obvious effects of aging (which are sometimes even impossible not to be heard and smelled) while staying happily relaxed, but rather want to fight aging with diets, cures, age appropriate exercise, prostheses, etc. Let us call this "anti-aging activities" (AAA). Aging as such, is for sure no disease, but an aging society like ours will set its sights on maintaining good health for as many of its members as possible and for as long as possible by using the appropriate anthropotechniques. This corresponds to the background premise of an ethics of a normative individualism: Every one of us only has one life and the quantity and quality of this life has the highest importance to (almost) all of us. It is taken for granted that no attempt to prolong life can ignore this quality.

Hence, AAA are not to be made fun of. Of course, the aim of AAA is not, not to age; as this is inevitable in complex organisms. So, "the ravages of time" go on, but the success of AAA is not measured by the abstract negation (no aging anymore) and its futuristic and trans-humanistic prospects, but by the delay of signs of aging and by the prolonging of the aging process itself. This will not give us eternal youth, but a prolonged second half of life. In the lifeworld, the idea of sprightliness presents itself best for this goal in the "second half of life". Old age is supposed to be lived in sprightliness as it used to be said about people who lived "sprightly until very old age".

In a cultural-comparative perspective, AAA lead to astonishing results. In World War I photographs many people look pretty old around the age of 40. In those times, old age already began at 60, whereas today it rather starts the life-phase of the "young oldies", for which the bell is ringed with the so-called restless retirement. Nutrition, hygiene, and not least medicine grant a third life-phase to a lot of people in the wealthy societies of the north, which can still be used and enjoyed; the price for this third phase, however, is the last phase of very old age, which is already set beyond the age of 80. AAA and IM are strongly connected. Because of its preventive and prophylactic character, IM is connectable to AAA. Hence, it fits well into our culture that is already streaked by AAA. IM could grasp the idea of "health up until old age" in a way that it would already start very early, like in dental hygiene. The idea of aging starting earlier than many people would think seems communicable on a health-political level. Of course, IM interprets the idea of health as something that not only the system of professional health care providers takes care and responsibility of, but so does every individual as well. As long as the access to this system is unrelated to individual behavior, there are no ethical constraints.

Whether accesses to the health system that are allowed by certain monetary payments (in the sense of Luhmann 1988) should be dependent or independent of individual behavior, is a health-political question like the one concerning the way in which a solidary or private insurance system should be regulated. An IM framing does not give clear-cut answers to health-political questions, even though short circuits and fallacies of ambiguous concepts are often made here. The concept of the individual does not speak for a system of private insurance as existing in the USA. Different health care systems are compatible with the IM framing and IM researchers should repudiate against being bogarted politically. Even in an IM framing, the inclusion of all citizens in a health care system is health-politically out of question. IM does not discount the egalitarian intuitions regarding the health care provision, which are widely spread among the German population. In that regard, IM can definitely be connected with general medical considerations. Elsewhere, it has been argued that many medical ethical discourses lead in to a multipolar normative area of tension (Ott 2008). This area of tension comprises the complex normative clusters (a) of the Hippocratic tradition, (b) of "informed consent", (c) of the freedom of research, and (d) of the question about the borderlines of the technologically possible (stem cell research, human cloning, etc.). This area of tension was reconstructed by normatively guided medical ethics; eudemonistic and health care political aspects were not discussed. This area of tension is not altered fundamentally by an IM framing. Those, who question the normative contents of these clusters in an IM framing, need to justify this.

How physicians would understand the Hippocratic tradition of healing and helping within an IM framing is open; however, this tradition cannot simply be abstractly negated. How medicine is supposed to deal with "the undiscerning" for example, is an old question within the Hippocratic tradition and which poses itself in a new way in IM, whereby we should know that lacking discernment and above all a lacking willpower correlate with the overall condition of the individual (e.g. phobias and addiction). Especially if IM pragmatically implies a holistic view to the human, phenomena like negligence, lack of willpower, impenitence, non-compliance etc. must not be understood intellectually and be moralized hastily.

Within an IM framing, the borders of freedom of research need to be respected too. The right not to be researched stays valid. All research that is done to and with human beings is subject to the criterion of well-informed consent, which can be withdrawn at any time and without indicating any reasons. The question of which research with people who are unable to consent is permissible (e.g. in research on dementia), needs to be discussed within an IM framing in the same way as outside of IM ("minimal risk, minimal burden"). The principles of autonomy and the maturity of patients, which are not to be taken for granted, but which rather are a medical task, could be taken seriously and unfolded within an IM framing.

In the following, we would like to advance an ethical proposal for the ethics of IM. The proposal wants to ethically underpin an IM framing onto three pillars: (1) informed consent (IC) *a fortiori*, (2) "*cura sui*", and (3) solidarity when dealing with risks. The three pillars can be connected with other features and the most important point is formed by the determination of IM as "Fröhliche Wissenschaft". In its ethical dimension, the IM framing should hence not be completely constituted deontologically, but should rather leave room for eudemonistic, virtue ethical and health political considerations.

8.4.2 Informed Consent Within IM Framing

The informed consent (IC) is before therapeutic as well as diagnostic measures, a deeply rooted normative principle (in ethical as well as juridical debates cf. Bunnik et al. 2013). From the juridical perspective, every intervention without IC is a criminal assault, as well as from the ethical perspective, every patient has the right to decide against medical recommendations. Since the IC principle is part of the central canon of principles of a post-paternalistic medical ethics and has proved to be a tower of strength in medical ethical debates, possible alterations in the understanding of IC need to be taken into account.

IC is some kind of individual defensive right, for whose utilization one does not need to give any reasons and which overrides collective goals (such as the goal of improving diagnostics). The IC principle is a "trump card", which individuals hold against collectives of physicians, researchers, co-patients etc. Part of the core of IC is that the consent must not be stimulated by massive positive or negative incentives, i.e. the informed decision must not be accompanied by severe disadvantages. The IC principle forbids consents "under constraint". Also, any consent may always be withdrawn. An interesting question is, however, which rather minimal incentives (nudging) are still compatible with the IC principle.

As far back as 1989, in his book "*Proceed with caution*", Holtzman issued the warning that an over-zealous promotion of genetic screening programs can take the personal freedom of choice away from the individual (Holtzman 1989). This could be accelerated, if systems of early diagnosis with approximately 100% sensitivity and specificity existed, which together with other negligible risks like over-diagnosis, negligible potential of harm ("minimal risk, minimal burden") thanks to early detection examinations, and the security against stigmatization or discrimination, raise the question of a possible "duty to consent". The Public Health Genetic debates about whether there might be cases, in which an initial moral obligation could be justified, in which preventive health care rules out the individual patient's will not to consent. IC will not directly be negated by IM; rather other principles and interpretations of IC could be strengthened in order to relativize IC in order to balance it politically with other aims and values. The principle of IC is likely to lose its dominant status over other collective aims. One approach is the "waiving" model introduced by Manson and O'Neill (2007). Manson and O'Neill presuppose a critic of the principle of patient's autonomy, which we do not share. Bullock (2010) clearly worked out the "waiving" model's weaknesses and dangers, whose main aspect is to motivate recruits, candidates and patients to waive their juridical documented rights, in order to legitimate certain interventions, which make sense, primarily from the perspective of physicians and researchers. This kind of "waiving" needs to be reached within the clinical setting and to be documented in a legally binding way. However, the "waiving" model is conceptually open to all forms of subtle manipulation and subreption, if a less demanding model of communication is presumed than by Manson and O'Neill. For an IM framing, the "waiving" model seems seductive, because it gives the "soft pressure to cooperate" a good conscience without completely invalidating the IC principle. Combinations of a meta-ethical emotivism and psychology of communication could support the "waiving" model in as much as many patients feel "safe" if they "trade" rights against the comfort of medical care. Those, who claim their rights, suddenly appear as egoists, who do not care about the benefit of research for the general public.

The suspicion that, especially in research, the functional demands of IM will lead to a soft pressure to cooperate cannot be denied. The cohorts need to be large enough to satisfy the epistemological criteria for scientific studies. Maybe from the medical perspective, this could be a healing form of pressure. The soft pressure to cooperate, even though paternalistic, can hence be justified: longer health and higher quality of life on average, better possibilities of treatments in the case of early diagnosis etc. In an IM framing, it can be assumed that medical results (studies etc.) will be taken into account in order to justify a "soft pressure". Such arguments could be presented: "The benefit of preventive-predictive IM for all (or indefinite many) is higher than the (possibly small) disadvantage, which only a few have to suffer from, because (for some random reason) they do not want to cooperate. The costs of not cooperating will then become too high for them, causing them to co-operate." Cooperation could be justified as an obligation of solidarity towards the general public. Wanting to flee from that could be declared as unreasonable and/or irresponsible. The burden of justification could be shifted in a way that making a draft on the IC principle actually needs to be justified. Having been a "trump card", the IC principle becomes an aspect which can be ruled out in the appreciation of values that are undertaken by physicians alone or in a committee.

Surely, it can be asked, whether the concept of IM necessarily leads to such shifts in the understanding of the principle of informed consent. Thus, a rejection of the demand "Consent must be specific!" with a shift towards a more contractualistic interpretation is discussed, because this seems unavoidable for the establishment of biobanks; since biobanks are supposed to render research possible, whose future hypotheses are yet unknown. The research-practical point of biobanks (and the effort of their establishment) would be thwarted if "Consent must be specific" were strictly valid. Hence, it seems ethically permissible, when mature persons give their consent to making pseudo- and anonymized data available to biomedical research, even though it is still unclear which research questions will be pursued in the future. Of course, this is a "blank check". Those, who are afraid of their data being used for military, eugenic or otherwise ethically questionable research purposes, should rather claim their right not to give their consent. It is important that no candidate or patient is put into a worse situation because of her refusal to consent. In these situations, the setting, in which declarations of consent are given, is very important. The clinical setting must not suggest that refusing to "waive rights" will turn an individual into a "second class" patient. But this problem seems principally solvable.

It would be strange, if the idea and the principle of IC would be kept up for therapies of patients, when the status of recruit and candidate (cf. sect. 8.2) were determined by other principles. Regarding the candidates and their dispositions and risks, it seems obvious to let themselves decide about the indicated intervention of "wait and see". A paternalism with candidates and a principle of maturity with patients seem conceptually inconsistent. Especially if, in an IM framing, individuals are stratified, examined, and treated predicatively-preventively in the various roles of recruits, candidates, patients, and convalescents over a long period of time, thereby, a sense of autonomy and responsibility could be strengthened on both sides.

People have non-medical life plans and goals, which lie outside of these roles and "careers", which get attributed within the medical system, and which physicians have to respect. Hence, in an IM framing, a lot speaks for a culture of consulting and shared decision making that is guided by good reasons. In an IM framing, the relation between physicians and those, who face them as individuals in their different roles with their own dispositions, risks, strengths, weaknesses, diseases and healths, could be humanized, if IC is not jeopardized for the sake of short term advantages in the recruiting process.

8.4.3 "Cura Sui"

This responsibility of one's own health can be understood in different ways: either as (a) a morally incomplete duty towards oneself, or (b) as a duty to good economy towards the solidary group, or (c) eudemonistically in the sense of a "*cura sui*" (Foucault 1993). These points are explained in more detail below:

a. In everyday life, we assume certain responsibilities for one's own health, which cannot be delegated. Usually, everyone has to brush their own teeth. In Kantian ethics, the health-related responsibility is grasped as an incomplete duty towards oneself, or to be more precise: towards humanity within one's own person. One's own body (German: Leib) is hence not, as John Locke (1988) argued, an estate, of which the person owns the "*ius usus sive abusus*". Everybody has duties towards one's own self in view of one's own body (Leib), which is part of one's "self". The self-reference, in the Kantian perspective, is not to one's own convenience. Here, we can find analogies of forms of external neglect and "letting oneself go", which are not to one's own convenience. To Kant (1968), these forms are degradations. Those who neglect their own appearance unnecessarily degrade themselves as persons. This can in a similar way be applied to neglecting one's own health. Surely, we assume a minimum of subsistence, which enables everyone to meet this responsibility. One cannot accuse a concentration camp inmate and a poorly supplied prisoner of war, or a neglected homeless person of not doing enough for their own health.

   However, these kinds of duties in view of one's own health would be incomplete duties, which cannot be fixed precisely. Brushing your teeth regularly is advised, however, not every single hour. Philosophical ethics cannot determine the exact extent of personal healthcare; anyway, such a general determination would also undermine the basic idea of IM. We just want to make the point that we all owe something to our own health, because we owe something to ourselves. Especially this abstract openness is the motivation for every individual to take care of their own health.

b. It is comparably less clear whether and, if so, to what extent an individual is responsible towards the solidary group of insured people regarding keeping their individual health care expenses to a minimum. Langanke and Fischer (2012) discuss a command of thrift and a prohibition of abusing money. Insurance companies assume that people generally involuntarily utilize health care services, as they still prefer to be healthy than sick. Using health care services in these cases is not a strategic "taking advantage of community money", but rather *stricto sensu* born out of necessity. Economic models, which transform individuals, who prefer to be healthy, into "*homines oeconomici*", who try to benefit most from their membership fees, overlook this. But do we get angry about our own health if we only need a few health care services, because we are rarely sick?

   IM could certainly lead to more of the following ways of thinking: The costs of a therapy could have been avoided by the individual's reasonable prophylactic behavior. So, why should I have to pay pro rata for avoidable costs? According to this thinking, people should be paying for everything that is avoidable according to the best medical knowledge themselves. However, it is questionable whether in an IM framing the avoidable and unavoidable can be distinguished clear enough in medical practice, i.e. also to stand up in court. Eventually, in many cases, IM upsets even those broad calculations. In the individual case, was diabetes, a stroke, an artificial heart valve or an artificial hip joint really avoidable or not? Regarding all the manifold risks and courses of diseases, informally introduced "red lists" and the recurring example of undiscerning smokers, will not be of much help. Moreover, these approaches do not take into account the individual configurations of "risks being given", "risks being taken" and "risks being imposed", which we will come back to later. At this point, we just want to hold onto the fact that in a probabilistic multi-case-multi-effect-approach the accountability of contra-factual avoidance behavior cannot be guided and is hence health politically eliminated.

c. Beyond duties of virtue towards oneself and the command of thrift towards the "fellows" of an insurance system, personal health care can also be understood as a meaningful, even enjoyable self-reference. To take care of one's own health becomes a "focal practice" (Borgmann 1984), which is done for its own sake, i.e. which we are happy to draw our attention to. "*Cura sui*" is the headline for such a "focal practice". This is not done as a duty but rather as a liking. It does not fill a life, but pleasantly permeates it. In a "*cura sui*", you "do something good to yourself". Thereby, possibilities arise to think about health care and preventive action with a life-world-attitude and within an IM framing, but without catalogs of duties and threats of penalties. It does not bother us that this kind of thinking is aligned with ancient tradition, because we have emphasized several times how IM does not need to be original.

An IM framing can also rely on our own intuitive-phenomenal feelings regarding our own well-being. Nietzsche called this feeling "the instinct of healing powers". We must not let this instinct be taken away from us, not even by a physician guided medicine—and that is exactly what "*cura sui*" draws on. Not following certain charts and measuring instruments, but rather instinctively, we know which kind of activities and diets are beneficial or harmful to us. This instinct is supposed to and can promote and strengthen IM in ourselves. In this sense, IM cannot only rely on the self-healing powers of a body, but also in the ability to reason and the willingness to cooperate of all those, who voluntarily want to take care of their own health. The idea of health is powerful and effective in human beings; but it is important to foster and promote it with the help of IM. That means, we plead for positive (rewarding) stimuli in order to take part in this kind of personal health care. In a certain way, the dechristianization of everyday life complies with us. Since many AAA-activities share neo-pagan aspects, IM as "*cura sui*" can draw on to the self-conception of many individuals, to whom diseases are no inquiries.

In view of the normative deficits of his concepts of biopower (Ott and Fischer 2012), Foucault's turn towards the conception of "*cura sui*" is less surprising than it seems. The path into eudemonistic ethics remains open even if the normative strategy of criticism can no longer free itself from its deficits and *aporiae*. Foucault took his final turn as an ethicist, not as an historian. The care for oneself, which precedes self-awareness, was understood as "*epimeleia*" in ancient times. To Foucault, this is an attitude, which was schematized in the Hellenistic period after the model of medicine and which can be determined as constant healthwise care (Foucault 1993).

The attitude of *epimeleia* implies that one becomes one's own physician (ibid.). This presupposes the perception of the healing arts as *techne* and also as *eupraxia*. Practicing the role of the good doctor with reference to oneself, i.e. becoming one's own therapist, requires certain techniques such as writing, *askesis*, temporary withdrawal from social life, but also dream interpretation and the cleansing of the conscience. In *epimeleia,* dietetics, hygiene and sexual education (not sexual "morals") are incorporated. In this sense, *epimeleia* can be described as "holistic". It is closer to today's human-ecological and alternative medical approaches than to orthodox medicine. *Epimeleia* must be exercised. Therefore, if there is always biopolitics being exercised on individual subjects, subjectivity emerges by *epimeleia* being exercised by the subjects themselves throughout their lives (Menke 2003).

To Foucault, the opposite to *epimeleia* is, in Latin wording, *stultitia*, folly. A fool is someone whose existence disperses in time, who does not care for anything, not even himself, and whose will is changeable and erratic. However, for Stoics, *epimeleia* always has the purpose of preparation for evils such as strokes of fate, diseases and, ultimately, death. Insofar, it contributes to a healthy life and to the preparation for diseases which shall be borne with indifference (*ataraxia*). The ideal of Stoics is the emotional calmness in the face of mundane incidents. Modern medicine would probably welcome stoic virtues in patients.

Epimeleia, as exercise, is self-care, which implies self-devotion. It is an attitude of attention and devotion towards oneself and others which is epitomised by certain practices. These are the so-called *Practices of Self*. These practices promote the attentive devotion to oneself without objectifying oneself. Foucault believes to find in *epimeleia* a form of care for oneself which is not committed to a Cartesian body-mind-dualism or a conception of self-disciplining. One can't exercise *epimeleia* out of the attitude of obedience to another person. In performing *epimeleia*, the concept of biopower becomes pointless. By such self-caring exercise, one might come to trust one's own vital body. This trusting is something other than permanent self-monitoring using apparatuses and devices. In the attitude of *epimeleia*, one does not monitor oneself in one's bodily functions but treats oneself in one's vital corporeality in a friendly and trusting way. The body appears trustworthy in its vital disposition to maintain and regain health.

What, then, is the relation between *epimeleia* and IM? To Lemke (2007, p. 130), IM is a regime of truth, a set of power strategies and a new context for practices of self. However, it is terminologically misguided to draw on the idea of power with reference to self-care. Self-care is no power relation in the sense defined by Foucault. It would be conceptually absurd and misleading to initially divide the person (for instance into the "good core" and the "weaker self") in order to then let that divided personality stage internal power games with "itself". This is neither the ancient sense of *epimeleia* nor its understanding in Foucault's work.

However, Lemke comes upon an important point, as IM could beguile into conceptualizing *epimeleia* as a permanent form of self-monitoring which has to be continually improved by technological means. There are indeed tendencies towards this, for instance the idea of the chip which is implanted under the skin and constantly transmits certain bodily results to a database.

However, *epimeleia*, understood as an attitude and as a practical exercise, is not self-monitoring, but rather awareness and trust in oneself. We understand *epimeleia* or self-care as the epitome of an ensemble of related therapeutic and transformative practices of attentive self-devotion. Self-care is "performative self-therapeutics". The "self" in this case is no independent substance ("Self") but only has the grammatical purpose of referring to one's own respective existence.

Every individual has this ensemble to choose from and it is unproblematic if he/she chooses the practices which he/she intuitively thinks (or "feels") will suit him/her. Practices of self are exercised voluntarily. IM could approve of and promote practices of self which are not to be reduced to the fastidious "scanning" of the body for signs of disease: The concentration in archery, the eroticism of Tango, the meditation techniques of yoga, the healthful relaxation while listening to music, the writing of a diary, the cleansing of the skin in a steam bath, contacts with nature, picnics with friends, the preparation of good food, therapeutic fasting, restful holidays etc. are "therapeutic practices" in a very broad sense which are obviously not medicinal in a strict sense but which could be concretized ("*concrescere*") by the possible combination of human ecology, alternative medicine and IM. We leave open the question of whether "good" sex should be considered as a therapeutic practice as well.

With respect to Foucault's philosophy, IM and *epimeleia* could be practiced complementarily: Preventive IM, which aims for well-tolerated and suitable therapies in a strict sense, and a careful and attentive self-devotion in self-therapeutic practices in a broad sense, do certainly not stand in opposition to one another and do thus not preclude one another logically, terminologically or in life practice. IM has to take a stance in the peculiar dialectics of subjects being individualized and disciplined by modern medicine on the one hand, while exercising and performing *epimeleia* on the other hand. To take a conceptual stance is logically independent from the empirical matter that only a few people are exercising *epimeleia*. An ethical culture of IM would not only tolerate and respect *epimeleia*, but would recommend and support it as both attitude and exercise. By doing so, it restricts a scientific-technological attitude from within. Dialectical relations can be specified to new and perhaps unusual combinations of IM and *epimeleia*. Such combinations are to be exercised in an experiential and experimental manner. We feel some initial sympathies with this perspective, but restrict ourselves to such outlook which must be substantiated within a comprehensive philosophy of IM.

Thus, we recommend to encapsulate "*cura sui*" in the framing of Individualized Medicine. "*Cura sui*" is clearly part of culture as a commonly shared set of practices. IM can contribute to a reform of somatic culture as it had been intended in the "Lebensreform" movement before 1914 (Heyll 2006). This rather middle-class movement was not ignorant against the more pressing needs of workers with respect to healthy conditions of labor, nourishment and housing.

8.4.4 Solidarity

In general, solidarity means that in the face of certain life situations and risks, people are willing to stand up for each other. The classic example was the class solidarity among laborers. The camaraderie among soldiers, which can go to the lengths of self-sacrifice, is also one form of solidarity. In anonymous systems of insurance, solidarity becomes more and more abstract and needs to be concluded via fees in the form of money. Thereby, it necessarily "cools down" and gives room to considerations, which put the individual benefit of insurance companies in the foreground. Private insurance companies take advantage of these considerations for their own sake. Solidary systems then turn into contracts between atomized individuals and enterprises, which calculate policies, which often motivate the policy holders to switch back to the shelter of the solidary system of insurances. Even IM is not capable of solving the problem of these dialectic contradictions of commercialized solidarity groups, which rely on the individual utility calculation. But IM can and should be aware of the roles it is intended to take over within the strategic games of insurance systems. Regarding the ethical contents of an IM framing, it is our aim to reanimate the idea of solidarity, which has become abstract by anonymity and commercialization. For that purpose we draw upon the debate of risks again (cf. sect. 8.2.3).

Health risks are partly constitutionally or genetically "given", partly "imposed" to individuals by others, and partly "taken" and accepted voluntarily. The genetic approach focuses on given risks, which individuals do not bear responsibility for, as nobody is responsible for their own genes. The socio-medical human-ecological approach focuses on risks, which are caused for individuals by conditions of living, working, environmental pollutants etc. There are debates on residuals of pesticides in groceries, on the harm of "electro smog" and about whether the sensitivity to chemicals should be considered to be a "real" disease or not (SRU 2004). The environmental movement made these environmental risks the subject of many political debates and achieved success in the reduction of many of them. At least, "hard" toxic substances like asbestos and DDT have been removed from the environment; now debates shift to address "soft" issues like noise. Interestingly, the post-war generation was exposed to many of these "imposed risks" (passive smoking, leaded gas, and the "fall-out" of over-ground atomic tests), whose long-term effects are still unknown or cannot be proved. IM is well compatible with the endeavors to minimize these imposed risks. Thus, cross connection to social medicine arise.

A lifeworld approach knows about all these risks, which people take in the various *nolens-volens* constellations in life. People ate too much candy as kids and repeatedly caught sunburn in adolescent years. While in school, people like to read a lot, but also had to read a lot, which, in the long run, might have harmed their eyes. For years, people did not sleep enough and did not take enough breaks and vacations in order to promote their careers. It is hard to compare the challenge, which one voluntarily takes on in demanding professions, with the stress of being forced to take on ever increasing workloads. Who would deny that drinks and tobacco played (or still play) the role of stimulants and/or tranquilizers for many at some point in their life. And who does not know the minor ailments that remain from former athletic efforts (e.g. as arthrosis)? And do we not have to take some risks in order to be able to reach a coherent feeling of our corporality over a period of life? Let him, who, in his own memory, appears to himself as the prophylactic prig, cast the first stone.

From its inner perspective, every experienced life shows a "*mixtum compositum*" of these three types of risks, which can be easily distinguished: risks being given, risks being imposed, and risks being taken. Analytically, these types can be separated from each other, in real life, however, every health risk is always a biographically concrete (*concrescere*: growing together) conglomeration. In this threefold sense, every risk profile is ultimately individual. For two risk classes ("given" and "imposed"), no one can be held responsible, unless one would want to try to justify a duty to avoid imposed risks wherever and whenever possible. Moreover, many of the taken risks are ambivalent, i.e. they can have health harming as well as health promoting effects, which, in retrospect, can hardly be distinguished. And of course, it always needs to be asked, which kind of knowledge of certain risks in earlier times can be presumed as obligatory afterwards. What should an educated lay man have been able to know about the connection between sun and skin cancer in 1975? On the one hand, we trace the cause of diseases down to individual genetics, and on the other hand, we have to acknowledge that a lot depends on individual behavior, if not on environmental conditions. The attempt to determine these heterogeneous and ambiguous types of risks in a certain individual with the aspiration of percental accuracy would require an escalating program of quantifications. And the uncertainties which are inevitably linked with probabilistic findings would multiply. An individualistic conception of responsibility would become an abstract moralizing, in so far as it would isolate specific momenta of the conglomeration and would directly associate these abstractions with concepts of morale and responsibility. Such moralizing is partly openly contradictory, because, from a medical perspective, some examples, like the abuse of alcohol are considered to be accepted diseases. To hold the alcoholic morally responsible for his or her addiction presumes that he or she could simply stop drinking—and this presumption contradicts all medical knowledge about addictive behavior. Addiction prevention is surely a meaningful goal; but it can rather be achieved by Scandinavian models like banning alcohol from supermarkets.

The solidarity within the health care system cannot be completely detached from other forms of solidarity. *Strictu sensu*, there is no human right to health, but there is a right to be materially capable to subsist, to take care of one's bodily hygiene, to move around, to rest etc. Hence, IM can go hand in hand with some insights of the old social-hygiene-movement regarding nutrition, living conditions, work place etc. An IM framing cannot and should not only address correlations between diseases, but correlations between diseases and life conditions as well. In so doing, forms of medicine, which deal with marginal groups of our society such as illegal immigrants, homeless people, drug addicts and immigrants among others, become relevant to IM. These groups of people, who are especially in need of assistance and help, are often isolated in an emphatic sense. In all its pride about clinical research, IM should not simply ignore these forms of medicine for the marginalized.

Of course, there are people, who accumulate health risks, who are called high risk candidates and to whom a change of lifestyle (a "*metanoia*") is medically highly recommended. These people disproportionally come from socio-economic backgrounds, which are disadvantaged in several regards. The empirical questions are, how large this group really is, what costs would normally be required for a group of that size and what the group's extra costs are? The ethical question that arises is, whether we do not owe solidarity especially to those who are or were behaving in ways detrimental to their health—and who, in several regards, are "punished enough". The popular "bashing" of people who are obese or who smoke is not completely motivated by medical reasons, but also consists of anti-pauperistic affections. It shares something pharisaical, namely the arrogant feeling of not being one of them. We do not say these groups of people behave in a way they should according to a *sub specie* well-meant interest in their own health. But we see more than just *akrasia* in the view of a critical social psychology, but self-destroying tendencies, which are rooted in a lack of self-respect and in missing recognition and respect in society. These tendencies will not disappear through intensified social-medical offers of support but would intensify them, because people would become dependent on support services. In the end, the question of the best strategies regarding high risk groups from low socio-economic backgrounds remains aporetic, in any case distressing.

As always, IM is generally to be connected with an attitude of solidarity. This means that the "faults" of patients' behavior should not be calculated and for these calculations to affect the costs of the therapies which they are entitled to receive. The reasons which speak against such a system are that it could become too complex, avoiding trouble, being discreet with our relationships with one another, generosity, lenience and sympathy. A health-political solidarity grows out of these latter virtues. The unending calculation of risks and costs along with the administrative bartering and the lawsuits and legal complaints will lead to pettiness, greed and hard-heartedness.

8.4.5   Conclusion

From all the debates on IM that have been going on over the last years, we have tried to pick up some components and to systemize them according to a hypothesis, which is both current and effective, with the idea of health in an IM framing that is still in *statu nascendi*. Certainly our one-sidedness and prejudices lie in the particular context of GANI_MED. We understand IM as a possible integrative health science in a highly modern society. Such a science is a "Fröhliche Wissenschaft", which is quasi transverse to various medical approaches. On the conceptual level, an IM framing should be able to integrate research on the biophily-hypothesis, on salutogenesis, alternative forms of healing, and last but not least, works of industrial and social medicine. It should ask what "vitality" is.

Now, we see that IM can also be connected with a cheerful calmness regarding the individual human mortality, with a portion of (self-)irony towards anti-aging activities, with a body-sensitive attitude of "*cura sui*" regarding one's own health, with a dialogical-consulting medicine, which upholds the principle of informed consent, and last but not least, with a solidary, and even big-hearted health care system. Responsibility for one's own health must and should not be negated in an abstract way; even independent of IM, that would be out of touch with everyday life. The epistemic grammar of IM is, as shown several times, flexible in regard to different health political options. IM enables us to design new scopes of medical practice. Nobody needs to be afraid of that. IM could just be as *discreet* as health itself. It could be inviting with all its offers and maybe be most beneficial when being an unobtrusive and inviting medicine of health. The physician's attitude of multiple offers of interventions of IM would be the one of "freigebende Fürsorge" ("releasing care") (Gadamer 2010, p. 141, translated by the authors).

IM can even capture the idea, which Nietzsche articulated to a "great" health. Even if Nietzsche's philosophy and ethics cannot be adopted *in toto*, his ideas on health gain more and more attention. A health is "great", when it is able to handle and digest. In this sense, health is always something extremely individual:

> For there is no health as such, and all attempts to define a thing that way have been wretched failures. Even the determination of what is healthy for your body depends on your goal, your horizon, your energies, your impulses, your errors, and above all on the ideals and phantasms of your soul (Nietzsche 1974, Aph. 120).

Acknowledgments The authors are grateful to Julia Engels for the translation of the article and Sally Werner for assisting with the references, as well as Veronika Surau-Ott and Dr. Astrid Lindberg for some inspiring ideas.

References

Antonovsky A (1987) Unraveling the mystery of health. Jossey-Bass, San Francisco
Anschütz F (1987) Ärtzliches Handeln. Grundlagen, Möglichkeiten, Grenzen, Widersprüche. Wissenschaftliche Buchgesellschaft, Darmstadt
Árnason V (2012) The personal is political: ethics and personalized medicine. Ethical Perspect 19(1):103–122
Aurenque D (2012) Personalized medicine as encouragement and discouragement of patient autonomy. In: Dabrock P, Braun M, Ried J (eds) Individualized medicine between hype and hope: exploring ethical and societal challenges for healthcare. LIT, Berlin, pp 33–50
Bartens W (2011a) Die Mogelpackung. http://www.sueddeutsche.de/wissen/personalisierte-medizin-die-mogelpackung-1.1121890. Accessed 6 May 2012
Bartens W (2011b) Wie fehlender Wohlstand die Lebenserwartung dämpft. http://www.sueddeutsche.de/leben/lebenserwartung-von-geringverdienern-nordic-walking-ersetzt-sozialpolitik-1.1236074. Accessed 6 May 2012
Beauchamp TL, Childress JF (2008) Principles of biomedical ethics. Oxford University, Oxford
Bloch E (1985) Das Prinzip Hoffnung. In: Werkausgabe, vol 5. Suhrkamp, Frankfurt a. M.

Boorse C (2004) On the distinction between disease and illness. In: Caplan AL, McCartney JJ, Sisti DA (eds) Health, disease and illness. Concepts in medicine. Georgetown University Press, Washington, D.C., p 77–89

Borgmann A (1984) Technology and the character of contemporary life. University of Chicago, Chicago

Bullock E (2010) Informed consent as waiver: the doctrine rethought. Ethical Perspect 7(4):529–555

Bunnik EM, de Jong A, Nijsingh N et al (2013) The new genetics and informed consent: differentiating choice to preserve autonomy. Bioethics 27(6):348–355

Buß M (2012) Zukunftsweisende Gesundheitsmodelle und ihre architektonische Interpretation. J Soc Manage 10(1):11–22

Buzzoni M (2003) Medicine as a human science between the singularity of the patient and technical scientific reproducibility. Poiesis Prax 1(3):171–184

Canguilhem G (2002) Écrits sur la médecine. Édition du Seuil, Paris

Danzer G, Rose M, Walter M et al (2002) On the theory of individual health. J Med Ethics 28(1):17–19

Düwell M (2008) Bioethik: Methoden, Theorien und Bereiche. Metzler, Stuttgart

Fischer T, Dörr M, Haring R et al (2015) Alarming symptoms of a paradigm shift? An approach to bridge the gap between hypothetical ethics and the current status of individualised medicine research. In: Vollmann J, Sandow V, Wäscher S et al (eds) Personalised medicine: ethical, medical, economic and legal critical perspectives. Furnham, Ashgate

Flessa S, Marschall P (2012) Individualisierte Medizin: vom Innovationskeimling zur Makroinnovation. Pharmacoeconomics 10(2):53–67

Foucault M (1973) Die Geburt der Klinik. Eine Archäologie des ärztlichen Blicks. Hanser, München

Foucault M (1985) Freiheit und Selbstsorge. Gespräch mit Michel Foucault am 20. Januar 1984. In: Becker H, Wolfstetter L, Gomez-Müller A et al (eds) Michel Foucault, Freiheit und Selbstsorge. Interview 1984 und Vorlesung 1982. Materialis, Frankfurt a. M., pp 7–28

Foucault M (1986) The care of the self—the history of sexuality, vol 3. Penguin, London

Foucault M (1990) The history of sexuality, volume one: an introduction. Vintage Books, New York

Foucault M (1993) Technologien des Selbst. In: Martin LH, Gutman H, Hutton PH (eds) Technologien des Selbst. Suhrkamp, Frankfurt a. M., pp 24–62

Franke A (2012) Modelle und Gesundheit von Krankheit. Hans Huber, Bern

Fuhr C (2013) Armut macht krank. http://www.aerztezeitung.de/politik_gesellschaft/versorgungsforschung/article/839622/top-thema-aerztetag-armut-macht-krank.html. Accessed 5 June 2013

Gadamer HG (2010) Über die Verborgenheit der Gesundheit. Suhrkamp, Frankfurt a. M.

Gadebusch Bondio M, Michl S (2010) Individualisierte Medizin. Die neue Medizin und ihre Versprechen. Dtsch Arztebl 107(21):A-1062/B-934/C-922

Gefenas E, Cekanauskaite A, Tuzaite E et al (2011) Does the "new philosophy" in predictive, preventive and personalised medicine require new ethics? EPMA J 2(2):141–147

Gert B, Clouser KD, Culver CM (1981) Malady: a new treatment of disease. Hastings Cent Rep 11(3):29–37

Golubnitschaja O (2009) Früherkennung ist oft nicht "früh" genug: Weshalb wir eine prädiktive Diagnostik brauchen. http://scienceblogs.de/vde-medtech/2009/09/11/fruherkennung-ist-oft-nicht-fruh-genug-weshalb-wir-eine-pradiktive-diagnostik-brauchen/. Accessed 18 June 2014

Griggs JJ (2009) Personalized medicine: a perk of privilege? Clin Pharmacol Ther 86(1):21–23

Grill M, Hackenbroch V (2011) Das große Versprechen. Der Spiegel 32:124–128

Habermas J (1981) Theorie des kommunikativen Handelns. Handlungsrationalität und gesellschaftliche Rationalisierung, vol 1. Suhrkamp, Frankfurt a. M.

Hegel GWF (1970) System der Wissenschaft. Erster Theil, die Phänomenologie des Geistes. In: Moldenhauer E, Michel KM (eds) Werke in 20 Bänden, vol 3. Suhrkamp, Frankfurt a. M.

Heidegger M (1977): Sein und Zeit. In: Herrmann, F-W von, Klostermann V (eds) Heidegger-Gesamtausgabe. Abteilung 1, Veröffentlichte Schriften 1914–1970, vol 3. Verlag Vittorio Klosermann, Frankfurt a. M.

Hempel U (2009) Personalisierte Medizin I. Keine Heilkunst mehr, sondern rationale molekulare Wissenschaft. Dtsch Arztebl 106(42):A2068-A2070

Heyll U (2006) Wasser, Fasten, Luft und Licht. Die Geschichte der Naturheilkunde in Deutschland. Campus, Frankfurt a. M.

Holtzman NA (1989) Proceed with caution: predicting genetic risks in the recombinant DNA era. John Hopkins University Press, Baltimore

Hucklenbroich P (2010) Der Krankheitsbegriff: Seine Grenzen und Ambivalenzen in der medizinethischen Diskussion. In: Höfner M, Schaede S, Thomas G (eds) Endliches Leben. Interdisziplinäre Zugänge zum Phänomen der Krankheit. Mohr Siebeck, Tübingen, pp 133–160

Hucklenbroich P (2013) Die wissenschaftstheoretische Struktur der medizinischen Krankheitslehre. In: Hucklenbroich P, Buyx A (eds) Wissenschaftstheoretische Aspekte des Krankheitsbegriffs. Mentis, Münster, pp 13–83

Hüsing B, Hartig J, Bührlen B et al (2008) Individualisierte Medizin und Gesundheitssystem. Zukunftsreport des Büros für Technikfolgen-Abschätzung beim Deutschen Bundestag. https://www.tab-beim-bundestag.de/de/pdf/publikationen/berichte/TAB-Arbeitsbericht-ab126.pdf. Accessed 17 June 2014

Hyland ME (2011) The origins of health and disease. Cambridge University Press, Cambridge

Juengst ET, Settersten RA Jr, Fishman JR et al (2012) After the revolution? Ethical and social challenges in 'personalized genomic medicine'. Per Med 9(4):429–439

Kant I (1968) Grundlegung zur Metaphysik der Sitten. In: Weischedel W (ed) Werke in zehn Bänden, vol 7. Wissenschaftliche Buchgesellschaft, Darmstadt

Knorr-Cetina K (1984) Die Fabrikation von Erkenntnis. Zur Anthropologie der Naturwissenschaft. Suhrkamp, Frankfurt a. M.

Kohane IS (2009) The twin questions of personalized medicine: who are you and whom do you most resemble? Genome Med 1(4). doi:10.1186/gm4

Kollek R, Lemke T (2008) Dimensionen genetischer Verantwortung. In: Der medizinische Blick in die Zukunft. Gesellschaftliche Implikationen prädiktiver Gentests. Campus, Frankfurt a. M., pp 223–287

Kuhn T (1962) Die Struktur wissenschaftlicher Revolutionen. Suhrkamp, Frankfurt a. M.

Langanke M, Fischer T (2012) Gesundheitsmanagement liegt mir im Blut, Individualisierte Medizin und gesundheitliche Eigenverantwortung. In: Gadebusch BM, Siebenpfeiffer H (eds) Konzepte des Humanen, Ethische und kulturelle Herausforderungen. Alber, Freiburg, pp 139–172

Langanke M, Fischer T, Brothers KB (2013a) Public health, it is running through my veins: personalized medicine and individual responsibility for health. In: Dabrock P, Braun M, Ried J (eds) Individualized medicine between hype and hope. Exploring ethical and societal challenges for healthcare. LIT, Berlin, pp 149–172

Langanke M, Fischer T, Erdmann P et al (2013b) Gesundheitliche Eigenverantwortung im Kontext Individualisierter Medizin. Ethik Medizin 25(3):243–250

Lemke T (2007) Biopolitik zur Einführung. Junius, Hamburg

Lemke T (2008) Gouvernementalität und Biopolitik. für Sozialwissenschaften, Wiesbaden

Locke J (1988) Two treatises of government. In: Laslett P (ed) Cambridge texts in the history of political thought. Cambridge University, Cambridge

Luhmann N (1988) Wirtschaft der Gesellschaft. Suhrkamp, Frankfurt a. M.

Maio G (2012) Chancen und Grenzen der personalisierten Medizin—eine ethische Betrachtung. GGW 12(1):15–19

Mancinelli L, Cronin M, Sadée W (2000) Pharmacogenetics: the promise of personalized medicine. AAPS PharmSci 2(1):E4

Manson NC, O'Neill O (2007) Rethinking informed consent in bioethics. Cambridge University, Cambridge

Marx K (1962) Das Kapital, vol 1. In: Marx-Engels-Werke (MEW), vol 23. Dietz Verlag, Berlin

Menke C (2003) Zweierlei Übung. Zum Verhältnis von sozialer Disziplinierung und ästhetischer Existenz. In: Honneth A, Saar M (eds) Michel Foucault. Zwischenbilanz einer Rezeption. Suhrkamp, Frankfurt a. M., pp 283–299

Nature Biotechnology Editorial (2012) What happened to personalized medicine? Nat Biotechnol 30(1). doi:10.1038/nbt.2096

Nietzsche F (1968) Twilight of the Idols, or, How to Philosophize with a Hammer. In: Kaufmann W (ed) The Portable Nietzsche. Viking Press, New York

Nietzsche F (1974) The gay science: with a prelude in rhymes and an appendix of songs (trans: Kaufmann W). Vintage Books, New York

Nietzsche F (1999a) Fröhliche Wissenschaft. In: Colli G, Montinari M (eds) Sämtliche Werke: kritische Studienausgabe in 15 Bänden, vol 3. De Gruyter, Berlin

Nietzsche F (1999b) Götzendämmerung und Ecce Homo. In: Colli G, Montinari M (eds) Sämtliche Werke: kritische Studienausgabe in 15 Bänden, vol 6. De Gruyter, Berlin

Ott K (2008) Diskursethik und die Grundzüge bioethischer Diskurse. In: Brand C, Engels EM, Ferrari A et al (eds) Wie funktioniert Bioethik. Mentis, Paderborn, pp 61–96

Ott K, Fischer T (2012) Can objections to individualized medicine be justified? In: Dabrock P, Braun M, Ried J (eds) Individualized medicine between hype and hope. Exploring ethical and societal challenges for healthcare. LIT, Berlin, pp 173–200

Plessner H (1975) Die Stufen des Organischen und der Menschen. De Gruyter, Berlin

Poste G (2011) Bring on the biomarkers. Nature 469:165–157. (doi:10.1038/469156a)

Rajan KS (2006) Biocapital. The constitution of postgenomic life. Durham, London

Ros A (1990) Begründung und Begriff. Wandlungen des Verständnisses begrifflicher Argumentation. Meiner, Hamburg

Rudnik-Schöneborn S, Langanke M, Erdmann P et al (2014) Ethische und rechtliche Aspekte im Umgang mit genetischen Zufallsbefunden—Herausforderungen und Lösungsansätze. Ethik Medizin 26(2):105–119

Schäfer L (1993) Das Bacon-Projekt. Von der Erkenntnis, Nutzung und Schonung der Natur. Suhrkamp, Frankfurt a. M.

Schleidgen S, Marckmann G (2013) Alter Wein in neuen Schläuchen? Ethische Implikationen der Individualisierten Medizin. Ethik Medizin 25(3):223–231

Snyderman R, Langheier J (2006) Prospective health care: the second transformation of medicine. Genome Biol 7(104). doi:10.1186/gb-2006-7-2-104

SRU (2004) Umweltgutachten 2004. Umweltpolitische Handlungsfähigkeiten sichern, p 89. http://www.umweltrat.de/cae/servlet/contentblob/465772/publicationFile/34305/2004_Umwelt-gutachten_Kurzfassung.pdf. Accessed 17 June 2014

Stegmaier W (2008) Philosophie der Orientierung. De Gruyter, Berlin

Vollman J (2013) Persönlicher—besser—kostengünstiger? Kritische medizinethische Anfragen an die "personalisierte Medizin". Ethik Medizin 25(3):233–241

Wagenaar C, Mens N (2010): Healthcare architecture in the Netherlands. Nai Publ, Rotterdam

Walach H, Meyer-Abich K-M (2005): Eigenverantwortung und gesunde Lebensweise. Thesen zu einer verantwortlichen Gesundheitspoltik für mündige Bürger. University Hospital Freiburg, Institute of Environmental Medicine and Hospital Epidemiology. http://www.psychologie.uni-oldenburg.de/wilfried.belschner/vl_gw_2005_walach.pdf. Accessed 5 Dec 2012

Watson JD, Cook-Deegan RM (1990) The human genome project and international health. JAMA 263(24):3322–3324

Weber M (1968) Der Sinn der 'Wertfreiheit' der soziologischen und ökonomischen Wissenschaften. In: Methodologische Schrifte, Studienausgabe. S. Fischer, Frankfurt a. M., pp 229–278

Wehkamp K (2012) Gesundheit als Potential. Warum der WHO-Gesundheitsbegriff verändert werden sollte! In: Gadebusch BM, Siebenpfeiffer H (eds) Konzepte des Humanen. Alber, Freiburg, pp 103–116

Wiesing U (1995) Kunst oder Wissenschaft. Konzeptionen der Medizin in der deutschen Romantik. Frommann-Holzboog, Stuttgart

Wiesing U (2004) Wer heilt, hat Recht? Über Pragmatik und Pluralität in der Medizin. Schattauer-Verlag, Stuttgart

Wikler D (1987) Who should be blamed for being sick? Health Educ Q 14(1):11–25

Windelband W (1907) Geschichte der Naturwissenschaften. In: Präludien, Aufsätze und Reden zur Einleitung in die Philosophie. J.C.B. Mohr, pp 355–379

Wittgenstein L (2009) Philosophical investigations (trans: Anscombe EM, Hacker PMS, Schulte J). Wiley-Blackwell, Oxford

World Health Organization (1946) Preamble to the constitution of the world health organization as adopted by the international health conference. http://www.who.int/about/definition/en/print.html. Accessed 18 June 2014

Chapter 9
The Concept of Disease in the Era of Prediction

Johann-Christian Põder and Heinrich Assel

Abstract Individualized Medicine is often accused of causing an extension of the concept of disease. It would thereby promote a pathologization of life, which is problematic in several respects. This critical query should be taken seriously. The aim of the innovative research in the field of biomarker-based predictive medicine is to increase the accuracy in predicting the onset of a disease. For those who are affected, the predictive knowledge often creates a state of uncertainty and stress, which could easily be construed as a disease. The rapidly growing amount of information about dispositions to disease creates new situations of decision-making and responsibility while also leading to the emergence of new personal and social patterns of identification (e.g., 'healthy ill' persons). The following section discusses the allegation of pathologization through an examination of the disease theories by Christopher Boorse, Peter Hucklenbroich, and Dirk Lanzerath. How is the relationship between disposition and disease interpreted in those representative theories? What are the consequences of this medical theory debate for the allegation that Individualized Medicine leads to pathologization of life?

Keywords Theory of Medicine · Individualized Medicine · Disease · Disposition · Prediction · Pathologization

9.1 Introduction

The aim of the following discussion is to ask whether the biomedical and biotechnological developments in the field of Individualized Medicine (IM) have any consequences for our perception and concept of disease. What happens with the concept of disease in the context of IM? The focus is on a specific question, which

J.-C. Põder (✉) · H. Assel
Theologische Fakultät, Lehrstuhl für Systematische Theologie,
Ernst-Moritz-Arndt-Universität Greifswald, Am Rubenowplatz 2–3,
17487 Greifswald, Germany
e-mail: johann-christian.poder@eelk.ee

H. Assel
e-mail: assel@uni-greifswald.de

© Springer International Publishing Switzerland 2015   165
T. Fischer et al. (eds.), *Individualized Medicine,* Advances in Predictive,
Preventive and Personalised Medicine 7, DOI 10.1007/978-3-319-11719-5_9

is of crucial importance for this subject matter: Does IM cause an extension of the concept of disease and a consequent pathologization of our lives? An answer to this question is sought through examination of three representative theories of disease (Hucklenbroich, Boorse, Lanzerath). However, the discussion should be started with a clarification of the notion of IM and identification of the aspects of IM, which could be of relevance for the concept of disease.

9.2 Individualized Medicine and the Concept of Disease

An objective discussion of the concept of disease in the context of IM requires sufficient clarity with regard to the concept and phenomenal content of IM. It could even be said that a public discourse about the opportunities, risks and limits of IM would be 'simply impossible' without such a clarity (Schleidgen et al. 2013, p. 1). Both Langanke et al. (2012a) and Schleidgen et al. (2013) acknowledge in their definitional analyses that there is currently a great semantic variability and vagueness when it comes to the meaning of 'personalized' or 'individualized' medicine. With critical and subversive opinions included, it is possible to identify at least five different main interpretations of IM: (a) IM is a 'misnomer' operating with empty promises; (b) IM cannot be seen as a novel concept as medicine has always been personalized; (c) IM means holistic medicine, focusing on individual needs of the patient; (d) the notion of IM covers methodologically heterogeneous concepts of personalization in medical technology, which have the goal of fine-tuning health care for the needs of the individual (unique therapeutic measures and stratification based on biomarkers); (e) IM is an approach in medical research and health care, distinguished methodologically by its reliance on biomarker-based stratification. For this study, the author has adopted the last interpretation, which has a clear conceptual definition and can be associated with the definitional clarification of IM proposed by Langanke et al. (2012a).

9.2.1 Elements of the Definition of IM—Biomarkers, Stratification, Prediction

According to Langanke et al. (2012a), IM includes, on the one hand, those approaches to medical research that aim to "*Identify and validate biomarkers and integrate them into clinical practice to enable better projection (prediction) of the onset or progression of diseases and/or of efficacy or adverse effects of treatments for specific groups of patients*". On the other hand, IM encompasses "*preventive, therapeutic and rehabilitative health care practices in which [...] biomarkers are used for systematic projection (prediction) of disease risks or progressions and/or prediction of efficacy or adverse effects of treatments*" (306–307; emphasis in original). This definition is based on the following interconnected elements: "biomarkers", "stratification" and "prediction".

a. "Biomarkers", used as key elements for making predictions in IM, are biological characteristics measured in specific laboratory tests. They can serve as indicators of physiological and pathological processes. Genetic biomarkers, in particular, were very important in the initial development of IM, especially in the rapidly developing field of pharmacogenetics. Today, the study of biomarkers in IM includes a wide range of -omics fields, such as genomics, transcriptomics, proteomics, and metabolomics. According to Langanke et al., IM can make use of both the genetic and epigenetic molecular biomarkers as well as non-molecular biomarkers (e.g., blood pressure, anthropometric data, data from medical imaging) and combinations of biomarkers from -omics fields and non-molecular tests (Langanke et al. 2012a, p. 309, for a different opinion see: Schleidgen et al. 2013, p. 9; cf. Hüsing et al. 2008, p. 12 and Langanke et al. chap. 2).

b. "Stratification" is crucial for the methodology of IM, because the high sensitivity and specificity of biomarkers enables a finer and improved division of individuals into sub-populations when predicting disease onset and/or therapeutic outcomes. Such stratification should improve the targeting of diagnostics and treatment and contribute to the development of individually tailored diagnostic and therapeutic approaches. In methodological terms, however, IM operates at the level of "groups of persons", not single individuals. For this reason, it is often suggested that 'stratified medicine' would be a more accurate term instead of IM (Schleidgen et al. 2013, p. 11; Dabrock 2011, p. 243).

c. "Prediction", specifically the predictive use of biomarkers, is the third main element of IM. Early identification of disease risks and determination of the best treatment options is the main goal of biomarker-based stratification. Early detection of disease risks, even before any clinical symptoms, should improve the efficacy of prevention. If a disease has nevertheless developed, the aim is to provide the patient with the treatment option, which has the best chances of success (Langanke et al. 2012b, p. 152; Karger and Hüsing 2011, p. 1–2).

9.2.2 The IM Concept of Disease—Reductionist, Instrumental and Expansive?

The lofty ambitions and visions, as well as the extensive, innovative research activities of such biomarker-based IM, which is geared toward prediction, also create a cause for concern and provoke critical questions. This opens up a varied and challenging field of ethical, legal and social problems, often leading to fundamental philosophical and anthropological questions. One such fundamental question concerns our understanding of disease and health. The question is, whether or how IM changes our perception of disease and health? The current discussion includes at least three interconnected arguments on the possibilities of how IM could influence our understanding of disease: the concept of disease in IM has been said to be "reductionist", "instrumental" and "expansive".

a. It is argued that the concept of disease in IM is "reductionist", because disease is interpreted primarily at the biological level, relying on biological data (biomarkers). According to Henke, IM is subject to 'biological reduction' as it 'inevitably ignores' the biographic and psychosomatic aspects (Henke 2011, p. 40). Woopen believes that in the course of development of IM, disease "will, even more than today, [...] become a phenomenon defined by its molecular characteristics and decoupled from subjective experience" (Woopen 2011, p. 101, cf. 105–106). The criticism of the reductionist 'scientistic orientation' (Bergdolt 2011, p. 27) of IM applies in particular to genetic research and its prominent position in IM (Kollek and Lemke 2008).

b. Secondly, it is argued that IM promotes an idea of "instrumental" feasibility and a potential for deliberate manipulation of disease and health. Predictive diagnostics, preventive care and innovative medical interventions can create the possibility of access and control in an area of life, which has previously had the character of an inevitable occurrence. This view is emphatically expressed by Kollek: "The individualization of medicine will lead to a fundamental shift in perception, i.e., health and disease would no longer be categories of chance or fate, but the objects and results of wilful action. 'If health is the result of the will, then disease would be the side effect of a lacking or misguided will'" (Hempel 2009, A-2070, cf. also Langanke et al. 2012b, p. 151). Such a shift would create a novel increase in individual responsibility for one's own health, but it could also mean a loss of solidarity in the provision of public health care.

c. Thirdly, IM is said to lead to an "expansive" extension of the concept of disease, or to an increasing pathologization of life. It is expected that, as a result of refinement and increasing utilization of predictive test methods, the number of 'diagnosable conditions' will increase significantly, consequently increasing the impact of medicine on our lives (Karger and Hüsing 2011, p. 3). The predictive knowledge about dispositions to disease would create new categories, such as 'healthy ill' or 'ill healthy' people (Stockter 2008, pp. 50–52; Bobbert 2012, p. 168; Zerres 2006). The findings (e.g., genetic peculiarities) obtained through a molecular procedure would then often be assigned the value of a disease, despite the absence of clinical symptoms or any experiences or suffering caused by the disease. This could lead, for instance, to a perception of "universal genetic morbidity" (Juengst 2009).

d. It is not possible here to present a detailed discussion of all three characteristics of the disease concept, allegedly derived from IM. As mentioned in the introduction, this study focuses on the third item, i.e., the question of possible expansive extension of the concept of disease. However, the accusation of a reductionist conceptualization of disease plays an important role in the context of this question as well (cf. the principal criticism of this accusation by Ott and Fischer sect. 8.3.3). With regard to the question of deliberate manipulation of disease and potential strengthening of individual responsibility, a reference can be made to the context-sensitive analysis by Langanke et al. (2012b).

9.2.3   IM and Extension of the Concept of Disease—Preliminary Remarks

Does the characteristic predictive, particularly genetic predictive, knowledge found in IM indeed cause an extension of the concept of disease, or maybe even a perception of "universal presymptomatic multimorbidity" (Bayertz 1998, p. 250)? Is it possible for a concept of disease to subsume the particular area of interest of IM— biomarker-based predictive knowledge? Before this question can be discussed in the context of selected contemporary theories of disease, some clarifying comments should be made.

Predictive knowledge is characterized by the ability to predict, with a greater or lesser degree of accuracy, whether a person will develop a particular medical condition in the future. The goal is to determine disease risks or a disposition to disease. Such knowledge is probabilistic by its nature; it operates with the statistical probability of development of a disease. Association analysis is the typical methodology employed in this context to determine whether a biomarker, e.g., a genetic mutation or a protein sequence, has any statistical relevance in relation to a disease. However, a completely accurate prediction is possible only in the case of a few, fully penetrative diseases (e.g., Huntington's chorea), while prediction becomes particularly difficult and scientifically challenging when it comes to complex, multi-factorial widespread diseases (e.g., cardiovascular conditions, diabetes, allergies). Hereditary breast cancer is a disease with a relatively high degree of probability of prediction. In the case of a BRCA1 mutation, the risk of disease is between 40 and 80% (German Medical Association).

Despite those margins, one can often encounter the view that predictive knowledge leads to a new category of people, the 'healthy ill'. This construct means an extension of the concept of disease to the disposition to, or the risk of, disease. A phenotypically healthy person without symptoms is regarded as having a disease if observed through the prism of the disposition to disease or disease risk. This dynamic has also been described as 'de-temporalization of disease': "*de-temporalization of disease* means an increasing detachment of the concept of disease from the symptoms (acute or chronic) and complaints occurring at certain points in time, and a 'forward displacement' of a diagnosis due to specific indications and 'risk factors'. The result is a 'healthy ill' person [...]" (Sauter and Gerlinger 2012, p. 205). Understandably, such an extension of the concept of disease would place the person concerned in an ambiguous position, which could be challenging both in psychological and social terms. Despite having no symptoms whatsoever, the person would be considered as being ill due to the disposition to disease (among others, this view could be adopted by social actors, such as employers or insurance companies). Due to this ambiguity, such a situation is often also seen as a third, paradoxical state *besides* disease and health (Lenk 2003, p. 177).

If no attempt is made in this situation to save some dispositions from pathologization by creating a classification of dispositions (e.g., based on the strength of correlation between a disposition and the manifestation of a disease, and the length of time to disease manifestation; cf. Bobbert 2012, pp. 179–189), the extension of the concept of disease to dispositions could easily lead to a "universal presymptomatic multimorbidity" (Bayertz). This is made possible by the refining of predictive methods, enabling to discover an increasing number of dispositions to disease. Research in human genetics, for instance, has revealed that every human being possesses a number of innate dispositions to disease: "We are all flawed mutants" (Collins 2011, p. 16). Every one of us has some genetic burdens or glitches, which are associated with disease risk. This enables Bayertz to conclude: "Everybody will be 'diseased'—but in a hidden and ambiguous sense" (Bayertz 1998, p. 250).

The extension of the disease concept to dispositions, including particularly the consequent universal pathologization, is in effect close to a disease concept, in which disease is defined at a molecular, e.g., genetic, level (cf. the accusation of biological reductionism 9.2.2/a). Here the constituent point of reference is no longer limited to a disease, which will become phenotypically manifest in the future (futuristic aspect), but it also encompasses the factual identification of a disease-associated molecular property with the disease itself (e.g., the association of the BRCA1 mutation with breast cancer). It is also possible that a phenotypically relevant risk of disease will be considered unimportant and a 'pathology' or a 'disease' will be defined by a mere deviation from a statistical 'standard state' of the genome. Either way, the disposition to disease in itself (*qua* disposition) is not included in the sphere of definition of disease and, instead, the concept of disease is defined at a molecular level. Even though this approach has several traits of reductionism, being based on biological determinism and inclining towards decoupling of the subjective experience of disease, it is possible to talk about the "extension" of the disease concept in this case as well—in the sense of expansive pathologization of life. Considering this, Magnus makes a critical comment in reference to a genetic definition of disease: "It is likely that we are all genetically diseased" (Magnus 2009, p. 241).

9.3   Predictive Knowledge and Theories of Disease

There are three main trends that emerge from the current literature on the philosophy of medicine: conflict, integration and ambiguity/lack of transparency (Põder and Assel 2014). Firstly, the debate has been, for many decades, shaped by the conflict between naturalistic and normative concepts of disease and health. On the one hand, modern medicine, and in its wake also the social perception is characterized by the dominance of the objectifying scientific perspective. The biomedical concept of disease and health makes the modern clinical practice verifiable and manageable. Medical judgment is guided and justified, to a significant degree, by biomedical knowledge (Paul 2006, p. 136). On the other hand, our experience of disease and health, as well as its conceptual articulation, is embedded in various normative—individual, social and cultural—contexts. Secondly, there are attempts to replace the (naturalistic or normative) reductionist view with a more complex, multidimensional analysis. This concerns both, an integrative perception of nature, person (existence) and culture, as well as a comprehension of the role of medical practice (doctor-patient relationship). This integrative trend expresses itself partially also in the so-called biopsychosocial model of disease, but it often lacks—at least in its strictly systems-theoretical versions—a consistent inclusion of the participant, or the first-person perspective. Generally, this second trend too is often marked by the naturalism-normativism-controversy as it poses the question of primacy among the different integrative elements. The third and final observation concerns a certain caginess and skepticism with regard to the possibility of a univocal definition of disease. The complex and difficult-to-grasp dynamic between empirical-scientific, existential and cultural dimensions of disease leads to conceptual ambiguity and lack of transparency (cf. the cluster-concept theory of disease, Stoecker 2009). It may also generate principal doubts about the necessity of a general definition of disease (Wiesing 1998, Hesslow 1993).

From this wide-ranging debate, three representative theories of disease have been selected for discussion on the following pages. Hopefully, this will contribute to the current debate by improving the understanding and appreciation of the relation between predictive knowledge and the disease concept, or the issue of potential extension of the disease concept to include dispositions. The first to consider is the naturalistic theory of Christopher Boorse. He has proposed one of the most prominent disease theories of the past decades and is a good representative of the antagonistic debate between naturalism and normativism. The second view to be discussed is an integrative position, which has a naturalistic emphasis as well: the reconstructive theory by Peter Hucklenbroich, based on the philosophy of science. An integrative viewpoint is also represented in the third position, but with a clear normative emphasis: the 'practical' disease theory by Dirk Lanzerath.

9.3.1   Biostatistical Theory of Disease (Christopher Boorse)

According to Boorse's biostatistical theory, disease is a deviation from the statistical normality of the functional capacity of an organism. He calculates the statistical normality, which is required for determination of a subnormal functional performance level, on the basis of the respective reference class (species, sex, age) and he understands the functional capacity in terms of survival and reproduction of the organism. Boorse assumes that every organism has a functional design or 'blueprint', in which all individual levels (e.g., organs, tissue, cells, genes) are aligned with the overarching functional goals (survival and reproduction). Consequently, not every statistical deviation can be classified as disease—the deviation should also involve a biological dysfunction of one or several subsystems of the organism.

This theoretical concept of disease is aimed at avoiding any reference to subjective or value-laden elements. Diseases are pathological processes or conditions, which can be described with scientific (biological) parameters. Consideration of the subjective disease experience (suffering, pain, impairment) is, therefore, not required for identification of a pathological dysfunction. The presence of clinical symptoms is, in principle, equally unimportant. Technically, every human being has some microdysfunctions, which are imperceptible for us (e.g., a dead neuron, Boorse 1997, pp. 50–51). Therefore is it not necessary to link, like Bayertz, the consequence of a universal pathologization to predictive knowledge and to the continuing discovery of new genetic dispositions to disease. Boorse stoically accepts Murphy's comment that, "a normal person is anyone who has not been sufficiently investigated" (Murphy 1976, p. 123; cf. Bayertz 1998, p. 249), as he states that: "It is certain in any case that to the pathologist no one is normal" (Boorse 1997, p. 50).

It seems, however, that a pathologization of "dispositions" (as an increased disease risk) is not compatible with Boorse's concept of disease. If disease is a dysfunction, it would be impossible to pass the judgment of morbidity on something, which has no functional nature in itself. Indeed, a disposition as an increased disease risk is not a biological function of the organism. Boorse himself points to the importance of differentiating between a disposition and a disease. Especially among physical states, it is easy "to confuse diseases with dispositions to become diseased under certain conditions" (Boorse 1977, p. 553). According to Boorse, such a confusion would result in a disintegration of the disease concept: "If whatever can cause disease were itself disease, everything would be a disease, since any causal connection is possible in a special environment" (ibid.). Admittedly, it is sometimes possible that a disposition is factually also a disease, e.g., AIDS as a disease also creates a disposition to fungal infections or certain types of cancer. However, AIDS does not become a disease by virtue of being a disposition but due to being itself a dysfunction. Indeed, many other dispositions—or, in other words, risk factors that combine to constitute the disposition to developing a disease under certain conditions—are not diseases (dysfunctions). In a more recent article, Boorse explicitly criticizes the tendency of increasingly treating risk-factor levels as diseases, as well as the concept of risk-based diseases. He makes the critical observation that there is a growing conceptual conflict between preventive and curative medicine, with an increasing confusion about their respective categories (Boorse 2012, unpublished).

9.3.2 Reconstructive Theory of Disease (Peter Hucklenbroich)

Hucklenbroich's theory of disease is aimed at creating a broad reconstruction of the disease concept of the modern scientific medicine in terms of the philosophy of science (Hucklenbroich 2013, pp. 18–25). Unlike Boorse, who had a similar reconstructive goal (Boorse 1977, p. 543), Hucklenbroich would like to do better justice to the biopsychosocial complexity of modern medicine. The core of Hucklenbroich's disease theory is based on five main criteria of pathologicity, with (a), (b) and (e) being the most important:

> (a) Shortening of lifetime expectancy, or immediate lethality, (b) Pain, and other specific somatic or vegetative complaints, (c) Infertility, i.e., inability of biological reproduction, (d) Inability or impairment of living together in human symbiotic communities, (e) Non-universal disposition of the organism to develop a condition that is pathological according to one or more of these criteria (this clause covers also conditions that are usually called risk/risk factor, disability, impairment, or handicap) (Hucklenbroich 2014, p. 1).

Unlike Boorse, Hucklenbroich attaches no importance to the concepts of function and statistical normality. As his reconstruction lacks a similar strong systematic foundation, the listed criteria (cf., e.g., Hucklenbroich 2013, p. 45), which are linked through an inclusive disjunction, could seem somewhat arbitrary or random, despite their alleged *prima facie* intuitiveness. However, he emphasizes the need for a detailed explication of those criteria in general pathology and nosology, as well as their shared characteristic of being 'threatening to life' in various ways (Hucklenbroich 2013, p. 51). Two additional conditions are important in this context: Firstly, it must be an involuntary process of the individual organisms. Secondly, there should be "at least one naturally occurring [...] alternative course, in which the process in question and its consequence (death, suffering, ...) does not occur" (Hucklenbroich 2013, p. 46). For Hucklenbroich, this last point represents an alternative to Boorse's (bio)statistical concept of normality (Hucklenbroich 2013, p. 72). Instead of a statistical comparison, we have a strategy which seems to resemble Popperian falsification: The hypothesis that all swans are white is falsified as soon as a black swan is found. Similarly, the hypothetical assumption that, for instance, cancer or stroke are healthy life processes (and not diseases) is falsified as soon as at least one 'better natural course' is found.

A more detailed discussion of this strategy is neither possible nor necessary in this context. What is interesting is the relation between disposition and disease according to Hucklenbroich. The main criterion (e) is naturally a conspicuous reference point. Hucklenbroich subsumes under the disease concept any conditions or events (dispositions), which increase the risk of becoming diseased in terms of the other listed criteria. He also explicitly extends the disease concept to dispositions. He consciously accepts the threat of universal morbidity, associated with such extension:

> The result of consistent application of this [...] criterion is a situation where a large number of people, possibly all of them, would have to be considered *ill*—in terms of being *carriers of one or several morbid dispositions* or *carriers of an (internal) risk of disease*, not in terms of the presence of actual complaints (Hucklenbroich 2010, p. 150; emphasis in original).

He refers, in particular, to the extremely high number and diversity of genetic dispositions. Despite complex intermediate forms (e.g., pleiotropy, in which a genetic arrangement can have both negative and positive consequences), we can assume that each of us carries at least one clearly negative disposition. This means that, according to Hucklenbroich, no human being can be regarded as being completely healthy from the perspective of a "strict medical disease concept" (Hucklenbroich 2010, p. 151).

Hucklenbroich's emphasis on etiopathogenesis is also important for the relationship between disposition and disease. Alongside nosology and disease qualification (the five primary disease criteria), it belongs to the three constitutive dimensions of the disease concept: Disease is, therefore, a phenomenon that is qualified as a disease, explained through etiopathological models, and systematized in nosological terms (Hucklenbroich 2013, p. 39). While Boorse's disease concept was characterized by an almost horizontal level of dysfunction, Hucklenbroich's concept is distinctive for the inclusion of a causal, etiopathogenic 'depth dimension'. However, this 'vertical' extension of the disease concept to various pathogenic causes and processes leads to a partial inclusion of conditions that would normally be classified as dispositions (e.g., co-causal pathogenic gene variants). Gottschalk-Mazouz notes in reference to the consequent threat of identifying disposition with disease: "Perhaps Hucklenbroich follows the medical parlance and thereby reflects [...] the trend of increasing extension of the disease concept, prevalent among scientists, physicians and pharmaceutical companies" (Gottschalk-Mazouz 2008, p. 71).

9.3.3   Practical Concept of Disease (Dirk Lanzerath)

Unlike Boorse and Hucklenbroich, Lanzerath takes the subject as the starting point for the determination of the disease concept. The subjective disease experience is the constitutive primary element: "*A given state is experienced as a state of health or disease only by virtue of the manner, in which we interpret it and accept it as a practical task*" (Lanzerath 2000, p. 223; emphasis in original). Based on that normative horizon of individual interpretation, or first-person perspective, will it be possible—in a secondary objectification—to talk about a 'pathological state' in a medical and scientific language (Lanzerath 2000, p. 255). According to Lanzerath, scientific parameters are quite relevant for the disease concept, but they can be identified as *constituent elements of disease* only through a subjective experience of disease.

This criterion—the constitutive priority of subjective experience—is also essential for Lanzerath's understanding of the relation between disposition and disease. The conditions that could develop into a disease but have no perceptible symptoms (e.g., hypertension) should not be identified as diseases. Lanzerath believes that this applies, in particular, to genetic causes of disease:

> There are no diseased genes and a carrier of a genetic disposition, which could be interpreted as the original reason for a disease, cannot be regarded as being ill simply by virtue of being a carrier of the disposition. Only a phenotypical manifestation [...] could be designated as a 'disease' and only a person in this state could be called an 'ill' person (Lanzerath 2000, p. 202).

If this was not the case, we all would always have some kind of disease. While Hucklenbroich accepted such a consequence, Lanzerath regards it as a dissolution of the meaning of the disease concept (ibid.). There is a danger of 'subtle elimination' of the boundary between disease and health (Lanzerath and Honnefelder 1998, p. 74). Furthermore, any genetic variants, which indicate a statistically determined or monogenically conditioned connection with a disease, cannot be identified with a disease itself. In order to avoid a lapse into 'geneticism' or 'reductionism', it is important to pay attention to the difference "between genotype and phenotype, between disposition and disease" (Lanzerath and Honnefelder 1998, p. 66).

However, the apparent clarity of this relation decreases after taking into consideration Lanzerath's further definition of the disease concept. On the one hand, he points to the socio-cultural setting of our individual interpretation and, on the other hand, he emphasizes the systematically central relationship between a doctor and a patient as the fundamental, practical and normative reference framework of the disease concept. In addition to scientific parameters, the subjective experience of disease is also linked with a multitude of social and cultural factors.

In this way, disease becomes an *interpretandum* with a cluster of premises, of which only a part can be raised by scientific means, while the other part develops in the practical relationship between a doctor and a patient under the conditions of the relevant socio-cultural environment (Lanzerath 2000, p. 260).

What is the relation between this interpretative openness regarding the *interpretandum* and the aforementioned clear boundary between disposition and disease? Could the latter still be disposable via interpretation? Lanzerath argues that the aforementioned openness could create a possibility for a productive 'ambiguity', enabling to 'hammer out' the disease concept in a practical doctor-patient relationship (Lanzerath and Honnefelder 1998, p. 60). However, one should also ask, how is it possible to prevent the extension of the disease concept to dispositions in the course of such 'hammering out'? According to Lanzerath, the doctor-patient relationship entails two constitutive, practical and normative elements: the subjective need for help and the provision of help by the doctor. The patient meets the doctor in a situation in which the patient experiences 'a subjective need for help' and 'a disturbance of wellness' (Lanzerath 2000, p. 259). However, could it not be possible that even a disposition, or a personally and socially perceived risk, can be experienced by a patient as a veritable, even massive 'disturbance of wellness'? (cf. Ott and Fischer sect. 8.2.3).

Lanzerath attempts to avoid certain extensions of the disease concept (e.g., to enhancement) by arguing that the wish of the patient should be critically corrected through medical indication. This should be handled by the medical profession, including professional standards for the objectives of healing, prevention and relief of suffering (Lanzerath 2000, p. 272). Langanke and Werner (2015) rightly note—at least as a tendency—the circularity of such a justification of action, as the indication relies on criteria that could itself be interpreted as a 'disturbance of wellness' (or the patient's wish). This criticism has the pathologization of old age in mind. The ground is even shakier when it comes to the extension of the disease concept to dispositions. Indeed, working with the dispositions to disease—even without a reference to a 'disturbance of wellness'—is fully consistent with the traditional understanding of one specific medical field of indication (prevention). The 'hammering out' of the disease concept to include dispositions remains inside the framework of the traditional medical indication, thereby making a smoother shift of meaning possible.

9.4   Does Individualized Medicine Lead to a Pathologization of Life?—A Perspective

After this review of three representative theories of disease, trying to determine whether they entail a pathologization of life—or, more precisely, pathologization of *dispositions*—the discussion can be closed by examining two questions. The first question is: What conclusions can be drawn from this review with regard to IM? The second question is short and to the point: Does IM lead to a pathologization of life?

9.4.1   Disease Theories, Pathologization of Dispositions, and IM

What is the result of the analysis of the disease theories of Boorse, Hucklenbroich and Lanzerath? It is notable that both Boorse and Lanzerath oppose a conceptual merger of preventive and curative medicine. According to Boorse, only a biological dysfunction, not merely a disposition to such dysfunction, can be identified as disease. For Lanzerath, a phenomenon can be 'interpreted' as disease only in the presence of subjective experience, reflecting a phenotypical, perceivable manifestation. Interestingly, this shows that the proponents of both concepts, one plainly naturalistic and the other clearly normative, try to argue against an extension of the disease concept to dispositions. However, both types of theories have their own specific problems.

As was hinted in the brief overview of the recent trends in the debate on medical theory (see the introduction to sect. 9.3), the disease concept of Boorse is increasingly viewed as being too narrow and reductionist. There have been many critical questions regarding the statistical normality as well as the concept of function. Furthermore, a theoretical concept that relies exclusively on biological factors fails to include the everyday and practical usages of the disease concept. Nevertheless, it could be seen that Boorse's concept lends itself well to maintaining a conceptual differentiation between disease and disposition. This is true even if we make the critical observation that an increasing knowledge of the functional significance of disease-associated genetic markers (as well as other biomarkers) could lead to a situation where many dispositions are exposed as (micro-)dysfunctions. This tendency towards pathologization is made logically and methodologically possible by the fact that Boorse generally argues for a physiopathologically extensive disease concept, which is decoupled from a phenotypical manifestation.

Lanzerath, in his integrative theory, does not want to neglect the biological dimension of the disease concept, but subjects it partially to the primacy of subjective individual interpretation and partially to the process of practical 'hammering out' in a doctor-patient relationship. This provides his theory with pragmatic and contextual flexibility and ensures an excellent integration of the practical aspect of the doctor-patient relationship, but it requires acceptance of conceptual ambiguity and openness. Instead of presenting a clearly defined concept of disease, he points to an *interpretandum*, a dynamic task of interpretation. As a result, it is not possible to exclude the prospect of interpreting a disposition (or the corresponding biological substrate) as a disease if there is a sufficiently strong disturbance of wellness.

Unlike Boorse and Lanzerath, Hucklenbroich presents a theory in which dispositions have been consciously subsumed under the disease concept. Hucklenbroich does not see the consequent pathologization as a problem. This does not mean that one attempts to eliminate "all morbid dispositions or all endogenous risk factors"; instead, their pathological nature should provide motivation for handling them in a sensible manner (Hucklenbroich 2010, p. 151; cf. Ott and Fischer sect. 8.2.3). Hucklenbroich's integrative theory of disease incorporates both the primary role of phenotype (suffering, pain, premature death) as well as the processes without any phenotypical effects, which are still seen as pathological in modern medicine.

However, compared to a normatively oriented theory (Lanzerath), his systems-theoretical view of the different levels of an organism (physiology, psyche, sociality) does not go beyond scientific objectification. It should also be noted that Hucklenbroich sets out to offer a *reconstruction* of a disease concept, which is already de facto guiding modern medicine, but his intention is not to make a decision about its validity. Therefore, his argument for the pathological character of dispositions could (but does not have to) be only a reflection of a factual "trend among scientists, physicians and pharmaceutical companies" (Gottschalk-Mazouz 2008, p. 71, cf. sect. 9.3.2 above).

What final conclusions can be drawn from this result with regard to IM? The theories of Boorse and Lanzerath are critical about the increasing tendency of the identification of disposition with disease, or the merging of preventive and curative medicine. This enables them to provide critical and corrective impulses for IM and for its self-conceptualizations. If such an identification, regarded from the perspective of IM, creates an intuitive feeling of discomfort or a need for theoretical reflection, these theories can serve as welcome points of reference. Unfortunately, their respective conceptual weaknesses have proven to be a significant hindrance to their use as tools for reflection. Nevertheless, Boorse and Lanzerath have presented two representative and well-defined disease concepts, which oppose the pathologization of dispositions, and maintain a high theoretical standard when discussing the related problems. However, the fact that one of the most ambitious modern reconstructive attempts, the theory of Hucklenbroich, argues for the pathologization of dispositions, represents a challenge for reflection. It prevents a reckless and too quick dismissal of this tendency as simply a—more or less problematic—expression of the spirit of our age.

9.4.2 Does IM Lead to a Pathologization of Life?

This question is formulated in a narrow and reductionist manner. However, the possibility that IM could play a significant role in a pathologization of life should be taken very seriously. The goal of innovations in the field of biomarker-based prediction (analysis of genetic information and omics data) is early detection of disease risks. This new predictive knowledge often creates a state of stress and uncertainty for the person concerned and this could easily be construed as a disease. The increasingly accurate and rapidly growing pool of information about dispositions to disease generates new situations of decision-making and responsibility, while also leading to the creation of new linguistic, individual and social identification patterns ('healthy ill' persons, 'gene carriers', etc.). It is not easy to assess the extent to which such 'technoscientific illness identities' (Sulik 2009, cf. Wehling 2012, pp. 16–17) will be adopted in the future as genuine options for the attribution of identity. Some previous studies (Wehling 2012, p. 17) have not confirmed an explosive trend in this direction. However, it should be remembered that even major developments in society can be subtle or indirect, being barely noticeable in the beginning (comparable, for example, with the creeping pollution of the environment). One instance of such indirect dynamic could be seen in the amalgamation of medicalization and pathologization: The refinement and increase of predictive knowledge is accompanied by the development of preventive strategies and interventions. This means that medical attention focuses increasingly on dispositions and risk factors, which are managed in a similar way to diseases. Such medicalization of dispositions seems to be in line with the conceptualization of dispositions as diseases (pathologization), whether we want it or not.

The development of linguistic praxis and the conceptualization of objects, methods and goals among the scientists, physicians and others involved in IM is naturally a very important element of this issue. The concept analyses of Langanke et al. (2012a) and Schleidgen et al. (2013) are particularly helpful with regard to the second aspect. These studies reveal a self-conceptualization of IM, operating with a clear differentiation between disposition (risk factors) and disease. The use of important future-oriented concepts, such as disease prediction, prevention and disease risks, seems to indicate a self-image, based on a conceptual difference between disposition and disease, between preventive and curative (therapeutic) medicine. A study of the actual language use in IM would probably produce mixed results. For instance, it is not rare for one to speak about diagnostics in the context of disease prediction and about *patient* populations in the context of prediction-related stratification (cf. Ott and Fischer sect. 8.2.3, who recommend the use of the term 'candidates' instead of 'patients'). Francis S. Collins, an enthusiastic proponent of IM, warned about such speech practices:

> If we're not careful, the increasingly accurate predictive power of personalized medicine could even begin to blur the concept of a diagnosis. Is an individual with a 60% risk of colon cancer already ill? […] No and no. We must seriously guard against that kind of slippage in semantics. Diagnoses should still be reserved for those who have developed actual symptoms of illness (Collins 2011, p. 17).

The philosophical (medicine theoretical) concepts of disease, which will have an impact on medical research and praxis, including IM, are among the important aspects to be considered when discussing these complex issues. Medical terms and concepts should not only be elaborated from a medical (or political and economic) perspective, but also from a *philosophical* perspective as an ongoing task and challenge. This aspect is also an element of the overall medical development of our time, which cannot be interpreted in merely deterministic terms. Most diseases involve a complex and multi-factorial dynamic with few linear causal relationships. Similarly, it could be said that IM does not automatically and deterministically lead to a specific outcome (such as a pathologization of life). Instead, it shapes our 'medical future' in a complex interaction with many other factors.

Acknowledgments The authors would like to thank Martin Langanke for inspiration and advice, and Alar Helstein for the translation of this text from German to English.

References

Bayertz K (1998) What's special with molecular genetic diagnostics? J Med Philos 23(3):247–254
Bergdolt K (2011) Individualisierte Medizin. Historische und aktuelle Aspekte. In: Schumpelick V, Vogel B (eds) Medizin nach Maß. Individualisierte Medizin—Wunsch und Wirklichkeit. Herder, Freiburg, pp 15–28
Bobbert M (2012) Krankheitsbegriff und prädiktive Gentests. In: Rothhaar M, Frewer A (eds) Das Gesunde, das Kranke und die Medizinethik. Moralische Implikationen des Krankheitsbegriffs. Franz Steiner, Stuttgart, pp 167–194
Boorse C (1977) Health as a theoretical concept. Phil Sci 44:542–573
Boorse C (1997) A rebuttal on health. In: Humber JM, Almeder RF (eds) What is disease? Humana, Totowa, pp 3–134
Boorse C (2012) Clinical Normality (unpublished draft). Paper presented at the international symposium on Christopher Boorse and the Philosophy of Medicine, University of Hamburg, 10-12 Sept 2012 http://www.philosophie.uni-hamburg.de/Schramme/Konferenzen_/ Boorse%20Clinical%20Normality.pdf. Accessed 16 April 2014
Collins FC (2011) The Language of Life. DNA and the revolution in personalized medicine. Harper Perennial, New York
Dabrock P (2011) Die konstruierte Realität der sog. individualisierten Medizin - sozialethische und theologische Anmerkungen. In: Schumpelick V, Vogel B (eds) Medizin nach Maß. Individualisierte Medizin—Wunsch und Wirklichkeit. Herder, Freiburg, pp 239–269
Gottschalk-Mazouz N (2008) Die Komplexität des Krankheitsbegriffs aus philosophischer Sicht: Theoretische und praktische, naturalistische und normative Aspekte. In: Zurhorst G, Gottschalk-Mazouz N (eds) Krankheit und Gesundheit. Philosophie und Psychologie im Dialog. Vandenhoeck & Ruprecht, Göttingen, pp 60–120
Hempel U (2009) Personalisierte Medizin I. Keine Heilkunst mehr, sondern rationale molekulare Wissenschaft. Dtsch Arztebl 106(42):A2068–A2070
Henke R (2011) Individualisierte Medizin heute. In: Schumpelick V, Vogel B (eds) Medizin nach Maß. Individualisierte Medizin—Wunsch und Wirklichkeit. Herder, Freiburg, pp 29–49
Hesslow G (1993) Do we need a concept of disease? Theor Med 14:1–14
Hucklenbroich P (2010) Der Krankheitsbegriff: Seine Grenzen und Ambivalenzen in der medizinethischen Diskussion. In: Höfner M, Schaede S, Thomas G (eds) Endliches Leben. Interdisziplinäre Zugänge zum Phänomen der Krankheit. Mohr Siebeck, Tübingen, pp 133–160
Hucklenbroich P (2013) Die wissenschaftstheoretische Struktur der medizinischen Krankheitslehre. In: Hucklenbroich P, Buyx A (eds) Wissenschaftstheoretische Aspekte des Krankheitsbegriffs. Mentis, Münster, pp 13–83
Hucklenbroich P (2014) Medical criteria of pathologicity and their role in scientific psychiatry—comments on the articles of Henrik Walter and Marco Stier. Front Psychol 5:128. doi:10.3389/fpsyg.2014.00130
Hüsing B, Hartig J, Bührlen B et al (2008) Individualisierte Medizin und Gesundheitssystem. TAB-Arbeitsberichte, vol 126. Büro für Technikfolgen-Abschätzung beim Deutschen Bundestag, Berlin
Juengst ET (2009) Concepts of Disease after the Human Genome Project. In: Caplan AL, McCartney JJ, Sisti DA (eds) Health, Disease and Illness. Concepts in Medicine. Georgetown University, Washington, pp 243–262
Karger CR, Hüsing B (2011) Personalisierte Medizin im Gesundheitssystem der Zukunft. Einflussfaktoren und Szenarien. Reihe Gesundheit, vol 44. Forschungszentrum Jülich, Jülich
Kollek R, Lemke T (2008) Der medizinische Blick in die Zukunft. Gesellschaftliche Implikationen prädiktiver Gentests. Campus, Frankfurt
Langanke M, Werner MH (2015) Der kranke Mensch. In: Oliver M, Giovanni M (eds) Orientierung am Menschen. Anthropologische Konzeptionen und normative Perspektiven. Wallstein, Göttingen (forthcoming)

Langanke M, Lieb W, Erdmann P et al (2012a) Was ist Individualisierte Medizin? Zur terminologischen Justierung eines schillernden Begriffs. Z Med Eth 58:259–314

Langanke M, Fischer T, Brothers K (2012b) Public Health—It is running through my veins: personalized medicine and individual responsibility for health. In: Dabrock P, Braun M, Ried J (eds) Individualized medicine between hype and hope. exploring ethical and societal challenges for healthcare. LIT, Berlin, 149–172

Lanzerath D (2000) Krankheit und ärztliches Handeln: Zur Funktion des Krankheitsbegriffs in der medizinischen Ethik. Karl Alber, Freiburg

Lanzerath D, Honnefelder L (1998) Krankheitsbegriff und ärztliche Anwendung der Humangenetik. In: Düwell M, Mieth D (eds) Ethik in der Humangenetik: die neuere Entwicklung der genetischen Frühdiagnostik aus ethischer Perspektive. Francke, Tübingen, pp 51–77

Lenk C (2003) Gesundheit in Zeiten des Humangenomprojekts: theoretische Konzepte, biotechnische Perspektiven, soziale Konsequenzen. In: Hornschuh T, Meyer K, Rüve G et al (eds) Schöne—gesunde—neue Welt? Das humangenetische Wissen und seine Anwendung aus philosophischer, soziologischer und historischer Perspektive, IWT-Paper vol 28. Universität Bielefeld, Bielefeld, pp 176–194

Magnus D (2009) The concept of genetic disease. In: Caplan AL, McCartney JJ, Sisti DA (eds) Health, disease and illness. Concepts in medicine. Georgetown University, Washington, pp 233–242

Murphy EA (1976) The logic of medicine. Johns Hopkins, Baltimore

Paul NW (2006) Gesundheit und Krankheit. In: Schulz S, Steigleder K, Fangerau H et al (eds) Geschichte, Theorie und Ethik der Medizin. Suhrkamp, Frankfurt, pp 131–142

Pöder J-C, Assel H (2014) Krankheit, Gesundheit und Gott. Verkündigung und Forschung 59(1):11–28

Sauter A, Gerlinger K (2012) Der pharmakologisch verbesserte Mensch. Leistungssteigernde Mittel als gesellschaftliche Herausforderung. Studien des Büros für Technikfolgen-Abschätzung, vol 34. Edition sigma, Berlin

Schleidgen S, Klingler C, Bertram T et al (2013) What is personalized medicine: sharpening a vague term based on a systematic literature review. BMC Med Ethics 14:55. doi:10.1186/1472-6939-14-55

Stockter U (2008) Präventivmedizin und Informed Consent - Zu den Anforderungen an die informierte Einwilligung in die Teilnahme an Screeningprogrammen. LIT, Berlin

Stoecker R (2009) Krankheit - ein gebrechlicher Begriff. In: Thomas G, Karle I (eds) Krankheitsdeutung in der postsäkularen Gesellschaft. Theologische Ansätze im interdisziplinären Gespräch. W. Kohlhammer, Stuttgart, pp 36–46

Sulik G (2009) Managing biomedical uncertainty: the technoscientific illness identity. Sociol Health Illn 31(7):1059–1076

Wehling P (2012) Die Medizin auf dem Weg zur Technowissenschaft? Technowissenschaftliche Krankheitsidentitäten und die Schwierigkeit der Technikfolgenabschätzung. Technikfolgenabschätzung—Theor Prax 21(2):15–21

Wiesing U (1998) Kann die Medizin als praktische Wissenschaft auf eine allgemeine Definition von Krankheit verzichten? Z Med Eth 44:83–97

Woopen C (2011) Individualisierte Medizin als zukunftsweisendes Leitbild. In: Schumpelick V, Vogel B (eds) Medizin nach Maß. Individualisierte Medizin—Wunsch und Wirklichkeit. Herder, Freiburg, pp 94–110

Zerres K (2006) Prädiktive Medizin: Der gesunde Kranke. In: Schumpelick V, Vogel B (eds) Arzt und Patient: Eine Beziehung im Wandel. Beiträge des Symposiums vom 15. bis 18. September 2005 in Cadenabbia. Herder, Freiburg, pp 554–564

Part V
Applied Research Ethics

Chapter 10
Informed Consent in GANI_MED—A Sectional Design for Clinical Epidemiological Studies Within Individualized Medicine

Martin Langanke, Jakob Fasold, Pia Erdmann, Roberto Lorbeer and Wenke Liedtke

Abstract Clinical epidemiological research is an important approach within Individualized Medicine. This research allows for the identification of possible biomarkers for specific diseases through the use of data from clinical treatments. From a research ethical point of view it needs to be noted that respective studies are mostly very complex. Therefore a variety of different normative requirements of both ethical and legal natures need to be met. On the one hand those studies aim for the scientific use of clinical records. On the other hand samples of biomaterial are to be collected, stored and analyzed using -omics methods. Sometimes there are also study examinations included in these studies which may produce incidental findings.

In this chapter a possible solution shall be demonstrated which meets the mentioned requirements: the informed consent process of the GANI_MED project. By presenting the consent documents design process and their different sections, it can be shown that a good ethical standard can be assured throughout the entire process while providing data from different sources to clinical epidemiological research. Furthermore the response patterns of the patients participating in the GANI_MED project will be presented. The data demonstrate that a possibility to deselect certain aspects of the study through the

---

Martin Langanke and Jakob Fasold contributed equally to this work.

M. Langanke (✉) · J. Fasold · P. Erdmann · W. Liedtke
Theologische Fakultät, Lehrstuhl für Systematische Theologie, Ernst-Moritz-Arndt-Universität Greifswald, Am Rubenowplatz 2–3, 17487 Greifswald, Germany
e-mail: langanke@uni-greifswald.de

J. Fasold
e-mail: Jakob.Fasold@uni-greifswald.de

P. Erdmann
e-mail: Pia.Erdmann@uni-greifswald.de

W. Liedtke
e-mail: wenke.liedtke@uni-greifswald.de

R. Lorbeer
Institute of Clinical Radiology, Klinikum der Ludwig-Maximilians-Universität, Pettenkoferstr, 8a, 80336 München, Germany
e-mail: Roberto.Lorbeer@med.uni-muenchen.de

© Springer International Publishing Switzerland 2015
T. Fischer et al. (eds.), *Individualized Medicine,* Advances in Predictive,
Preventive and Personalised Medicine 7, DOI 10.1007/978-3-319-11719-5_10

sectional consent documents is actually used by the participants. The data also show that the patients of different morbidity cohorts answer in a different way.

Keywords Informed consent · Consent form · Research ethics · Contractualism · Clinic epidemiologic research · GANI_MED · Incidental findings · Response rates · Consent patterns

10.1   Introduction

Clinical epidemiological research is an important approach within Individualized Medicine. This research allows the identification of possible biomarkers for specific diseases through the use of data from clinical treatments. From a research ethical point of view it needs to be noted that respective studies are mostly shaped in a very complex design. Therefore a variety of different normative requirements of both ethical and legal natures need to be met. On the one hand those studies aim for the scientific use of clinical records. On the other hand samples of biomaterial are to be collected, stored and analyzed using the -omics-methods. Sometimes there are also examinations included in these studies which may produce incidental findings.

All of the mentioned aspects of this research need to be explained during the informed consent process. The patient needs to be enabled to decide competently on his or her participation and to deselect those aspects which are not wanted if possible.

In this chapter a feasible solution to the problems depicted above shall be presented with the informed consent process of the GANI_MED project.

After a brief presentation of the GANI_MED project itself and its structure the essential ethical implications are discussed and assessed.

Then the respective legal framework of the German law is described with the necessary focus on the laws of data protection. On the basis of a contractualistic understanding of the informed consent process three research ethical minimum requirements are discussed to which physicians and biomedical researchers involved in GANI_MED agree to. These three ethical principles are (a) a commitment that the agreements concluded have to be met, (b) a transparency requirement and (c) the obligation to minimize the anticipated stress of the participants.

The application of these three principles may show that a moderate research ethical contractualism is an incomplete normative model for clinic epidemiological research and therefore an extension through stronger norms could be discussed.

The informed consent documents are described and the basic design is explained. It is intended to demonstrate in which way the documents have to be formally designed to meet the transparency requirement.

Finally the first results of the quantitative analyses of the response patterns of the GANI_MED participants will be presented and discussed from a research ethical perspective. It can be shown that the participant's possibility to deselect certain aspects of the different studies of the GANI_MED project through the informed consent documents is used. These patterns vary between the different patient cohorts. And yet another conclusion can be drawn from the response patterns: Despite the implementation of staff trainings, conversation guidelines and other instruments to standardize the informed consent process, the response patterns differ significantly between several different interviewers in the same cohort.

10.2 GANI_MED—Ethical Implications of the Study Design

GANI_MED—Greifswald Approach to Individualized Medicine is an interdisciplinary research consortium which aims to connect Individualized Medicine with clinical applications (Grabe et al. 2014). It is located at Greifswald University, Germany, Mecklenburg Pomerania. The project's aims are to integrate specific results and methods that originate from Individualized Medicine into the daily clinical routine. Individualized Medicine can be understood as a stratifying medicine (Hüsing et al. 2008; Langanke et al. 2011; Schleidgen et al. 2013). This scientific approach intends to achieve a better prediction of the risk to develop a disease and also a better prediction of whether or not a certain therapy is effective for a patient (for more details cf. Langanke et al. chap. 2). For that purpose different biomarkers have to be identified. The relevant information is gained amongst others from the so-called -omics methods (e.g. genomics, transcriptomics, metabolomics, proteomics etc.). Hence biomarkers can be used to subdivide patient populations into either (high) risk populations or into groups of patients which respond better or worse to certain therapies. If such biomarkers are available which are diagnostically conclusive and clinically validated they are to be implemented into the daily clinical routine. This may be achieved if biomarkers are embedded into medical scores which enable the physician to distinguish between high-risk or low-risk patients. Also -omics based tests could help to determine the best suited therapy for a specific patient.

10.2.1 Structure and Organization

In order to achieve the scientific objectives of identifying and validating biomarkers and using them to improve clinical diagnostics and therapies, an interdisciplinary research consortium was founded. The GANI_MED-project is a cooperation of different institutes and faculties of Greifswald University including sub-projects such as ethics and economics (for more details cf. Grabe chap. 3). To better organize a project of this size, GANI_MED is partitioned into two major layers (for the following see Fig. 10.1).

Layer 1 centers on the structural aspects of the project. This division's tasks includes establishing and operating a biobank, a bioinformatics related working group, an IT-infrastructure (research platform), an education program as well as the verification of quality standards and the development of concepts to secure sustainability.

Layer 2 bundles all activities, which aim to the formation of patient cohorts. Besides this the second layer focuses on the identification of possible biomarkers

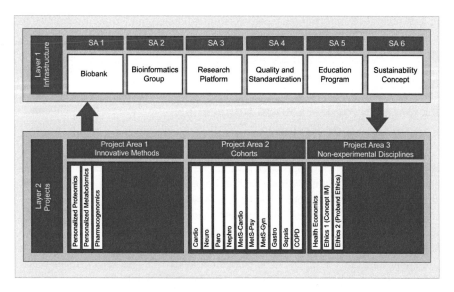

Fig. 10.1 GANI_MED—Project structure

through the basic research in the fields of medicine and human biology. Furthermore layer 2 includes ethical and health economical analyses and studies.

The centerpiece of layer 1 is the combination of the biobank with the research platform. This research platform is supplied with data from the clinical routine of patients who take part in GANI_MED. These data are made available by the participating clinics of the university hospital (Langanke et al. 2012).

The biobank on the other hand will be supplied with human biological material like blood, urine, saliva and other extra material donated by the patients participating in GANI_MED. The biobank will store these materials and make them accessible to researchers who will especially use methods originating from the so-called -omics field.

Layer 2 aims, among other goals, to the development of patient cohorts. To do so patients of the university hospital who fit a certain profile are asked to participate in GANI_MED. If a patient suffers from a specific disease and shows certain symptoms he or she is asked to take part in GANI_MED. These specific illnesses include widespread diseases like stroke or cardiac insufficiency and the so-called "emerging diseases" such as fatty liver (Langanke et al. 2012).

The ethical sub-projects in layer 2 focus mainly on conceptual and regulatory aspects of GANI_MED. This includes the design and implementation of the informed consent process with all associated documents and ensuring an allover excellent ethical standard. The health economical sub-project centers on the financial aspects of Individualized Medicine in combination with the analysis of the sustainability of GANI_MED.

10.2.2 Research with Patient's Data

The GANI_MED project as depicted above raises many research ethical challenges especially through the integration of the different research fields of medicine, natural science and health economics into one joint project. Aside from the many methodological challenges that become apparent in the interaction of the mentioned research fields, concrete procedural, ethical and legal questions need to be answered. Since GANI_MED has to include patients into the research project, all relevant laws and guidelines need to be taken into account. However not all aspects (e.g. the regulatory framework of the biobank) are specifically managed in the German law. Therefore normative frameworks need to be developed which do not contradict current laws and enable the researchers within GANI_MED to achieve the scientific goals and which allow for long-term planning. One of these goals is the identification of suitable biomarkers for the therapeutic and diagnostic use. The methods of choice are statistical association analyses, which identify the statistical correlation of various medical parameters (e.g. anthropometrical data, laboratory values, nucleotide polymorphisms) with certain diseases as well as the progression of this illness and their symptoms. These methods are also used to determine whether or not a therapy is effective for a specific group of patients.

The data used for the association analyses are gathered from different sources. Besides the data originating from the clinical routine, information on comorbidities is received from the specific GANI_MED anamnesis and further questionnaires. These data are accompanied by analyses from the biomaterial stored within the biobank. In some cases data from specific examination which are not part of the clinical routine are gathered as well. These examinations are called "study examinations" and they do not have a diagnostic or therapeutic value. These study examinations serve the sole purpose of gaining scientific knowledge on a certain disease.

Furthermore, data are gathered over a long period of time on the course of the illness and possible secondary diseases. For that reason the GANI_MED project cooperates with several non-clinical institutions, and even some non-medical institutions. Such institutions are resident doctors, certain disease registers and also local residents' registration offices. Within the GANI_MED project it is possible to re-contact patients for further surveys or follow-up studies. Besides the medical studies, the health economical analyses require certain data from the health insurance companies and the Association of Statutory Health Insurance Physicians (ASHIP). The ASHIP is a German association of medical doctors and psychotherapists which are accredited for the ambulant treatment of statutory insured patients.

10.3   Regulatory Demands and the Research Ethical Background

In Germany the goals of the GANI_MED project cannot be achieved without an "opt-in" informed consent process if they should be in accordance with German law, regulations of data security and good ethical practice (Langanke et al. 2011). This is due to the specific aims the GANI_MED project has set itself. At certain points of the research procedure, the participating patient needs to be re-identified after his or her data and biomaterial has been encoded in order to suit the demands of the national as well as the local law of data security. Therefore the collected data have to be pseudonymized which allows a re-identification of the patients. In contrast, data which cannot be matched with the donating patient are anonymized.

10.3.1   Protection of Data and Privacy

The German National Law of Data Security (germ. Bundesdatenschutzgesetz, abr. BDSG) as well as the relevant Law of Data Protection of the state Mecklenburg Pomerania (abr. DSG-MV) distinguishes between anonymization and pseudonymization with reference to the term of personal data (Bundesdatenschutzgesetz 2009; Gesetz zum Schutz des Bürgers 2011). The law defines "personal data" as "a single information on individual or material aspects of a certain or determinable person." Therefore the data collected by the GANI_MED project has to be considered as personal data because the scientific relevant information are connected with identifying data (e.g. name, address, date of birth, etc.). If the identifying data are removed, a set of anonymized data are created. Yet, if the identifying data are replaced with a code number which allows a reconnection of the number and the patient, pseudonymized data are obtained. The reconnection itself can be accomplished by an original list which contains the code numbers and the identifying data. But this original list is not accessible to the researchers. Therefore the data can be considered anonymized for the daily use but a reconnection is possible through the original list. In other words, it does not matter in which form the collected data are presented to the researchers. If in principle a reconnection of code numbers and identifying data is possible, these data are legally considered pseudonymized regardless in which form they are used. Only if a reconnection is not possible the data are anonymized. Hence pseudonymized data takes an intermediate position between not encoded personal data and anonymized data.

According to the legal definition of anonymization and pseudonymization it becomes apparent that the GANI_MED project cannot use anonymized data if the scientific goals are to be achieved. On the one hand re-contacting patients and follow-up studies are integral parts of the project, on the other hand, patients need to be identifiable if they visit the medical center a second time or if data are required which are stored in other institutions than the university medical center. Besides this a re-identification is necessary since data from different sources are combined into one set of data. In order to combine this information, it needs to be known how these data correspond. Nevertheless not every scientist has to know the identity of the participant, for example researchers focusing on the "-omics" methods. The demand of data minimization and data avoidance which results from the BDSG § 3a need to be met within GANI_MED (Bundesdatenschutzgesetz 2009). Therefore an economical data management was installed which ensures through a graded access authorization that only those data are accessible which are actually required by the particular scientist. Also a structural information technological separation of the personal and scientific data is required.

Since pseudonymized data are analog to personal data these encoded yet re-identifiable data are considered as data worth being protected. This finds its expression in the right of informational self-determination. This right is a derivation of the constitutional protection (Grundgesetz 2012) of the human dignity (germ. "Schutz der

Menschenwürde", see "Grundgesetz" Art. 1) and the right of the development of one's personality (see "Grundgesetz" Art. 2). The juristic expression of this constitutional right is found in the national law of data security (BSDG § 4) which states that personal data must not be used without the explicit consent of the person concerned (Bundesdatenschutzgesetz 2009). Apart from some legal exceptions, this has to be applied to all personal data regardless of the nature of the data. Hence, the GANI_MED project is obliged to implement an informed consent process in order to use the personal data and biomaterial of the participating patients. Every patient has to give his or her consent for the collection, processing, storage, usage and cession of his or her data. If a patient is not able to give his or her consent e.g. in case of a mental impairment or a coma, a legal representative may give the consent in the participant's stead. However this is only possible if a representative was appointed before the ability to consent was lost. All consents need to be given in writing with the signature of the patient. This may also include signatures made on screen surfaces of suitable terminals.

The informed consent documents given to the participant-to-be has to explicitly explain the purpose of the intended research. This is also a legal requirement of the BDSG, the earmarking of a purpose. This means that only those scientific goals may be pursued with the use of the collected data and biomaterial to which the donating patient consented to. If the data or the biomaterial should be used for a different scientific purpose a respective consent is needed.

Due to the legal demands of data minimization and data avoidance, an information technological separation of the scientific data and the personal data of the participating patients is required (Reng et al. 2006).

However this separation raises the question of how to re-identify a certain person for example if a patient needs to be re-contacted but without compromising the data security of the personal data. The separation in IT of scientific and personal data and their possible re-connection was confided to a Trusted Third Party which was installed within GANI_MED. Such a Trusted Third Party allows the legal requirements of data security to be met, concerning the pseudonymization of personal data and their storage. Within the GANI_MED project this central trust office oversees all re-identifications of the entire project to increase efficiency and effectiveness. Since the members of the trust office are not further involved in the research of the GANI_MED project the threat of the misuse of data is minimized. Therefore it is only possible to re-identify a certain patient if the trust office conducts the re-identification of the pseudonomized data. Hence, no researcher may directly access personal data of the participating patients as long as he or she is not also directly involved in their clinical treatment. But even if a researcher is simultaneously the attending physician of the participating patient, the doctor may only access the data supplied by the clinical database, not the research database. To access these information the authorization of the trust office is needed.

10.3.2 Research Ethics—A Moderate Contractual Approach

In the following section the ethical set of norms are presented which formed the guidelines used during the development of the GANI_MED informed consent process.

The main purpose of explaining the research ethical norms and their context is not to completely contribute new insights to the scientific discussion. This explanation is intended to present a normative ethical theoretical frame of the informed consent process and its deduction. Theories which derive from internationally recognized codes often lack a thorough deduction of their ethical norms (Langanke and Erdmann 2011; Erdmann 2014).

The normative theoretical frame which is to be presented originates from two main aspects. The first aspect is the consideration that medical research as conducted by the GANI_MED project is performed in a given legal framework. The second aspect focuses on the prerequisites which are needed to formulate the basic norms and two additional fairness requirements. These ethical premises are not very strong so that the greatest possible consensus is achieved. This small set of norms is the centerpiece of this theoretical frame. This set of norms can also apply to researchers who developed their own ethical theory from the international codes (Langanke and Erdmann 2011; Erdmann 2014).

It needs to be said that the theoretical frame which will be presented does not claim to give all the answers to all the problems of biomedical research. The argumentation focuses mainly on non-interventional studies especially on epidemiological and clinical epidemiological studies.

At first two fundamental premises are depicted which are needed to establish and apply the entire moderate contractual model. Furthermore normative implications are analyzed which can be drawn from the basic normative premise of this model that the consent of a patient to participate in a medical study is equivalent to a contract. In the next step we propose to take two additional fairness requirements into account. We will finally critically examine whether the presented contractual model is adequate for a clinical epidemiological study like GANI_MED.

10.3.2.1 Two Premises of a Moderate Contractual Model

1. Compliance with the law:

The moderate contractual model relies on being compliant to the applicable law. This simple requirement has to be stated since all the medical research conducted in Germany has to be performed within the legal frameworks. Therefore applied research ethics can never recommend solutions which contradict the law. Recommendations can use the preference rule that positive law has a higher degree of commitment than other norms (Erdmann 2014).

By referring to the legal obligations many controversies about the handling of participants of medical studies can be avoided. For example, if a study is to be compliant to the law, then a non-disclosure strategy without exceptions concerning the handling of incidental findings cannot be designed, if the type of scientific data analysis allows the detection of incidental findings instantly or at least in time. Besides the fact that such a model would contradict many of the international codes, an invariable non-disclosure can result in cases of failure to render assistance which can be deducted from the criminal code § 323c, especially if assistance is necessary and reasonable (Strafgesetzbuch 2014). In such a case it is irrelevant whether the conducting researchers are part of the medical staff or not. The only important requisites are that findings may occur in the study which require immediate treatment and that the ability is given to recognize the necessity of such findings (Erdmann 2014).

From the prerequisite of the compliance with the law it can also be deducted that the consent form for the participation in a study like GANI_MED cannot contain elements which are either illegal or violate moral principles, or compromise rights which are guaranteed by the constitution through the right of the protection of human dignity (Grundgesetz 2012), or even compromise laws of data protection (Erdmann 2014).

2. Compliance with scientific standards:

The moderate contractual model can only be applied on studies which follow the principles of good scientific standards by using the methods which are best suited for the respective study and are also accepted by the scientific community. Methodologically insufficient research is—in accordance with the important research ethical codes—*per se* unethical (Erdmann 2014).

It needs to be mentioned that in case of long term epidemiological studies conflicts may occur between the obligation of keeping the highest methodological standards and the ethical demand to disclose at least some incidental findings (Hoffmann and Schmücker 2010). This is due to the fact that the disclosure of incidental findings may lead to a self-induced bias in the sample which may have a negative effect on the methodical quality of the results.

10.3.2.2 The Contract-like Character of the Consent

The participation in the GANI_MED project is of course voluntary. Patients—or non-patients, as is the case wth three of the cohorts—who match specific criteria for inclusion are asked to participate in GANI_MED on certain conditions. These conditions are explained through the information material and in information meetings. If the patient agrees to participate under these conditions the institution responsible for the study and the autonomously deciding person enter into a special contract (Rudnik-Schöneborn et al. 2014; Ehling and Vogeler 2008).

Therefore the participants may legitimately expect not only that the confirmed agreements are met but are met to the conditions to which were agreed upon. This seems to be relevant especially with regards to cohorts in which incidental findings can occur. In GANI_MED incidental findings can arise in one pure non-patient-cohort (Paro, see Fig. 10.1) and two other cohorts which partially include non-patients (Nephro and MetS-Gyn, see Fig. 10.1).

The ethically important aspect is that the commitments which were made by the institution conducting the study are binding. The participants need to rely on the fact that the agreements which were made are to be kept.

Two aspects can be deducted from the fact that the relationship between the participant and GANI_MED is a contract. The first one is the fact that the participant has no legal claim to participate in GANI_MED. The second aspect is that the responsible institution cannot insist on the examinations if the patient withdraws his or her consent even after he or she signed the consent form. This is due to the fact that an obligatory right of withdrawal is a standard part of the consent form used in GANI_MED. Regarding the first aspect, GANI_MED explicitly reserves the right to exclude persons willing to participate from certain or all medical studies. As the consent documents declare, this is especially the case if certain examinations would negatively influence the person's medical condition.

When the informed consent process is understood as a contract and the prerequisites of the compliance with the law and scientific standards are met, four additional general ethical conclusions can be drawn:

1. The people responsible for the GANI_MED study are in principle entitled by the research interests to set the conditions for conducting the study as long as they are in accordance with the law.
2. It cannot be deducted from the presented model that an epidemiological or clinical epidemiological study needs to be designed in a way in which incidental findings are delivered promptly or at all. Also concepts of non-disclosure can be implemented in this contractual model such as a "data quarantine" which stores the data for a long period of time before the analysis, as discussed in Puls et al. (2010). It is only important that the participant is made aware of the conditions during the informed consent process.
3. A non-interventional study which may generate incidental findings can choose a way of disclosure which does not sabotage the scientific aims through a self-inflicted bias. This risk of a methodological self-created influence can occur if too many incidental findings are disclosed, as depicted in Hoffmann and Schmücker (2010). This applies especially on long term non-interventional studies with several points of data collections. Even by keeping in compliance with the criminal code several possibilities are available to disclose an incidental finding. The most freedom is given by the handling of "conspicuities" whose pathological value is not clear, or by dealing with findings which, according to the prevailing medical doctrine, do not cause any need for treatment.
4. The test persons have to rely on the institution concerning the keeping of the promises made.

10.3.2.3 Two Fairness Requirements

Further ethical demands become apparent when the informed consent process, its content, its presentation and its associated processes are linked with certain ethical norms like the concept of fairness.

In regard to the presented research ethical contractualism, epidemiological and clinical epidemiological studies are "measured" here by two fundamental fairness requirements. This is done without giving any further ethical demands, e.g. care requirements which would only derive from a Hippocratic ethos. This decision was made because the research subjects of GANI_MED have either—if they are also patients—the status of participating patients or they take part as non-patients (Langanke and Erdmann 2011; Rudnik-Schöneborn et al. 2014).

The first norm ("transparency norm") claims a basic quality standard of the notification process and the associated documents. The second norm ("requirement of minimizing anticipated stress") demands basic conditions concerning the form of the notification process in case of incidental findings within medical studies. Both norms are connected to research ethical ideals which are mostly indisputable, or at least are not being questioned.

1. Transparency

To guarantee a valid informed consent (IC), the test persons and patients who participate in scientific studies, should have understood (a) under which conditions they participate and (b) which anticipated consequences in terms of decision-making are connected to their participation (WMA Declaration of Helsinki 2013, Art. 26). Both aspects have to be clearly distinguished. On the one hand, transparency is demanded concerning the (e.g. oral) form of presentation of the clarification of content. On the other hand, it is also demanded that possible consequences for decision making are emphasized in the context of clarification.

If one agrees on these two aspects of transparency, one will also concede that the informed consent documents have to be prepared as transparently as possible. Having said this, it should be noted that transparency is an ideal and its realization cannot be simply claimed. However, the degree of realization can be distinguished (an informed consent document may be more or less transparent), so that, for example, a better transparency could be claimed in a concrete case. The demand for transparency calls us to strive for the ideal that all possibilities to optimize transparency should be used.

This means that known difficulties, e.g. with the formulation of informed consent documents, should be avoided and that there is an attempt to optimize all of the processes involved. This also means that Diagnostic or Therapeutic Misconceptions need to be taken into account.

Phenomena such as Diagnostic or Therapeutic Misconception are unquestionably a valid explanation as to why transparency in medical studies cannot be established to a desirable extent through the informed consent processes. However, Therapeutic Misconceptions do not give any justification for moving away from the aim of achieving maximum transparency and to just accept the shortcomings in the realization that transparency cannot be achieved to a desirable extent (for the terms "Therapeutic Misconception" and "Diagnostic Misconception" see Appelbaum et al. 2004 and Heinrichs 2011). From a genuine ethical perspective it can be requested that all means are used to decrease Diagnostic and Therapeutic Misconceptions. Therefore, this transparency norm adheres to the fact that the informed consent process enables a person to make an autonomous decision about whether to participate in medical research studies or not. This can only be done if the participant knows the aims of the study as well as the conditions and the, possibly unfavorable, consequences which are linked to the participation. This also applies to the possible case that a legal guardian is appointed for an incapacitated person who gives the consent representatively (Erdmann and Langanke 2011; Rudnik-Schöneborn et al. 2014).

Taking this ideal into account, any signals supporting Diagnostic or Therapeutic Misconceptions should be avoided wherever possible. Clinical epidemiologic projects like GANI_MED are easily endangered to fall prey to such misconceptions since these projects mainly recruit participants in hospitals. This means that the impression might be given to a patient that participating in the study may have a positive effect on the treatment. But also from the perspective of the recruiting physicians and medical staff, uncertainties concerning their own role might occur whether they are medical personnel, medical researchers or simply researchers. This can be another reason why Diagnostic and Therapeutic Misconceptions take place.

The transparency requirement becomes more important because consequences of decisions can be imagined which would greatly affect the life situations of people. Incidental findings and the stress connected to these can be classified as very serious. This is why the participants in the Paro-cohort (see Fig. 10.1) must be made perfectly aware of the conditions and of possible consequences which are linked to the participation. This is due to the fact that in this cohort a MRI of the head is undertaken. The participant has to be made aware of the conditions under which incidental findings are disclosed and of the types of findings which will not be returned. Therefore, the test persons require the knowledge concerning the expected "rewarding", the conditions of result disclosure, and the possible stress which results from the notification process if imaging techniques are used.

In the case of the MRI of the head of the Paro-cohort there was a vigilant attempt to eliminate all hints that this MRI has the same diagnostic value as a MRI made within a treatment in order to satisfy the transparency requirement.

2. Requirement to minimize anticipated stress

The transparency requirement requests, on the textual level, to communicate consequences of decisions clearly whereas this requirement requests the minimization of anticipated stress within the study process (e.g. the notification process) as well as potential consequences of decisions for the participants (with regard to the given constraints, e.g. limited staff, time and money in medical studies) (Erdmann and Langanke 2011; Rudnik-Schöneborn et al. 2014).

Anticipated stress includes all circumstances which are known from the context of medical research. For those circumstances it can be predicted—either through a scientific analysis or everyday experience—that they can possibly cause frustration, fear or physical suffering (criterion 1).

Moreover, situations are meant of which the potentially stressing character is already proven in the appropriate and accessible medical, psychological and research ethical literature and other similar sources (criterion 2).

Lastly, anticipated stress can also be predicted by medical, psychological and/or nursing expertise if the situation caused by the study is known from the (daily) practice (criterion 3).

Regarding non-interventional studies, the requirement to minimize anticipated stress shall be explained concerning the disclosure of health-related findings or incidental findings (for the term "incidental findings" see Wolf 2008; Heinrichs 2011; Erdmann 2014 and Erdmann chap. 12).

When considering a variety of psychological and medical studies it is well known that the information about a serious pathological finding, e.g. a tumor, can cause great fear, stress, even extreme stress situations for the people affected. This can also be accompanied by all known medical complications. Furthermore, it could be assumed that this stress could be dealt with in a much better way in a talk with a doctor than to get a written result. This is the reason why findings are only disclosed in personal conversation within the clinical context. The six-step SPIKES algorithm is a protocol to deliver bad news which was developed to meet the high demands of the field of oncology. Although epidemiological studies normally "only" produce findings which need to be assessed further, such an algorithm may also prove to be useful in a scientific context (for the SPIKES- Algorithm see Baile et al. 2000, for the discussion on the communication standards for delivering bad news see also Kurtz 2002; Fallowfield et al. 2002; Fallowfield and Jenkins 2004; Back et al. 2005).

Hence it can be anticipated that a written disclosure of findings with a health-relevant potential can cause stress, at least occasionally (Erdmann 2014). Such stressful experiences can be avoided by implementing algorithms similar to the SPIKES algorithm to disclose the findings.

If this is true, the requirement to minimize anticipated stress demands wherever possible to abstain from written-only notification of health-related findings. This also applies if the disclosure by a personal conversation is more costly and time consuming than a written notification. This is the reason why any incidental findings which occur in the Paro-cohort of the GANI_MED project are disclosed in personal conversation between the participant and the senior physician of the Clinic and Polyclinic for Neurology. This is a definite decision against the option to transmit the findings to the participant's general practitioner or the attending medical specialist and not to the participant personally. As experience shows, younger participants often do not have a general practitioner at all and participants in general tend to give false details about their physicians.

The same concept was implemented in the GANI_MED informed consent process regarding health-related information which is generated by future research on the biomaterial using the -omics methods. If studies identify and clinically validate certain biomarkers from the biomaterial of a GANI_MED participant, these participants would be contacted again and invited to personal briefing. This of course can only happen if the participant consented to a possible re-contacting and a disclosure of results from the -omics research in the consent. The most important criterion whether or not health-related information is to be disclosed to the participant is the choice of the participant documented in the consent form. This applies to possible incidental findings from the MRI of the head of the Paro-cohort as well as to results from the analyses of the biomaterials.

The consent patterns in the section of the consent form which deals with the disclosure of health-related results from the analyses of the biomaterial vary considerably between the different GANI_MED cohorts. It seems that age and membership of a certain cohort are the key determinants affecting the consent patterns which range from approximately 85 to 100 % (see Table 10.2 in sect. 10.5).

10.3.2.4 Clinical Epidemiologial Research as an Ethical Special Case

In the case of the GANI_MED project the presented set of contractual norms has proven to be a common denominator between the physicians and researchers involved on the one hand and the ethicists which developed the informed consent process on the other. Yet from an ethical perspective this is not a strong argument. Also it cannot be deducted that all non-interventional studies similar to GANI_MED are bound to contractual concepts. The point can be raised that clinical epidemiological studies are an ethical special case since they share research ethical relevant aspects with clinical studies. This assessment can be supported by the fact previously mentioned that clinical epidemiological studies assign their participants double roles of patients and research subjects. The same concept applies on the conducting researchers who are mostly physicians or medical staff in general. As a possible result Diagnostic and Therapeutic Misconceptions can have a strong effect on clinical epidemiological studies. Another argument can be made to assess clinical epidemiological studies in a different way as pure epidemiological studies. It cannot be expected from a participant of a clinical epidemiological study—who is also a patient—to distinguish between those two roles completely. Hence it may not be clear for them that they find themselves in different circumstances with different rights and obligations than patients.

Taking all objections into account it needs to be said that the presented set of norms is open to logical consistent extensions, especially on the level of the fairness requirements. This may imply further and also stronger norms. Hence the presented contractual approach does not claim to be a complete and concluded research ethical theory. This model tries to work out the basic ethical minimum requests towards clinical epidemiological and pure epidemiological studies if the two following norms are to be accepted: On the one hand the participation should not only be viewed as formally voluntary. This voluntariness is to be translated into actual processes (transparency). On the other hand it should be assumed that humans are sentient beings and that they should not suffer avoidable harm or to speak with Kamlah 1973 to respect the needs of other people (minimizing the anticipated stress).

10.4   The GANI_MED Informed Consent

After depicting the theoretical and conceptual requirement and ethical considerations of a good ethical standard of a clinical epidemiological study we want to present some informed consent processes and documents which are in use in the GANI_MED project.

GANI_MED is an interdisciplinary research association but it was decided that every study of each subproject which affects the participating patients are to be disclosed and need to be approved by the ethics committee and—of course—by the participant. This includes those studies which normally do not need an ethics committee vote such as health economical studies. Also the participants are informed on all the sub-projects which would be working with their data or samples, even in some cases in an anonymized way.

10.4.1   The Informed Consent Documents

The complexity of the GANI_MED project and its scientific goals are the main factors to restrict the transparency of the informed consent documents. Therefore the information material was divided into three documents closely linked to one another. The informed consent process was accordingly structured to clearly separate the information and the consent. The documents forming the centerpiece of the GANI_MED informed consent process are the information booklet (IB), the consent form (CF) and the authorization form (AF).

The information booklet is specially shaped according to the respective cohorts. In other words there is not one in use but ten information booklets were prepared and everyone is customized to describe the respective cohort, their scientific aims and how the study itself is shaped. Its purpose is to supply the patient with all the information needed.

The consent and the authorization form are documents split into certain cohort-specific sections but the same consent and authorization forms are mostly used by all cohorts in GANI_MED. In both the consent and authorization forms the patient is asked to check the respective box whether he or she is willing to participate in GANI_MED and in which sub-projects. The separation of consent form and authorization form results from the different focus and further use of the documents. In the consent form the patient can state whether or not he or she will take part in the study, even down to the specific sub-projects. Hence this document focuses on the patient, the consent and a possible withdrawal. In the authorization form on the other hand the patient is asked whether the staff members of the study can contact the general practitioner of the patient, the health insurance company and also disease registers. Therefore the authorization form will be sent to different institutions to prove the consent of the patient that certain data may be accessed for scientific use. This means that copies from the authorization form will be sent to those institutions and since this document is only one and a half pages long, the handling is simplified. Through the separation of consent form and authorization form it hopefully becomes obvious to the patient that these documents have different focuses so he or she is not obliged to check every box.

If a patient fits the desired profile in the participating clinics or wards he or she is asked to participate and all of these three documents are given to the patient. After some time to consider—ideally 24 h—the patient is briefed directly on the ward. The patients make their decision on the spot on a portable tablet PC. Since

Matrix of Cohorts					
Cohort	Medical Research Studies	Risk of Incidental Findings in Medical Research Studies	Information Booklet (IB)	CF-Sections	AF-Sections
Cardio	No	Not applicable	X	1a-b, 3, 4, 5, 6, 7	1, 2, 3, 4, 5
Neuro	Yes	No	X	1a-b-c, 3, 4, 5, 6, 7	1, 2, 3, 4, 5
Paro*	Yes	Yes	X	1a-b-c, 2, 3, 4, 5, 6, 7	1, 2, 3, 4, 5
Nephro*	Yes	Yes	X	1a-b-c, 2, 3, 4, 5, 6, 7	1, 2, 3, 4, 5
MetS-Card	Yes	No	X	1a-b-c, 3, 4, 5, 6, 7	1, 2, 3, 4, 5
MetS-Psy	Yes	No	X	1a-b-c, 3, 4, 5, 6, 7	1, 2, 3, 4, 5
MetS-Gyn*	Yes	Yes	X	1a-b-c, 2, 3, 4, 5, 6, 7	1, 2, 3, 4, 5
Gastro	No	Not applicable	X	1a-b, 3, 4, 5, 6, 7	1, 2, 3, 4, 5
Sepsis	No	Not applicable	X	1a-b, 3, 4, 5, 6, 7	1, 2, 3, 4, 5
COPD	No	Not applicable	X	1a-b, 3, 4, 5, 6, 7	1, 2, 3, 4, 5

Not actively managed by IT   Actively managed by IT, Selection and signature on Tablet PC   * Includes extern probands

Consent Form (CF)			
Section		Description of Section	Jumps
1	a	Approval of scientific use of the clinical data	
	b	Approval of implementation in health economical studies	
	c	Approval of implementation in medical research studies	
2		Communication of incidental findings from medical research studies?	
3		Approval to the storage of biosamples	
4		Communication of health related results of future research on the biomaterial?	"Yes" only if "Yes" in 5
5		Approval of renewed contact	
6		Information on right to withdrawal the consent	
7		Final remarks	

Authorization Form (AF)	
Section	Description of Section
1	Approval to contact general practitioner/ attending physician + release from discretion
2	Approval to contact disease related registers
3	Approval to request an extract of the extended register of residents
4	Approval to request fee related data from the health insurance fund (or nursing care insurance fund)
5	Approval to request fee related data from the Association of Statutory Health Insurance Physicians

Basemodule: "No" = Total refusal   Offered in all GANI_MED cohorts   Offered only in individual cohorts

Fig. 10.2 Cohort specific risk of incidental findings and sections of the used informed consent documents in GANI_MED

the information booklet does not need to be actively administered, both the consent form and the authorization form are presented digitally. If the patient is willing to participate he or she checks the respective boxes and signs the document directly on the tablet PC. The informed consent process is completed after a final talk with a physician who countersigns the consent form and the authorization form.

10.4.2 The Design of the Consent Form (CF)

A detailed description of the design of the informed consent procedure and its implementation into the IT system is found in Bahls et al. (chapter 11). Therefore a short explanation will suffice at this point (see Fig. 10.2).

The consent form and the authorization form have been designed with several sections. This means that the patient may consent to a full participation or refuse to participate, but he or she may also rule out certain aspects of the study. Depending on the respective sub-project the patient has to check four or five sections in the consent form. Additionally this document contains two sections which do not need to be actively selected but provide information on withdrawal and some final remarks.

The first section includes the consent to the scientific use of the patient's clinical data and the consent for the health economical study. Certain sub-projects have implemented study examinations. This means that certain examinations are performed which are normally not part of the clinical routine of the respective medical discipline. The use of a MRI at the dental ward may serve as an example of a study examination. If a study examination is implemented in a sub-project the consent for the conduct is requested in the first section of the consent form.

The second section focuses on potential results from study examinations. Within the second section, it is asked whether or not findings of therapeutic value which arise from the study examinations should be disclosed (see Fig. 10.2).

The third section requests the storage of biomaterial.

The fourth section is the analog to the second section. Here it is asked whether potential future results of therapeutic value which came to light from the -omics analysis of the biomaterials should be disclosed (see Fig. 10.2).

The fifth section asks for the consent of re-contacting. One can only agree to the fourth section if the consent is given in the fifth section. Should someone agree to the fourth section but do not to the fifth, the agreement to the fourth section becomes void (see Fig. 10.2).

The sixth and seventh sections are the information sections mentioned above (see Fig. 10.2).

10.4.3   The Design of the Authorization Form (AF)

The authorization form consists of five sections which do not vary between the different sub-projects. If the patient consents to the first section he or she authorizes the study staff to contact the general practitioner of the patient and also authorizes the general practitioner to provide selected medical information about the patient to the project. This is done to understand the medical history of the patients participating. Hence it is a release from the doctor-patient confidentiality (see Fig. 10.2).

In the second section the patient's permission is asked for to contact the respective disease registers (see Fig. 10.2).

The third section would enable the study staff to collect certain information from the register of residents. Among other reasons, this information is needed to learn whether a patient has deceased (see Fig. 10.2).

If the patient consents to the fourth section of the authorization form he or she allows the staff of the study to contact and gather information from the patient's health insurance company. This information is mainly cost-related in order to conduct the health economical sub-project (see Fig. 10.2).

A similar consent is asked for in the final section (see Fig. 10.2) to contact the Association of Statutory Health Insurance Physicians (ASHIP), as mentioned in paragraph 10.2.2.

10.5 Response Patterns in GANI_MED—Empirical Data

With the help of the sophisticated IT system which is used to register, process, encode and monitor the participating patients, it was possible to show the specific response patterns of all the participants in the GANI_MED project. The following four tables seek to focus on ethical relevant aspects of the informed consent process.

Table 10.1 shows the response patterns of all patients *willing to participate* in the GANI_MED project. This is imperative to mention since the exact number of patients who refused to be included into GANI_MED is unknown, although a non-responder questionnaire was implemented which allows the interviewers to anonymously assess patients regarding the reasons for their refusal. This questionnaire was completed in 193 cases, but due to practical reasons and communicative difficulties it was not regularly used. Furthermore one has to take into account that patients have the right to refuse to answer the questions of this questionnaire. In consequence the number of total refusals in GANI_MED cannot be calculated. Therefore the given percentages in Table 10.1 represent only those patients who agreed in principle to participate in the GANI_MED project, more concrete: who agreed at least to section one of the consent form (scientific use of clinical data plus—in certain cohorts—study examinations). Although the percentages seem to be high, it is important to see that about ten percent of the participating patients use

Table 10.1 Full and partial consent to the consent form (CF) and authorization form (AF)

Cohort	Number of patients	Consent to consent form (CF) and authorization form (AF)			
		Full consent (%)		Partial consent (%)	
		CF	AF	CF	AF
Cardio	908	89.76	93.39	10.24	6.61
Neuro	437	85.81	80.55	14.19	19.45
Paro	630	96.35	92.70	3.65	7.30
Nephro	364	84.89	68.32	15.11	31.68
MetS-Gyn	102	86.27	77.45	13.73	22.55
MetS-Card	452	92.70	95.35	7.30	4.65
MetS-Psy	248	93.55	81.85	6.45	18.15
Gastro	255	98.43	99.61	1.57	0.39
Sepsis	22	72.73	66.67	27.27	33.33
COPD	105	75.24	97.14	24.76	2.86
Total	3,523	90.58	88.47	9.42	11.53

State: 03.03.2014

Table 10.2 Consent to the individual sections of the consent form (CF)

Cohort	Number of patients	Consent to the individual sections of the CF (%)				
		Section 1	Section 3	Section 4	Section 5	Section 2
Cardio	908	100	99.34	90.20	92.51	–
Neuro	437	100	99.77	86.04	89.02	–
Paro	630	100	99.84	96.98	99.68	99.20
Nephro	364	100	98.9	85.99	95.6	92.31
MetS-Gyn	102	100	99.02	86.27	98.04	98.00
MetS-Card	452	100	99.12	93.36	97.12	–
MetS-Psy	248	100	96.37	93.95	95.97	–
Gastro	255	100	98.43	98.82	99.22	–
Sepsis	22	100	95.45	72.73	72.73	–
COPD	105	100	100	75.24	87.62	–
Total	3,523	100	99.12	91.09	94.89	98.02

State: 03.03.2014

the possibility for a partial consent. This means that having the consent form and the authorization form split into sections enables the patient to rule out certain aspects of the study and that this possibility is used by every tenth participant. If such a possibility was not implemented in our case either three hundred patients would have not participated at all or would have given their consent without being fully content with it. It can be said that having the informed consent documents split into sections is worth pursuing since patients make use of it.

Table 10.2 shows the respective percentages of consent for every section in which consent is asked. Section 2 focuses on the disclosure of incidental findings which can arise from study examinations in some cohorts. Since only three of the cohorts have implemented study examinations going hand in hand with the risk of incidental findings, the second section only appears in the consent form of these cohorts. It is also important to state that sect. 4 and 5 are linked both logically and through the IT system. In sect. 5 it is asked whether the patient may be re-contacted by the study staff, while sect. 4 deals with the disclosure of results which eventually arise from the -omics analysis of the biomaterial. Hence if a patient does not agree to be re-contacted, possible health related results from -omics research cannot be disclosed. Since the CF is filled out on a tablet PC, the logical link between sect. 5 and 4 is also fixed into the program of the tablet PC, meaning that sect. 4 is deselected automatically if sect. 5 is rejected. If a "classical" paper document is used, the use of such backwards links is not recommendable since they may cause confusion.

Furthermore, if a patient does not agree to the first section, he or she is not included in the GANI_MED project and is therefore not recorded, hence the given percentage.

It is obvious that the patterns differ between the cohorts. Hence it can be said that the demands and concerns of the patients do not only differ individually but they also differ between the types of participants. Age and the severity of the disease seem to have an influence on the decisions of the patients. This was confirmed by the study staff, e.g. for the patients of the Nephro cohort. These patients are often elderly persons who suffer from severe forms of chronic kidney disease which reduce their life expectancy. Therefore they often see for themselves no benefit from the disclosure of potential health relevant information from the -omics analyses in the future (in 5 or 10 years time). In consequence only 85.99 % of these Nephro patients agreed on sect. 4 of the consent form. Yet it is not possible to find a mono-causal connection between the patient populations and their response patterns. Ethically relevant is the fact that different patients have different demands and needs. Hence this variety needs to be reflected in the informed consent documents. Having the consent documents split into sections can serve as a realization of the demand to fulfill or at least recognize the variety of needs.

In Table 10.3 the consent pattern of the individual sections of the authorization form are presented. It is noteworthy that there is no clear correlation between the response patterns of the different cohorts. The pattern appears to be more dependent on the belonging to a certain cohort than on the sections themselves. It seems that not always the same aspect is equally relevant to different types of patients and/ or cohorts, which sounds trivial at first. Yet it is remarkable that for example only 84.44 % of the patients on the neurological ward are willing to release their general practitioner from the doctor-patient confidentiality while a significant portion of Nephro participants do not want to grant access to their health insurance data. Since this article focuses on general ethical principles it would go beyond the scope of this text to deeper evaluate the reasons for this pattern. As discussed below with regards

Table 10.3 Consent to the individual sections of the authorization form (AF)

Cohort	Number of Patients	Consent to the individual section of the AF (%)				
		Section 1	Section 2	Section 3	Section 4	Section 5
Cardio	908	96.15	98.68	98.68	98.35	98.46
Neuro	437	84.44	95.19	97.94	95.88	95.88
Paro	630	96.51	98.25	97.94	96.83	96.67
Nephro	364	78.79	92.56	88.15	83.75	84.02
MetS-Gyn	102	96.08	89.22	98.04	86.27	91.18
MetS-Card	452	97.79	98.23	98.45	98.01	98.23
MetS-Psy	248	91.13	88.71	88.71	89.11	89.52
Gastro	255	100	100	100	99.61	99.61
Sepsis	22	66.67	71.43	66.67	66.67	66.67
COPD	105	99.05	98.1	98.1	98.1	98.1
Total	3,523	93.01	96.42	96.51	95.12	95.34

State: 03.03.2014

Table 10.4 Interviewer dependence of response rates to consent form (CF) and authorization form (AF)

Interviewer	Consent patterns			
	Full consent (%)		Partial consent (%)	
	CF	AF	CF	AF
Interviewer 1 of cohort A	98.56	98.24	1.44	1.76
Interviewer 2 of cohort A	68.42	79.47	31.58	20.53
Interviewer 1 of cohort B	96.40	95.60	3.6	4.40
Interviewer 2 of cohort B	97.56	87.20	2.44	12.80
Interviewer 3 of cohort B	96.20	94.02	3.8	5.98

to Table 10.4 the personal attitudes of the interviewers and their individual "style" of clarification seem to have a significant influence on the participant's response. Nevertheless it can be stated that patients make use of the different sections of the authorization form. Since the patients take the opportunity to use the sections even more than within the consent form having different sections seems to be justified for projects like GANI_MED.

Table 10.4 shows the response rates to the consent and the authorization form but focuses on individual interviewers who conduct the informed consent interviews and answer patient's questions. Although the entire informed consent process is highly standardized and the interviews are performed using guidelines, the patterns of the different interviewers scatter broadly. It is imperative to say that it would be a critical mistake to assume interviewer 1 of cohort A did "a better job" than interviewer 2 since he or she got the higher response rates. A high response rate naturally means the interviewer acquired more participants but it does not say whether the informed consent process is done properly. It is very likely that the interviewer who receives lower consent ratings invests more time and empathy into the informed consent process by taking uncertainties and fears of the patients seriously. Therefore a normative statement cannot be given whether high consent rates are better or worse than lower rates. It is important to note that the informed consent process—regardless of the basic protocol—has a strong empathic aspect which cannot be neglected. These results may indicate that coaching the conversational skills of the staff members may be as important as the generation of guidelines.

10.6   Conclusions and Recommendations

The informed consent process of the GANI_MED project as depicted in this chapter may serve as an example of how to meet the ethical and legal challenges of including patients into a clinical epidemiological study in a responsible way. Clinical epidemiological studies are essential approaches for Individualized Medicine, especially in regard of identifying biomarkers for certain diseases. Therefore it seems reasonable to sum up the main results and recommendations which can be drawn from the practical experiences made during the GANI_MED project.

1. Clinical epidemiological studies which generate hypotheses in the field of Individualized Medicine, and are therefore similar to the GANI_MED project, typically try to pursue several scientific goals in one project. This involves the access of clinical data for scientific analyses as well as data from third parties, the conducting of additional study examinations and the collecting and storing of biomaterials which are to be analyzed using the -omic methods.

2. The inclusion of patients in such sophisticated studies without an explicit consent (opt-in) is not justifiable for the following reasons:

   - This kind of research is only possible with pseudonymized data which means not anonymized and therefore re-identifiable. Hence personal rights are affected especially in regard to data security.
   - The pseudonymization also affects data which are generated during the storage and analyses of biosamples.
   - The analyses of biomaterial and the conducting of study examinations may produce health-related information which is worth being disclosed to the participants. It is not possible to handle this risk in an ethically responsible way without asking the participants regarding their preferences.

3. Since studies like GANI_MED need to collect data in a pseudonymized form, respective processes have to be developed, which meet the necessity to thoroughly protect these person-related data. This implies the IT procedural separation of person-related data and scientific data, the implementation of a graded access authorization and the implementation of a Trusted Third Party. This Trusted Third Party can decode the encrypted data and relink it with the person-related data when it is necessary (e.g. re-contacting) (see Bahls et al. chap. 11).

4. It is suggested that the consent in epidemiological and clinical epidemiological projects should be designed on the basis of a research ethical model which is "in touch" with legal demands. In the case of the GANI_MED project the physicians and researchers involved and the ethicists who developed the informed consent process have agreed on a contractual model. This model needs to meet the following three requirements:

   - A contractual base norm: The consent documents form a contract. Hence the concluded agreements are to be understood as binding for both parties. The right of withdrawal of the participant is a part of this contract.
   - Transparency requirement: To create an ethically valid informed consent the participating patient or non-patient needs to understand the conditions under which he or she participates. Also the possible consequences of the participation which can be anticipated need to be understood. Both aspects have to be clearly distinguished: The presentation of any information—written or spoken—needs to be as transparent as possible. Furthermore, the possible consequences linked with the participation need to be addressed explicitly during the informed consent process. From a methodical point of view the requested transparency is an ideal which demands the identification and use of potentials to optimize the informed consent process and all related documents.

- Requirement to minimize anticipated stress: This requirement demands that by designing potentially stressful studies, anticipated stress is to be avoided or at least minimized. This includes the disclosure of incidental findings. This demand should not be weighed up against pragmatic constrains such as limited staff and limited financial recourses. The actual applied meaning of "anticipated stress" can be specified by three criteria of anticipation.

5. The presented model of a moderate contractualism can be expanded by other ethical norms if they are at least logically consistent. The necessity of such expansions can be discussed regarding clinical epidemiological studies since they share several aspects of clinical studies.
6. In order to meet the requirements of the transparency requirement it is recommendable to design the informed consent documents with different sections. It is also advisable to divide the relevant content between two documents, the consent form and the authorization form.
7. The consent form of a clinical epidemiological study similar to GANI_MED should contain the following sections:

   - Section 1: Consent to the scientific use of clinical data. This section is at the same time the consent to the participation in GANI_MED itself. If the consent is denied here, the person does not participate in GANI_MED at all. This section can contain, if applicable, the consent to conduct study examinations which may not be diagnostically required.
   - Section 2: If the study examinations can lead to incidental findings and if these findings are to be disclosed promptly, the participant is asked whether or not such notification is wanted.
   - Section 3: The consent to the storage and analyses of biomaterials.
   - Section 4: If the potential biomarkers should be clinically validated, health-related information can be generated which have a personal relevance to the donor of the analyzed biomaterials. In this section it is asked whether or not such results of future research are to be disclosed.
   - Section 5: The consent is asked that the study staff may re-contact the participant. Section 5 and 4 are logically linked. An approval to section 4 requires an approval to section 5. A notification about results from the analyses of the biomaterials is only possible if the participant wants to be re-contacted.
   - Further paragraphs should contain information on the right of withdrawal and references to conditions of the participation if necessary.

8. The consent to access data of third parties is ideally asked in a separate authorization form. This document should be brief so it can easily be faxed to the respective third parties. The following institutions are interesting for clinical epidemiological studies especially if health economical studies are included:

   - General practitioners and tending specialists
   - Disease register
   - Register of residents (tracking in case of relocation, mortality follow-ups)
   - Health insurance companies
   - Association of Statutory Health Insurance Physicians

9. In order to meet the requirement to minimize anticipated stress the notification process should be designed very carefully especially those which disclose incidental findings or results from the analyses of biomaterials. A variety of notification algorithms is available and should be chosen accordingly. A simple written notification is not recommended. The delivery of a written notification of health-related findings always needs to be accompanied by personal conversation with a (study) physician.

10. Through the quantitative analyses of the consent patterns in GANI_MED three main conclusions can be drawn:

   – If the consent form and authorization form are divided into sections, allowing participants to deselect certain aspects of the study, the participants do use this option. It can be assumed that through the different sections the participant can express his or her consent more specifically, empowering the voluntariness of the participant.

   – The consent patterns are disease-cohort-related. Different types of patients attach importance to different aspects. On the one hand this shows that different populations of patients lay down different requirements towards a medical study. On the other hand this is another argument for the informed consent documents to be split into sections since the response patterns of the different cohorts show only little similarities.

   – Despite a high level of standardization as well as a constant training of the interviewers, the response rates of different staff members of the same cohort differ significantly. Although a qualitative assessment regarding the performance of those staff members cannot be drawn from this data, it shows clearly that the consent briefings and informational conversations are very character-dependent. In the light of this it is recommendable to implement staff training focusing on conversational skills.

References

Appelbaum PS, Lidz CW, Grisso T (2004) Therapeutic misconception in clinical research: frequency and risk factors. IRB: Ethics & Human Research 26(2):1–8

Back AL, Arnold RM, Baile WF et al (2005) Approaching difficult communication tasks in oncology. CA Cancer J Clin 55:164–177

Baile WF, Buckmann R, Lenzi R et al (2000) SPIKES—a six-step protocol for delivering bad news: application to the patient with cancer. The Oncologist 5:302–311

Bundesdatenschutzgesetz in der Fassung der Bekanntmachung vom 14. Januar 2003 (BGBl. I S. 66), das zuletzt durch Artikel 1 des Gesetzes vom 14. August 2009 (BGBl. I S. 2814) geändert worden ist (2009). http://www.gesetze-im-internet.de/bundesrecht/bdsg_1990/gesamt.pdf. Accessed 22 May 2014

Ehling J, Vogeler M (2008) Der Probandenvertrag. MedR 26:273–281

Erdmann P (2014) Zufallsbefunde aus bildgebenden Verfahren in populationsbasierter Forschung: Eine empirisch-ethische Untersuchung. Mentis, Münster (forthcoming)

Fallowfield L, Jenkins V (2004) Communicating sad, bad, and difficult news in medicine. The Lancet 363:312–319

Fallowfield L, Jenkins V, Farewell V et al (2002) Efficacy of a Cancer Research UK communication skills training model for oncologists: a randomized controlled trial. The Lancet 359:650–656

Gesetz zum Schutz des Bürgers bei der Verarbeitung seiner Daten (Landesdatenschutzgesetz—DSG M-V) Vom 28. März 2002 Mehrfach geändert durch Artikel 2 des Gesetzes vom 20. Mai 2011 (GVOBl. M-V S. 277, 278) (2011). http://www.landesrecht-mv.de/jportal/portal/page/bsmvprod.psml?showdoccase=1&doc.id=jlr-DSGMVrahmen&doc.part=X&doc.origin=bs. Accessed 22 May 2014

Grabe HJ, Assel H, Bahls T et al (2014) Cohort profile: Greifswald Approach to Individualized Medicine (GANI_MED). J Transl Med 12(144)

Grundgesetz für die Bundesrepublik Deutschland vom 23. Mai 1949 (BGBl. S. 1), zuletzt geändert durch das Gesetz vom 11. Juli 2012 (BGBl. I S. 1478) (2012). http://www.bundestag.de/bundestag/aufgaben/rechtsgrundlagen/grundgesetz/. Accessed 22 May 2014

Heinrichs B (2011) A new challenge for research ethics: Incidental findings in neuroimaging. Bioeth Inquiry 8:59–65

Hoffmann M, Schmücker R (2010) Die ethische Problematik der Zufallsbefunde. In: Puls R, Hosten N (eds) Ganzkörper-MRT-Screening. Befunde und Zufallsbefunde. ABW Wissenschaftsverlag, Berlin, pp 1–16

Hüsing B, Hartig J, Bührlen B et al (2008) Individualisierte Medizin und Gesundheitssystem, TAB-Arbeitsbericht Nr. 126. https://www.tab-beim-bundestag.de/de/pdf/publikationen/berichte/TAB-Arbeitsbericht-ab126.pdf. Accessed 27 Juni 2014

Kamlah W (1973) Philosophische Anthropologie. Sprachkritische Grundlegung und Ethik. Bibliographisches Institut, Mannheim

Kurtz SM (2002) Doctor-patient communication: principles and practices. Can J Neurol Sci 29 (suppl. 2):23–29

Langanke M, Erdmann P (2011) Die MRT als wissenschaftliche Studienuntersuchung und das Problem der Mitteilung von Zufallsbefunden: Probandenethische Herausforderungen. In: Theissen H, Langanke M (eds) Tragfähige Rede von Gott. Festgabe für Heinrich Assel zum 50. Geburtstag 9. Februar 2011. Verlag Dr. Kovač, Hamburg, pp 198–240

Langanke M, Brothers KB, Erdmann P et al (2011) Comparing different scientific approaches to personalized medicine: research ethics and privacy protection. Personalized Med 8(4):437–444

Langanke M, Erdmann P, Dörr M et al (2012) Gesundheitsökonomische Forschung im Kontext Individualisierter Medizin: Forschungsethische und datenschutzrechtliche Aspekte am Beispiel des GANI_MED-Projekts. Pharmacoeconomics Ger Res Articles 10(2):105–121

Puls R, Hamm B, Hosten, N (2010) MRT ohne Radiologen—ethische Aspekte bei bevölkerungsbasierten Studien mit MR-Untersuchung. Fortschr Röntgenstr 182:469–471

Reng CM, Debold P, Specker C et al (2006) Generische Lösungen der TMF zum Datenschutz für die Forschungsnetze in der Medizin. Im Auftrag des Koordinierungsrates der Telematikplattform für Medizinische Forschungsnetze. Medizinisch Wissenschaftliche Verlagsgesellschaft, Berlin

Rudnik-Schöneborn S, Langanke M, Erdmann P et al (2014) Ethische und rechtliche Aspekte im Umgang mit genetischen Zufallsbefunden: Herausforderungen und Lösungsansätze. Ethik in der Medizin 26(2):105–119

Schleidgen S, Klingler C, Bertram T et al (2013) What is personalized medicine: sharpening a vague term based on a systematic literature review. BMC Medical Ethics 14:55

Strafgesetzbuch in der Fassung der Bekanntmachung vom 13. November 1998 (BGBl. I S. 3322), das zuletzt durch Artikel 1 des Gesetzes vom 23. April 2014 (BGBl. I S. 410) geändert worden ist (2014). http://www.gesetze-im-internet.de/bundesrecht/stgb/gesamt.pdf. Accessed 22 May 2014

WMA Declaration of Helsinki—Ethical Principles for Medical Research Involving Human Subjects. Adopted by the 18th WMA General Assembly, Helsinki, Finland, June 1964 and amended by the: 29th WMA General Assembly, Tokyo, Japan, October 1975, 35th WMA General Assembly, Venice, Italy, October 1983, 41st WMA General Assembly, Hong Kong, September 1989, 48th WMA General Assembly, Somerset West, Republic of South Africa, October 1996, 52nd WMA General Assembly, Edinburgh, Scotland, October 2000, 53rd WMA General Assembly, Washington DC, USA, October 2002 (Note of Clarification added), 55th WMA General Assembly, Tokyo, Japan, October 2004 (Note of Clarification added), 59th WMA General Assembly, Seoul, Republic of Korea, October 2008, 64th WMA General Assembly, Fortaleza, Brazil, October 2013 (2013). http://www.wma.net/en/30publications/10policies/b3/. Accessed 22 May 2014

Wolf SM (2008) The challenge of incidental findings. J Law Med Ethics 36:216–218

Chapter 11
Ethics Meets IT: Aspects and Elements of Computer-based Informed Consent Processing

Thomas Bahls, Wenke Liedtke, Lars Geidel and Martin Langanke

Abstract Traditionally, informed consent documents are a set of paper sheets signed by a participant. This approach is not appropriate for today's requirements and it does not provide for instant access or for the automated processing of its contents, especially when using data and biomaterials for specific research purposes. Storing a scan of the paper document does not facilitate automated processing in complex, large-scale study environments. Once paper is archived and all other informed consent processing is done in electronic form, then ethics really meets IT. This chapter will explain selected technical and ethical aspects on the subject of processing data items that belong to persons who have agreed to take part in medical research projects using their medical or identifying data such as name, date of birth, etc. Approaches and mechanisms to manage informed consent are explained, discussing the GANI_MED solution as an example project. Requirements with regards to an automated processing by an IT platform are summarized. Quality criteria for informed consent documents are named and discussed as the quality of informed consent documents have a significant impact to the precision an IT platform can ensure when determining scope and contents of an informed consent. Ways to determine the scope of an informed consent and its automatic application during a data use and access process are presented. A data model to manage the scope challenge for the example of a hospital-wide broad consent is

T. Bahls (✉) · L. Geidel
Institut für Community Medicine Abt. VC, Universitätsmedizin Greifswald, Ellernholzstr. 1–2, 17489 Greifswald, Germany
e-mail: thomas.bahls@uni-greifswald.de

L. Geidel
e-mail: lars.geidel@uni-greifswald.de

W. Liedtke · M. Langanke
Theologische Fakultät, Lehrstuhl für Systematische Theologie, Ernst-Moritz-Arndt-Universität Greifswald, Am Rubenowplatz 2–3, 17487 Greifswald, Germany
e-mail: wenke.liedtke@uni-greifswald.de

M. Langanke
e-mail: langanke@uni-greifswald.de

© Springer International Publishing Switzerland 2015
T. Fischer et al. (eds.), *Individualized Medicine,* Advances in Predictive,
Preventive and Personalised Medicine 7, DOI 10.1007/978-3-319-11719-5_11

introduced. To conclude, implementation aspects of such IT-based platform managing consent are discussed.

Keywords Informed consent · Consent processing · Consent management · Consent quality · Consent scope · Broad consent · GANI_MED · Use & access · Trusted Third Party

11.1   Introduction

Traditionally, informed consent documents are a set of paper sheets signed by a patient or healthy volunteer. This approach is not appropriate for today's requirements and it does not provide for instant access or for the automated processing of its contents, especially when using data and biomaterials for specific research purposes. Storing a scan of the paper document does not facilitate automated processing in complex, large-scale study environments. Once paper is archived and all other informed consent processing is done in electronic form, then ethics really meets IT. This chapter will explain selected technical and ethical aspects on the subject of processing data items that belong to persons who have agreed to take part in medical research projects using their medical or identifying data such as names, date of birth, etc. One of the larger medical research projects the authors of this chapter are engaged with is the GANI_MED project (Langanke et al. 2011; Langanke et al. 2012; Grabe et al. 2014). Examples and insights the authors share are partly based on the works with GANI_MED at the University Medicine Greifswald, Germany.

Current medical research, including those aiming at individualization of therapy, often relies on a number of different data types from various sources (e.g., case report forms authorized from medical staff, questionnaires a patient filled in himself, medical devices such as an ergometer, or MRI images). These different data types and sources need to be combined "somehow" (i.e., using IT tools and knowledge) in order to work on a research topic. This refers to two additional aspects: access to data and use of it. Research networks such as Community Medicine (2014) have often agreed to a common set of use and access policies, thereby enabling a steering board to control which data, including biomaterials, is used and for what purpose. But use and access are not only subject to these policies: First and foremost, legal regulations and ethical standards have to be observed. Legal regulations in Germany, for instance, comprise of federal and state laws that regulate data and information privacy. Personal data such as name, address, date of birth, sex, race, etc. (identifying data, IDAT) are considered subject to special protection measures (Langanke et al. 2011; Langanke et al. 2012) compared to data a person has, for instance, submitted by answering a questionnaire (medical data, MDAT). Typical measures taken to observe these regulations sufficiently are technical as well as organizational separation of these personal or identifying data from all other (medical) collected data and are used for research purposes. Both partial data sets can be linked by identifiers generated, so-called pseudonyms. Pseudonyms allow for re-identifying a person but provide no hints to researchers about a particular person's identity when working with their medical data. Anonymization does not provide for this option (Langanke et al. 2011; Langanke et al. 2012), all identifying data are removed and a link cannot be reconstructed again. Care should be taken that not only name, address, or date of birth must be considered personal data but also a number could be classified as personal data. An example for such a number could be a case ID in a hospital information system (HIS). Research staff who are also being active in medical treatment in the hospital have access to the HIS as well and could easily resolve the link and get knowledge about a person's identity. If data sets used for specific research purposes would contain such numbers, data privacy measures would be managed insufficiently. A correct measure would have been to replace the original case ID value by some pseudonym value as well. Data privacy as briefly referenced here is closely linked to ethics but will not be within primary focus in the remainder of this chapter. More details are available in data privacy concepts, e.g. Reng et al. (2006). However we will consider four fundamental aspects:

1. Current medical research relies upon data of various types from different sources and comprises of both identifying and medical data. Combination of these data is often an algorithmic challenge. Furthermore, the amount of data does often no longer allow for manual processing. Spreadsheet usage is neither an option in many cases as a spreadsheet application is not a good choice to combine large amounts of text, numbers, images, lab analytes, etc. from different sources. Because of the necessity of complex algorithms to combine all these data the use of Information Technology is inevitable.
2. Various regulations apply to data access and usage. Research networks may have their specific regulations; but first and foremost legal regulations apply. Among other aspects, legal regulations in Germany stipulate a person's explicit agreement to processing their data in the case that personal data are involved.
3. Such agreement may comprise certain authorizations and is called informed consent. As the informed consent is a normative item, it must be part of all data processing, including collection, storage, usage, and transfer. The agreement names a purpose that can be more specific or broader. In either case it sets a scope for the data processing agreed. Thereby the informed consent is a central ethical as well as legal element for all data processing using IT.
4. All IT-based data processing must be carried out in accordance to legal and ethical regulations, including a person's informed consent. As the term "processing" refers to data collection and storage as well as usage, computer algorithms could only tell if something is allowed if the informed consent is not monolithic but provides specific sections (policies) thereby allowing to check if the action finds a match with an agreed policy.

As the informed consent is one of the fundamental elements of IT-based data processing in the context of medical research, the following paragraphs will highlight some aspects in more detail.

11.2 Requirements of IT-Based Storage and Processing of Informed Consent

Making personalized medical data available for medical research presumes a person's agreement. Persons could be either patients or healthy volunteers. The contents of the agreements and data are closely related. Given the fact that usually more than one research project is carried out based on the same data, it is desirable that the informed consent documents are as uniform and comparable as possible, and thereby aiming at a maximum of synergies for later automated processing. Furthermore, a uniform and comparable structure of such documents precludes human error. Recording and storing the informed consent and data should be bound together as closely as possible.

Because of the mandatory requirement to process only data that have been agreed to become processed, IT must have knowledge about the respective informed consent contents, its link to all data covered by it, and IT must be able to process the contents. An adequate consent form does not allow for interpretation but is strict in determining the scope, purpose, and actions that are agreed upon. Only with such strict implementation IT can check and tell in an automated way whether usage of particular data for export and transfer, or whether their use is for research-orientated analysis and that reporting has been agreed on.

Ideally, the recording of a person's agreement is done using IT mechanisms from the beginning. This provides for a minimum rate of mistakes as no paper document is transcribed into an electronic format but the latter is used instantly. An electronic format is a prerequisite for later IT-based automated processing. If an informed consent will be declared and recorded in writing, it is important that only personnel, who has received training with distinct qualification for this, transcribes the informed consent contents into the electronic format. Furthermore, a scan of the original document should be attached to the electronic format for later reference as well as quality assurance.

A number of challenges with significant ethical reach have to be met when designing and implementing processes related to informed consent in an IT platform. The challenges appear manageable once the platform is dedicated to a specific project as portions of the processes and mechanisms could be tailored to the specifics of the project. In case the IT platform covers a multitude of studies and research projects like, for example, the entire HIS, the structure and contents of the (study-specific) consent form become the crucial factor for a successful implementation. Challenges with significant ethical reach that have to be met comprise of:

1. Ethical deficits in the consent form remain as long as the the participant's legitimate expectations on the commitment cannot be mapped confidently to algorithms that enable IT to process an informed consent and decide upon the compliant use of data. Withdrawals of consent must be processed reliably and instantly. This refers particularly to:
   - the scope of examinations that is consented or not (scope with regards to contents), for example in case of studies with patients that use data from regular treatment;

- the scope with regards to the time period an agreement covers, for example in long running observation studies with patients; and
- the interactions of multiple informed consent documents in the case that a person has been included in more than one study, contents of an examination changes during the course of a study, or separately consented third-party data sets.

The quality of a consent form—and thereby its ethical validity—has significant impact to the ability to process it, which is often underestimated. Therefore, logical structure of consent forms and the interface between them and the IT processes should be focused on more intensively than is usually the case.

2. Platforms that are expected to service a multitude of studies and research projects face another usually underestimated challenge. Several studies of different types have diverse and partly divergent requirements, some of them may be the result of legal regulations such as the Medical Device Directive (MDD) of the EU or local embodiments such as the German Medizinproduktegesetz (2013). If these diverse and partly divergent requirements are managed by stand-alone solutions with regards to structure and contents of the respective consent forms, the rate of mistakes will increase, which is considered as an ethical risk.

It is therefore desirable both from a processing as well as an ethical perspective to supply research with a homogeneous and uniform sectional informed consent architecture that facilitates to manage different study types with adequate documents which

- use a uniform structure that recurs across documents and can be recognized as such,
- cover identical requirements of several studies with identical text modules, and
- implement requirements depending on the type of a study with specific text modules that are, however, compatible with the overall informed consent framework.

11.3   Ethics Meets IT—The GANI_MED Example

GANI_MED is a clinic-epidemiologic research project that aims at the identification of approaches to individualize therapies, primarily based on biomarkers. Persons taking part in GANI_MED are recruited in cohorts, most of them being cohorts of patients while one cohort works with healthy volunteers. Ten cohorts have been defined and approximately 4,500 persons will be recruited by the end of the project (Langanke et al. 2011; Langanke et al. 2012; Grabe et al. 2014). In contrast to pure epidemiological studies, GANI_MED has no fixed and predefined examination schedule. Data from clinical routine devices such as ergometers, MRI, and the hospital information system are collected in addition to cohort-specific examinations. As the vast majority of participants come as patients to the hospital seeking medical treatment, inclusion into GANI_MED is not the first measure taken. The data set, however, should be as complete as possible. To meet this allegedly simple

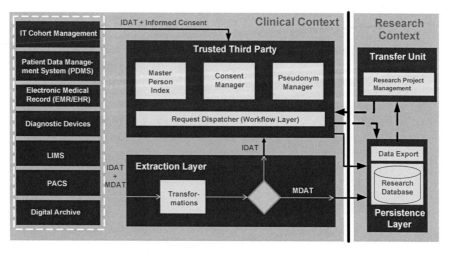

Fig. 11.1 High-level architecture of the IT platform in GANI_MED

requirement, the interaction between the agreement signed at some point later in time during a person's treatment and the fact that data might be generated prior to this signature must be carefully considered when designing and implementing the informed consent and the respective processes. The IT platform must be capable to process and cache data upon their generation until it can reliably determine whether or not these data can be processed within the scope of an informed consent.

Figure 11.1 illustrates the high-level architecture of the IT platform developed in GANI_MED. It consists of five building blocks; each of them shares specific aspects with ethics:

- a number of data sources (diagnostic devices, hospital-wide infrastructure systems such as LIMS or PACS, and the IT Cohort Management),
- an Extraction Layer (EL), which works as an interface and cache between the data sources and the research database with persisted medical data (MDAT), and also implements mechanisms to separate identifying data (IDAT) and MDAT received from diagnostic devices and other hospital-wide infrastructure systems,
- a Trusted Third Party, which works as an organizational unit which implements measures for data privacy, and as a technical instrument to supply pseudonyms and to keep the actual as well as the historic versions of all persons' informed consents and making them technically available to the remaining IT platform for automated data processing,
- a Persistence Layer primarily permanently stores all consented MDAT together with the respective pseudonyms and makes them available by technical interfaces, and
- a Transfer Unit (TU), which works as an organizational unit managing the use and access regulations of a project or research network.

Data, which are collected for GANI_MED, originates from various types of source systems, some of them "broadcasting" their information once (this could be

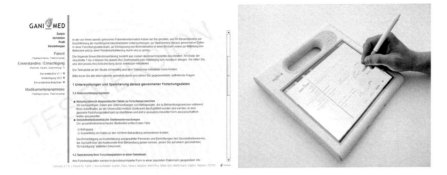

Fig. 11.2 The recording of informed consent in GANI_MED

considered as push style) while others export data upon administrative and often manual actions (this could be considered as pull style). Self-answered questionnaires are used as well, most of them being in paper form requiring later transcription into electronic form. Electronic Case Report Forms (eCRF) are used as well. These eCRFs are completed by trained medical staff on so-called Mobile Clinical Assistants—a hospital-compatible form of tablet computers with WiFi, barcode scanner, docking station, hot-swappable batteries, and a water-resistant design with the option of a disinfectable touch screen. Figure 11.2 shows a 2013 hardware model of these tablet computers, together with screenshots of an example of the consent form and the signature area of it.

Besides completing eCRFs, these tablet computers are also used to record personal data such as name, address, date of birth, etc. for all persons taking part in GANI_MED. Furthermore their informed consent is recorded in electronic form. Prior to being able to enter any medical data, a person must complete two informed consent documents comprising of a consent form and an authorization form. These are displayed and have to be signed at the end by both the person now included in GANI_MED and the physician, together with the date and name in readable form. Both signatures are given on the touch screen, both hardware models with resistive screen technology (i.e., a regular pen could be used) and capacitive screen technology working with specific pens exist. As primarily patients and other participants might not be used to signing a document this way, an exercise area is provided to make them familiar with the technology (while leaving the consent form shown in the background), prior to giving their signature. Signature pads are expected to become a future solution as they provide for additional records such as biometric parameters like writing pressure.

The most prominent advantage of this approach is that the contents of the signed consent form are available to the IT platform instantaneously (upon synchronization with the server application). No media transcription is necessary; mistakes that could easily occur during transcriptions are avoided. Instant availability of the consent provides for as-early-as-possible processing of data originating from diagnostic devices or analyts from the LIMS that are "pushed" from those systems.

Matrix of Cohorts				
Cohort	Medical Research Studies	Information Booklet (IB)	CF-Sections	AF-Sections
Cardio	No	X	1a-b, 3, 4, 5, 6, 7	1, 2, 3, 4, 5
Neuro	Yes	X	1a-b-c, 3, 4, 5, 6, 7	1, 2, 3, 4, 5
Paro*	Yes	X	1a-b-c, 2, 3, 4, 5, 6, 7	1, 2, 3, 4, 5
Nephro*	Yes	X	1a-b-c, 2, 3, 4, 5, 6, 7	1, 2, 3, 4, 5
MetS-Card	Yes	X	1a-b-c, 3, 4, 5, 6, 7	1, 2, 3, 4, 5
MetS-Psy	Yes	X	1a-b-c, 3, 4, 5, 6, 7	1, 2, 3, 4, 5
MetS-Gyn*	Yes	X	1a-b-c, 2, 3, 4, 5, 6, 7	1, 2, 3, 4, 5
Gastro	No	X	1a-b, 3, 4, 5, 6, 7	1, 2, 3, 4, 5
Sepsis	No	X	1a-b, 3, 4, 5, 6, 7	1, 2, 3, 4, 5
COPD	No	X	1a-b, 3, 4, 5, 6, 7	1, 2, 3, 4, 5

Not actively managed by IT   Actively managed by IT, Selection and signature on Tablet PC   * Includes extern probands

Consent Form (CF)			
Section		Description of Section	Jumps
1	a	Approval of scientific use of the clinical data	
	b	Approval of implementation in health economical studies	
	c	Approval of implementation in medical research studies	
2		Communication of incidental findings from medical research studies?	
3		Approval to the storage of biosamples	
4		Communication of health related results of future research on the biomaterial?	"Yes" only if "Yes" in 5
5		Approval of renewed contact	
6		Information on right to withdrawal the consent	
7		Final remarks	

Authorization Form (AF)	
Section	Description of Section
1	Approval to contact general practitioner/ attending physician + release from discretion
2	Approval to contact disease related registers
3	Approval to request an extract of the extended register of residents
4	Approval to request fee related data from the health insurance fund (or nursing care insurance fund)
5	Approval to request fee related data from the Association of Statutory Health Insurance Physicians

Basemodule: "No" = Total refusal   Offered in all GANI_MED cohorts   Offered only in individual cohorts

Fig. 11.3 Sectional informed consent architecture in GANI_MED

The informed consent documents displayed present separate consent and authorization sections to be checked. Figure 11.3 illustrates the sectional design of the informed consent in GANI_MED which is 1:1 implemented by the electronic representation. Each section and, thereby, each "Yes" or "No" of a respective person is implemented as a policy value in the IT platform. Policy values are used later to decide whether or not a given use of data is permissible. Assuming a high-quality and specific design and verbalization of the consent form, there is no need to "map" or "interpret" content for or, even worse, by the IT platform (Langanke et al. 2012).

A consent form sets a scope to the agreement: with regards to time as well as to contents. Both need to be considered when collecting, storing and using data. Therefore, data must be collected and stored together with some meta information that allows the scope to be checked in an automated way. Almost all examinations, results, and other data in a hospital environment are associated with some case ID. Such case IDs have begin and end dates attached and can be used in a number of ways: they can denote a single exercise, all data belonging to a person in a specific study or study episode, all data belonging to a person's treatment for a specific diagnosis etc. Case IDs provide a flexible means to group data. Therefore, an informed consent could be modelled as one further item associated to a specific case or number of case IDs. This way, there is always a limited scope, which is more desirable than an all-scope ("all my past and future data this hospital has knowledge of"), which can be ethically problematic with regards to the transparency principle (for this ethical principle see Langanke and Erdmann 2011; Erdmann 2014).

The Extraction Layer is another architectural IT component with a close relationship to ethics. It works as an interface and cache to all data sources connected to the IT platform. For each and every value that passes the Extraction Layer, it needs to decide whether or not the item falls under the patient's consent and may, therefore, become processed further or has to be discarded. Each value is first assigned to a person's identity. Furthermore, origination of the value has to be determined: A diagnostic device may or may not be used exclusively for one project or study. Then, the device together with the person's identity determines the necessary information to verify the current consent status. In case a device is a shared resource among more than one project, study, organizational unit of the hospital, or exercise, then some additional information regarding the scope of the value is required. Such could be a flag, a separate identifier, a study-specific case ID, and so on. The same logic applies for other data sources. The Extraction Layer caches all data for which it cannot yet decide whether a valid consent status applies. Such may be the case with a stroke patient who is not able to agree upon arrival in the hospital but agrees one or 2 weeks after his arrival, having received treatment in the meantime. The period the Extraction Layer or similar architectural components cache such data may vary depending on study types, use cases, and the concept coordinated with the officer for data protection and privacy.

The role of a Trusted Third Party in such IT architecture is apparent: it holds and manages all personal data, including the consent forms. The association between consent forms and identities of persons is known. The Trusted Third Party as an IT component provides technical interfaces for other platform services that could be used to determine whether a person exists, which identifier it has been assigned (for some specific scope, e.g., a study), and whether a positive consent has been recorded. In order to determine the consent status, more parameters such as study or project, case ID, or the relevant informed consent section (policy) must usually be verified. The Trusted Third Party is also consulted for supporting actions such as to contact a person again in the case of incidental findings or to check a person's vital status with a population register (for the ethical problem of managing incidental findings from imaging and genetics see: Wolf 2008; Booth et al. 2012; Brothers et al. 2013; Rudnik-Schöneborn et al. 2014; Erdmann 2014). Furthermore, the Trusted Third Party is consulted prior to the use of any data and the current agreement status is checked.

The Transfer Unit manages the specific regulations of a research network, for instance. A Transfer Unit is rather an organizational instrument with people working according to specific processes than a technical component of some IT architecture. Transfer Unit staff ensure that data are used according to the specific regulations as well as per the agreements and authorizations recorded in the IC document. For that, it interacts technically with the Trusted Third Party. All data which is to be used are typically extracted from a quality-assured research database. The Transfer Unit also ensures that only as much data as necessary and only those that are necessary for a given purpose are transferred. All transferred data sets are stored for future reference, thereby having a copy that can be used to reproduce the exact transfer independent from any evolution of the original database. Both the Trusted Third Party as well as the Transfer Unit have very limited options to verify the purpose of data use automatically.

11.4 Managing the Scope Challenge

The scope with regards to content, that is, the type and amount of data or materials to be processed in accordance with the consent given, is determined by the consent form. An IT platform should be able to manage various consent forms at the same time, thereby allowing for situations in which a person has signed multiple consent forms (i.e., the person is a member in multiple studies, research projects, etc.). The scope and span of these documents could vary significantly. Examples range between epidemiological studies, patient-based studies with cohorts, clinical trials and hospital-wide broad consent-based research. The span with regards to the amount of data and the themes covered by a consent form will vary significantly between several documents as well. Consequently, a temporal course within one document set has to be considered. Therefore, a consent form should not be bound to persons, which would imply severe modeling and decision problems when it comes to usage of data and material. Instead, real or artificial case IDs within a hospital information system provide a proper solution to this problem.

The model chosen for the scope with regards to the time period has a direct impact to the consent management process when bringing the multiple layers together (e.g., person, case, data covered by a specific informed consent document, context data has been collected in, validity period of a person's consent and authorization forms), to determine a simple "Yes" or "No" to a complex question. Furthermore, the information process of the recruitment phase is directly impacted as well: Simple approaches with a clear logic can be explained and agreed upon with manageable effort and a balanced rate of mistakes. Additionally, simple approaches with a clear logic can be represented in computer algorithms leading to automatic and likewise correct decisions.

Experience with the GANI_MED project results in a scope decision model for hospital-wide broad consent-based research as shown in Fig. 11.4. It could perhaps be easily adapted to other models with a more limited span.

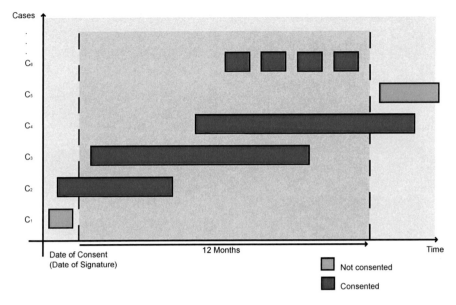

Fig. 11.4 A scope decision model with case-based data management

The scope with regards to the time period of a consent form is determined by the date the document has been signed. Date and signature work as follows in a confining manner:

- The consent form is valid for exactly 12 months, starting at the date of signature. All data collected in the context of consented case IDs within this period count as consented. Use of data collected within these 12 months is not limited.
- Persons not being able to agree at the time they arrive at the hospital may lead to situation C_2. The case ID has been generated, and perhaps data has been recorded, prior to the consent form being signed. The case ID is, however, valid even after the consent form has been finally signed. Such case ID counts as consented in a way that data belonging to this case ID can be used under the consent, including those data having been collected or generated prior to the signature.
- An analog logic applies for situation C_4: This case ID has still been created within the 12 month corridor while all data belonging to this case count as consented, including those that have been collected or generated after the 12 month period is over.
- Situation C_6 illustrates the option with fine-grained case ID models. All those case IDs within the corridor of 12 months count as agreed. The logic of C_2 and C_4 applies analogously.
- Persons can be expected to receive treatment in the hospital repeatedly. Therefore, the IT platform must be able to check whether a still-active consent form exists for the person, i.e., the current date is within an active 12 month period. No new consent form need be signed in these situations.

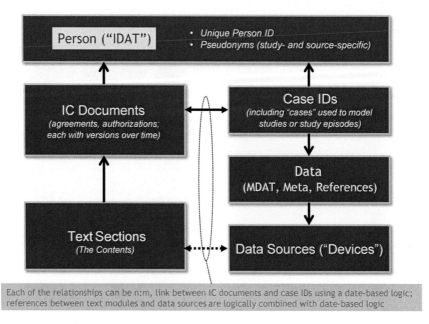

Fig. 11.5 Data model for a case-based data management

- Finally, an option should also be conceptualized to record the disagreement of a person, thereby allowing to offer those persons a new consent form at some later time knowing that they had disagreed at some earlier opportunity.

Besides the scope-related remarks made, modeling of interactions between data collected and stored, the informed consent (IC) presents another relevant aspect of a practical implementation in an IT platform. Fig. 11.5 suggests one suitable model.

A person, represented by their IDAT, has additional parameters: a unique person identifier as well as a number of specific pseudonyms. A number of consent forms signed by the person with a date of signature as well as a number of case IDs are also assigned to that person. Data sources, data, metadata, and references (e.g., PACS, LIMS, or biobank identifiers or links) are assigned to case IDs. Consent forms and case IDs can be linked and combined by a date-based logic as explained above. A case ID that is linked with said data can be referenced by multiple consent forms, and a consent form can cover multiple case IDs. The consent form is split into sections such as a general consent to collect, store, process, and use data; consent with regards to biomaterials, authorization for the release of medical records from the general practitioner, etc. These sections could be assigned to case IDs and/or data sources. Interesting details comprise of the following:

- Sections such as the authorization for the release of medical information from the general physician may not be linked with any data stored in an IT platform while the authorization must be part of the consent form record kept by a Trusted Third Party.

- A disagreement for taking biomaterials should be linked with the "data source" biobank to allow for an automatic decision or check when transferring or registering materials in a biobank system.

11.5 Implementation

11.5.1 IT Implementation Model of an Informed Consent

Due to the electronic ascertainment or processing of measurements and the resulting increase of collected data as well as the rising amount of participants of surveys, earlier paper form-based administration of consent documents is no longer feasible.

An electronic management of consent forms must enable the researcher to find out whether participants agreed to a specified target use or not. Considering this stringent requirement it is necessary for the consent form to have a very formal layout. Coiera and Clarke (2004) define four types of consent: (1) General consent, which allows every possible kind of data use; (2) General consent with disagreement to particular aspects (policies); (3) General denial with agreement to particular aspects; (4) General denial, which prohibits any further use of data.

All these types of consent can be represented with the mentioned generic sectional consent. At first, all possible uses which one can consent to, are collected. Each of these atomic units is reflected by a policy. Many policies are summarized to a consent template with an optional header and footer. The policies can and should be reused in different consent templates, so that this system forms a kind of construction kit. This architecture takes into account that different consents share common policies like "consent to store personal data" or "consent on re-contacting".

This formal procedure provides the basis for a later on automatic—and thus as much as possible free of interaction—check of the consent state and avoids any likely additional work for staff (Coiera and Clarke 2004), causing problems of tolerance.

In the case of long-term surveys, the probability of changes to the consent rises. Reasons for such necessary modifications range from over-specific phrasing of the original consent, legal changes or the implementation of additional measurements into a study design. Consequently one needs a model to administrate, as well as a concept to request different versions of the consent. So it is necessary to provide single policies with a kind of version number. In this case one has to consider that this version number shows different degrees of meaning in view of the content. Under most simple conditions it may be a consecutive number. However it may also be composed in a specific schematic (e.g. major.minor—1.0, 1.1, …). Thus single parts of version number could have different meanings—in the case of the example: a change of the part of the version number in front of the dot corresponds to legal or content changes of policy. In contrast to this, a change of version number behind the dot represents a so-called "cosmetic" change of policy (e.g. an orthographic adjustment). Practically, those persons who develop and authorize the consent should decide and indicate to IT whether a change is to be reflected in its version as a major or a minor change.

As a consequence of this possible meaning of parts of the version number it has to be feasible to inquire of the consent state for a range of the version number. Thus an intern, ideally configurable rule based system is necessary which is able to handle all possible occurring cases of version conflicts (e.g. a consent signed by one participant in different versions with varying consent states).

11.5.2   Improvements to the Basic Implementation Model

So far a system which is capable of storing and retrieving single policies in different versions has to be constructed. These policies are, together with an optional header and footer, combined to form a so-called consent template, which is versioned as well. So such a consent template correlates to an empty paper based consent document. Version numbers and the rule based query system to handle version conflicts should be configurable.

From a technical point of view a consent consists of at least a reference to a consent template and therefore a list of policies, a corresponding amount of consent status, a reference both to the consenting person and the physician which informs him and finally the consent date. A consent state is always connected to a policy and can be set to different values e.g. "accepted", "declined" and "unknown". The list of possible consent states should be configurable as well, although its impact on the rule based query system has to be noted. Since an identity management system should be provided by other means, the references to both the participant and physician should consist of some sort of ID. It is possible to store multiple synonym IDs, which eases the query process in the case that different systems know different IDs (e.g. pseudonyms). This, however, may lead to data duplication with all its implications and an accumulation of identity data which is questionable from a data privacy point of view.

In order to map a broader range of real world consents a more generic model could be achieved by adding free text fields to a consent template. This opens the possibility to store any additional information connected to a consent. Since the data content of such free text fields is not well defined, it is difficult to process it further.

Another useful option is the possibility to store the scanned consent document or the scanned or captured signature or even a qualified electronic signature. These scans could be used to provide a descriptive view of the stored consent information or may even replace the original paper-based consent document, although the pure digital storage of consent is not in compliance with regulations in Germany at the time of writing.

Figure 11.6 shows the model developed so far, on the left the common paper based consent; the right side contains a schematic view of the corresponding structures of the electronic consent management.

In addition to the discussed structures necessary to store the consent, some components which help to improve the collection of consent data and thus prevent

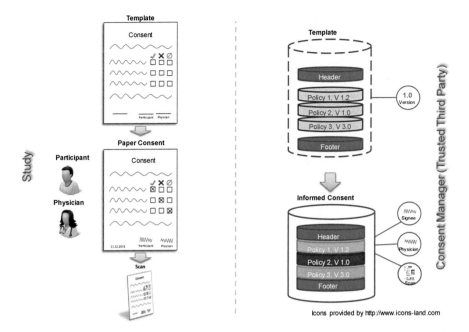

Fig. 11.6 A consent management model with paper and electronic representation

inconsistencies are appropriate. One of these is the possibility to flag a policy as mandatory whenever a denial of such policy would lead to an exclusion from the study or survey (e.g. "consent to store personal data"). It still is possible to store a consent where the consent state for a mandatory policy is "declined", but subsequent queries for any policy within this consent should lead to a denial, regardless of their consent state. Another option is a default state which could be set for any policy within a consent template. This value is meant as a hint for the consent input system and should never be set by the consent manager once a non-default value had been recorded. Finally, structures to declare dependencies between policies should be provided (a "consent to use some kind of biosamples" needs only to be given if a preceding "consent to retrieve samples" is agreed upon).

11.6 Additional Aspects

11.6.1 Quality Assurance—References and Transcribed Information

In an ideal IT architecture implementation, everything is connected. Therefore, a LIMS would know whether a person has agreed to take biosamples and which analytics or biobanking options have been agreed upon. It would only process those materials that have clearly been agreed upon. In practice, there are many differences from theory: technical measures such as interfaces, for example, may be a limiting factor in a specific platform implementation, or a study design may not allow for such strict workflow. Existence of biosamples is often referenced in some eCRF sections, together with a lab or sample number. However, samples could be discarded for several reasons during lab analytics, and sample numbers could be mistyped or mixed. Check digits and barcode scanners could help to lower the number of mistakes. Quality assurance must therefore also cover such consistency checks with external devices and information sources and must correct possible discrepancies together with feedback to medical staff accounting for those discrepancies in order to foster a training effect. A similar aspect applies in the case that the informed consent is declared and signed on paper. As has been explained earlier, a transcription into electronic format is indispensable. Such transcription will never be free of mistakes, which is why quality assurance must verify whether the electronic informed consent representations are complete and correct. Double-blinded input would be an appropriate though expensive measure. Random verifications are the minimum level of quality measures to be taken.

11.6.2   Informed Consent Documents' Quality

Importance of clarity and specificity of the informed consent document text have been emphasized a couple of times. This is, on the one hand, motivated by ethical standards, and on the other hand by the fact that when software algorithms have to decide upon the contents of an informed consent document, decisions will be correct if the underlying text is clear and specific. Otherwise the software would need to overcome a lack of definition by either automated interpretation or ignorance, both being unacceptable from legal and ethical points of view. One worst-case measure would be to supplement a mapping matrix that adds the necessary clarity and specificity the consent form lacks and maps such supplemented information to the sections/policies a consent manager operates upon.

Furthermore, consent forms may consist of multiple documents—often an information booklet (IB, see Fig. 11.3), for example, explaining study motivation, setup, and exercises; and the sectional consent form text, which can be kept short by having an accompanying information booklet. Normative information that need be used during decision making should be bundled in the consent form only. Then, the informed consent database may hold a versioned reference to the information booklet but algorithms could work on the normative consent form only. Assignment and update mistakes when revising these documents are avoided; specificity issues due to the structure of the document set used can easily be prevented.

Another quality-related issue with informed consent documents is the number of versions of each informed consent document set to be managed during the course of a study or project. Experience tells us that aspects of the study change and, as a consequence, the informed consent documents change over time. Typical examples of changes during the course of a study are sample sizes or examinations conducted.

Versions of informed consent documents need be managed adequately by IT measures, with the effect that a Transfer Unit, for instance, would need to understand about the several policies and their changes over time to formulate its inquiry towards the Trusted Third Party properly. The complexity of these inquiries and an increased mistake rate are the trade-off for having multiple informed consent versions. It is advisable that the sections and structure of the informed consent document should be as standardized and harmonized as possible, including the same paragraphs for the same facts, in order to minimize such trade-offs.

11.6.3 Purpose-Related Decisions

The purpose of data usage is typically part of a data usage application or an inquiry between the Transfer Unit and a Trusted Third Party. The purpose in a consent form cannot usually be a specific term but will allow for interpretation ("medical research in the field of cardiology"). Whether a research project originating a data usage application falls under this purpose, cannot be determined automatically by computer algorithms. Determining whether or not an application is within the range of purposes agreed must be done by humans. If there is a board whose members discuss and decide upon data usage applications, this would be one suitable panel to make such purpose-related decisions. In problematic cases we recommend a consultation with the responsible ethics commission.

11.6.4 Free Text Fields

Another problematic area from a computer algorithm's perspective are text fields in forms that could be used for writing additional text. Experience tells us that such text fields will sooner or later be abused for virtually any purpose. Withdrawals of consent, changes of address, additional medical indications, or corrections of input mistakes are noted herein. People expect those free text fields to be read and interpreted by humans, and expect them to act upon these "instructions"—while in practice these texts are usually just stored permanently within a database. Many staff do not acknowledge the purpose of such fields for commentary, optional additional information, and other non-binding information. As a consequence, free text fields should be avoided wherever possible. They may be better presented interactively if there is a known reason to add additional information.

11.6.5 Withdrawal

Every participant of a study has to have the possibility to withdraw his given consent without any restrictions. There are no further requirements on how this withdrawal is expressed and the form of the withdrawal is not specified. It can be expressed by

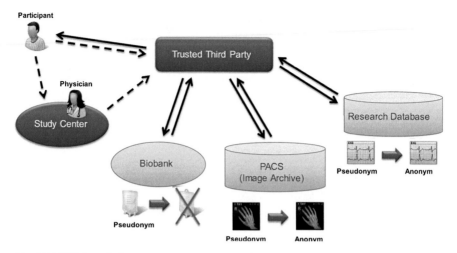

Fig. 11.7 Withdrawal of an agreement (starts and ends at the participant)

telephone, fax, e-mail, or paper mail. The participant is informed at the time of his/her consent which organization unit is responsible for the receipt of the withdrawal.

In contrast to the many and unspecific options to declare a withdrawal, the necessary structures and processes to handle such revocation of consent need to be strictly defined. It is important who receives the withdrawal, which organization units need to be informed about the withdrawal and who is in charge of informing the relevant organization units such as the biobank. In general, the first address to send a withdrawal would be the place where the participant signed the consent. There the withdrawal is documented. Another reasonable choice is the Trusted Third Party, which is also responsible for the consent management. This or similar central units assume the task to apprise all organizations which need to take actions on the withdrawal. These actions of course need to be defined too. They reach from simply "nothing" to "pseudonymization" or "anonymization" to "delete data" and/or "destroy samples". After handling the defined task, the central unit should be informed, which in turn informs the participant about the completed process. This final notification is the reason why the personal data of the participant need to be pseudonymized or deleted last (Pommerening et al. 2014). Figure 11.7 shows the entire process of a withdrawal.

11.7 Conclusions and Recommendations

This chapter explained selected technical and ethical aspects on the subject of processing data items that belong to study participants of epidemiologic and clinic-epidemiologic studies, which are enrolled to identify predictive biomarkers in the context of Individualized Medicine. As the informed consent is one of the fundamental elements of IT-based data from such studies, the following aspects have to be high-lighted.

1. Because of the mandatory requirement to process only data that have been agreed to become processed, IT must have knowledge about the respective informed consent contents, its link to all data covered by it, and IT must be able to process the contents. An adequate consent form does not allow for interpretation but is strict in determining the scope, purpose, and actions that are agreed upon.

2. The quality of a consent form—and thereby its ethical validity—has significant impact to the ability to process it, which is often underestimated. Therefore, the logical structure of consent forms and the interface between them and the IT processes should be focused on more intensively than usually done. This is, on the one hand, motivated by ethical standards, and on the other hand by the fact that when software algorithms have to decide upon the contents of an informed consent document, decisions will be correct if the underlying text is clear and specific. Otherwise software would need to overcome a lack of definition by either automated interpretation or ignorance, both being inacceptable from legal and ethical points of view.

3. It is desirable both from a processing as well as an ethical perspective to supply research with a homogeneous and uniform sectional IC architecture that facili-tates to manage different study types with adequate documents which use a uni-form structure, cover identical requirements with identical sections of text and implement requirements, depending on the type of a study, with specific sections of text.

4. Based on the example of the GANI_MED project this chapter presents a high-level architecture of an IT platform consisting of five building blocks:
   • data sources such as diagnostic devices, hospital-wide infrastructure systems and the IT Cohort Management,
   • an Extraction Layer working as an interface and cache between the data sources and the research database,
   • a Trusted Third Party as an organizational unit implementing measures for data privacy as well as a technical instrument to supply pseudonyms and to keep all relevant given consents of the participants available,
   • a Persistence Layer primarily storing all consented medical data together with the respective pseudonyms and
   • a Transfer Unit as an organizational unit managing the use and access regula-tions.

5. The scope of an agreement with regards to the time period is determined by the consent form. The scope decision model has a direct impact on the consent man-agement process. It should enable transparent and accurate decisions to be made about which data from which period of time are available for medical research.

6. Due to the electronic ascertainment or processing of data and the resulting increase of collected data as well as the rising amount of participants of surveys, the previous form of the paper-based administration of consents is no longer feasible.

7. Quality assurance must also cover such consistency checks with external devices and information sources and must correct possible discrepancies together with feedback to medical staff accounting for those discrepancies in order to foster a training effect.
8. Another quality-related issue with IC documents is the number of versions of each IC document set to be managed during the course of a study or project.
9. Which purpose a research project originating a data usage application falls under cannot be determined automatically by computer algorithms, but must be done by humans.
10. Free text fields in consent forms should be avoided wherever possible.
11. Structures and processes to handle withdrawals of consent need to be defined. It is important who receives the withdrawal, which organization units need to be informed and who is in charge of informing the relevant organization units.

References

Booth TC, Waldman AD, Wardlaw JM, Taylor SA, Jackson A (2012) Management of incidental findings during imaging research in "healthy" volunteers: current UK practice. Br J Radiol 85:11–21

Brothers KB, Langanke M, Erdmann P (2013) The implications of the incidentalome for clinical pharmacogenomics. Pharmacogenomics 14:1353–1362

Coiera E, Clarke R (2004) e-Consent: the design and implementation of consumer consent mechanisms in an electronic environment. Jam Med Inform Assoc 11(2):129–140

Community Medicine (2014) http://www.community-medicine.de. Accessed 12 May 2014

Erdmann P (2014) Zufallsbefunde aus bildgebenden Verfahren in populationsbasierter Forschung: Eine empirisch-ethische Untersuchung zur Ganzkörper-MRT in SHIP. Mentis, Münster (forthcoming)

Grabe HJ, Assel H, Bahls T et al (2014) Cohort profile: Greifswald Approach to Individualized Medicine (GANI_MED). J Transl Med 12:44

Langanke M, Erdmann P (2011) Die MRT als wissenschaftliche Studienuntersuchung und das Problem der Mitteilung von Zufallsbefunden: Probandenethische Herausforderungen. In: Theissen H, Langanke M (eds) Tragfähige Rede von Gott. Festgabe für Heinrich Assel zum 50. Geburtstag 9. Februar 2011. Verlag Dr. Kovač, Hamburg, pp 198–240

Langanke M, Brothers KB, Erdmann P et al (2011) Comparing different scientific approaches to personalized medicine: research ethics and privacy protection. Per Med 8(4):437–444

Langanke M, Erdmann P, Dörr M et al (2012) Gesundheitsökonomische Forschung im Kontext Individualisierter Medizin: Forschungsethische und datenschutzrechtliche Aspekte am Beispiel des GANI_MED-Projekts. Pharmacoeconomics German Research Articles 10(2):105–121

Medizinproduktegesetz in der Fassung der Bekanntmachung vom 7. August 2002 (BGBl. I S. 3146), das zuletzt durch Artikel 4 Absatz 62 des Gesetzes vom 7. August 2013 (BGBl. I S. 3154) geändert worden ist (2013) http://www.gesetze-im-internet.de/bundesrecht/mpg/gesamt.pdf. Accessed 13 May 2014

Pommerening K, Drepper J, Helbing K et al (2014) Leitfaden zum Datenschutz in medizinischen Forschungsprojekten—Generische Lösungen der TMF—Version 2. http://www.tmf-ev.de/. Accessed 26 June 2014

Reng CM, Debold P, Specker C, Pommerening K (2006) Generische Lösungen der TMF zum Datenschutz für die Forschungsnetze in der Medizin. Im Auftrag des Koordinierungsrates der Telematikplattform für Medizinische Forschungsnetze. Medizinisch Wissenschaftliche Verlagsgesellschaft, Berlin

Rudnik-Schöneborn S, Langanke M, Erdmann P et al (2014) Ethische und rechtliche Aspekte im Umgang mit genetischen Zufallsbefunden: Herausforderungen und Lösungsansätze. Ethik in der Medizin 26(2):105–119

Wolf SM (2008) The challenge of incidental findings. J Law Med Ethics 36:216–218

Chapter 12
Handling Incidental Findings from Imaging Within IM Related Research—Results from an Empirical-Ethical Study

Pia Erdmann

Abstract This chapter gives a summary of an empirical-ethical study which was carried out as a sub-project of GANI_MED-Greifswald. The aim of this sub-project was to investigate the reaction of participants, caused by incidental findings from a whole-body MRI in the context of SHIP, an epidemiological study conducted in Western Pomerania, Germany. As well as studying the potentially negative reactions of the participants after being informed of incidental findings, it was examined whether the participant's subjective perception of health is influenced and whether this leads to a changed risk-benefit evaluation concerning the participation in a MRI within the medical research context. In order to investigate this, an analysis of the currently most important codices and relevant literature were taken into consideration to define the subject of discussion and regulation and thus, to define the subject of research in detail. Afterwards the study was carried out according to a mixed-methods design which showed firstly which kind of stress is indeed a result of participating in the MRI examination and which factors cause this. Furthermore, it was shown that the risk-benefit evaluation and thus, the willingness to participate, had not changed significantly after the whole-body MRI examination. However, this might be due to the fact that the research subjects often misjudged the conditions for participation by overestimating the benefits and underestimating the risks. The chapter ends with a short summary of the deductions which were drawn from this study in order to be implemented into the procedures of GANI_MED.

Keywords Incidental findings · Whole-body MRI · Empirical ethics · Therapeutic misconception · Diagnostic misconception · Breaking serious news algorithms · GANI_MED

P. Erdmann (✉)
Lehrstuhl für Systematische Theologie, Theologische Fakultät, Ernst-Moritz-Arndt-Universität Greifswald, Am Rubenowplatz 2–3, 17487 Greifswald, Germany
e-mail: pia.erdmann@uni-greifswald.de

© Springer International Publishing Switzerland 2015
T. Fischer et al. (eds.), *Individualized Medicine,* Advances in Predictive,
Preventive and Personalised Medicine 7, DOI 10.1007/978-3-319-11719-5_12

12.1   Introduction

This chapter is a summary of my Phd thesis which was accredited by the University
Medicine Greifswald in 2014 (Erdmann 2014). It gives insights into an empirical-
ethical study which was carried out as a sub-project of GANI_MED–Greifswald
Approach to Individualized Medicine (Grabe et al. 2014). The aim of this sub-
project was to investigate the reaction of participants caused by incidental findings
from a whole-body MRI in the context of SHIP, an epidemiological study conducted
in Western Pomerania, Germany (Völzke et al. 2011). As well as the possibly negative reactions of being informed of incidental findings, it was examined whether the
participant's subjective perception of health is influenced and whether this leads to
a changed risk-benefit evaluation concerning the participation in a MRI within the
medical research context. In order to investigate this, an analysis of the currently
most important codices and relevant literature was taken into consideration to define the subject of discussion and regulation and thus, to define the subject of research in detail. Afterwards the study was carried out according to a mixed-methods
design which showed firstly which kind of stress is indeed a result of participating
in the MRI examination and what factors cause this. Furthermore, it was shown that
the risk-benefit evaluation and thus, the willingness to participate, had not changed
significantly after the whole-body MRI examination. However, this might be due to
the fact that the research subjects often misjudged the conditions for participation
by overestimating the benefits and underestimating the risks. The conclusion drawn
from this study is firstly that information has to be given to the participants with
great care, as in so many other clinical studies, in order to avoid phenomena such as
diagnostic and therapeutic misconception and secondly that factors causing stress
can be avoided by adjusting the conditions of participation, for example by changing the mode of informing research subjects of the results. The chapter ends with a
short summary of the deductions which were drawn from this study in order to be
implemented into the procedures of GANI_MED.

12.2   The Whole-Body MRI Within SHIP

The "Study of Health in Pomerania" (SHIP) researches the frequency of diseases
and their relevant risk factors for people who live in Western Pomerania, Germany.
(Völzke et al. 2011). In order to do this several basic and special examinations are
carried out (e.g. basic lab program, basic medical check-ups, examination of skin,
sleep clinic). Since 2008 it is possible to participate in a whole-body MRI examination in the context of SHIP. The participants who are interested in this are informed
in advance (in written and oral form) that they will be contacted only when there is
a relevant result and that there can be false-positive or false-negative findings which
should be checked again by a general practitioner or specialist. The participants are
also informed that the examinations do not correspond with a health screening. The
question of whether the diagnosis is relevant is clarified with the help of a complex

finding and notification procedure. For this purpose, initially a consultation with the participant takes place immediately after the examination. The participants will be informed about so-called ad-hoc findings which need acute treatment and therefore hospital admission is necessary. However, this happened very rarely so far ($\leq 1\%$). After the ad-hoc findings, a procedure with second and, if necessary, third opinions follows. In the case of unclear results (varying opinions or the lack of a precedent case) an advisory board is consulted which is an interdisciplinary committee which will decide on the question of whether the participant will be notified.

12.3 Ethical Analysis and Development of Research-Leading Questions

Giving an answer to whether the consequences of incidental findings from a whole-body MRI are ethically reasonable is methodologically done in several steps: Firstly, it is important to move from the question of simple reasonableness towards refined research-leading questions. In order to do so, the whole issue complex was structured into three ethical problem areas—informed consent, dealing with incidental findings, balancing risks and benefits. In a second step it was examined which norms are suggested and discussed with regard to these three problem areas in current bioethical regulations and the research-ethical discourse.

Which conditions have to be met in order to get a valid informed consent? Do binding conditions exist for dealing with incidental findings, and in particular concerning the question of whether incidental findings will be reported, and if so, what kind of findings? Concerning the weighing up of risks and benefits it was clarified in which form and to what extent stress caused by the incidental findings was considered reasonable and how respondents, who think about participating, weigh up the expected risks and benefits and thus evaluate for themselves whether they are willing to take the risk of stress?

For this analysis the following set of rules were examined:

- WMA Declaration of Helsinki (2013)—the internationally most famous set of rules
- The relevant guidelines by the Council for International Organizations of Medical Sciences (CIOMS), particularly:
  a. International Ethical Guidelines for Biomedical Research Involving Human Subjects (2002)
  b. International Ethical Guidelines for Epidemiological Studies 2009
  c. International Guidelines for Ethical Review of Epidemiological Studies (1991)
- Convention for the Protection of Human Rights and Dignity of the Human Being with regard to the Application of Biology and Medicine: Convention on Human Rights and Biomedicine, Council of Europe (1997)
- Additional Protocol to the Convention on Human Rights and Biomedicine Concerning Biomedical Research (2005)

- (Muster-)Berufsordnung für die in Deutschland tätigen Ärztinnen und Ärzte (2011)—the professional code of conduct for doctors by the German Medical Association
- Stellungnahme der Deutschen Gesellschaft für Humangenetik zu genetischen Zusatzbefunden in Diagnostik und Forschung (2013)—a statement of the German Society of Human Genetics on additional genetic findings in diagnosis and research
- Leitlinien der Bundesärztekammer zur Qualitätssicherung der Magnet-Resonanz-Tomographie (2000)—guidelines of the German Medical Association for quality management of MRI which include relevant comments and recommendations.

In a short summary (a detailed presentation of the regulation status will follow later in the context of the presentation of the results) the analysis of the codices and the relevant research-ethical discussion showed that the conditions for a valid informed consent within the codices are defined but—according to the current discourse—these conditions are not always met, which often causes therapeutic misconception in clinical studies. "Therapeutic misconception" is a phenomenon indicating that subjects who participate in research studies misjudge the character of research, i.e. they overestimate the benefit of participating in the study and underestimate the risks (first: Appelbaum et al. 1987).

Regarding the dealing with incidental findings it was emphasized that no (or only very controversial) discussions exist concerning all aspects, especially the following questions:

- What is considered as an incidental finding?
- Should incidental findings be disclosed?
- If a notification about findings takes place, what will be reported and how?

The norms concerning balancing risks and benefits are formulated in the codices in a vague way. It is often demanded to avoid risks, and if this is not possible it is desirable to minimize them at least. Moreover, they should be in proportion to the expected research benefit (e.g. International Ethical Guidelines for Epidemiological Studies 2009, p. 28). However, it remains unclear what is reasonable or proportionate. This is due to the fact that so far no binding methodology for the evaluation of risk/benefit exists. There are currently only suggestions concerning the risk-assessment (compare e.g. Rid et al. 2010 or Raspe and Hüppe 2012).

In a third step, research-leading questions highlighting the problem areas are differentiated further.

Informed consent: What do the research subjects know exactly about the conditions of participating with regard to the reported results, the time span and the mode of notification, as well as the potential risks and the potential benefits of their participation?

Attitudes of the participants regarding the dealing with incidental findings, independent of the real notification procedure in SHIP: What are their opinions on a strict non-disclosure strategy? Which kind of incidental findings do the they want to know about? How would they like to be notified? Do the participants want a notification saying that there is nothing relevant to report?

Balancing risks and benefits: Will the notification about an incidental finding after a whole-body MRI in the context of SHIP lead to stress for the people concerned, and if so, to what extent? Is the stress caused justifiable? Which factors are likely to cause stress? And on the side of the benefits: Do the participants for example feel a greater assurance in the healthy state of their body?

The analysis and development of research-leading questions is followed by assessing these questions in a practical way. A triangulation procedure will be used for the study design, as shown in the following section.

12.4   Methods

The research design of this study is a mixed-methods approach; in particular it is a "sequential explanatory mixed methods design" (Plano Clark and Creswell 2008, p. 466). More precisely this means that a two-stage triangulation procedure was chosen which combines quantitative and qualitative surveys.

The quantitative survey was undertaken by using two questionnaires: the first one was given to the participants immediately after their examination, the second one was mailed to them, either four weeks after a notification about an incidental finding was sent to them, or after it was confirmed that there were no results relevant for reporting. A follow-up interview was carried out with 20 respondents who stated either a high level of stress or doubted another participation in a whole-body MRI in the second questionnaire. The research sample consisted of all SHIP participants who had an appointment for a whole-body MRI between March 3rd and July 23rd 2010 ($N=549$). A questionnaire was given to 454 participants because some missed the appointment or the examination could not take place due to reasons such as metal found in the body or claustrophobia. 439 participants filled out the first questionnaire, 409 filled out the second one and sent it back to the MRI center. 131 out of 409 participants who filled out the second questionnaire received a notification about an incidental finding.

For the qualitative part of the study there were respondents whose answers met a specific pattern in the second questionnaire, and therefore the sampling was based on the model of "theoretical sampling" (Lamnek 2010, p. 95). These included that the persons concerned (a) indicated either very severe stress, severe stress or dissatisfaction or (b) refused to participate in a further MRI scan or (c) had expressed stress results (at the time of filling out the second questionnaire) which strongly differed from what they expected when they filled out the first questionnaire.

The two-stage procedure was efficient because firstly, questions which arose from the results of the quantitative survey were clarified in the face-to-face interview and secondly, opinions and attitudes could be investigated which are harder to evaluate by quantitative surveys.

12.5   Results

The presentation of the results will be done in three sections which, following the analysis of the ethical codices, have the titles "Informed Consent", "Dealing with incidental findings" and "Balancing Risks and Benefits". The results of both the quantitative and qualitative surveys are taken into consideration.

12.5.1   Informed Consent

One of the most important prerequisites for informed consent is the understanding of the content of information by the people who were informed. This includes for the present case of a MRI examination in the context of SHIP the understanding that it is a participation in an epidemiological study, not an examination within a health screening. Thus, in the context of the question of understanding the content, it is interesting whether the mode of participation has been understood. This can also be deduced from the reasons of participation.

Several different factors regarding the motivation of the subjects to participate in the MRI examination were found: Questionnaires and interviews suggest that a major motivating factor is the desire to know more about one's health. Moreover, some participants expected to have certain medical conditions clarified. Other participants thought that the MRI examination can be used as a medical check-up, e.g. for early detection of cancer or relapses. Despite these high expectations regarding the benefit, most of the participants continue to attend regular check-ups. The fact that participating at the same time serves scientific needs is another important factor. Other reasons for participating include recommendation from third parties (treating physician, relatives or friends), or a technical-scientific interest in the MRI, even though this is no decisive factor for participation.

Thus the analysis of the reasons of participation shows that the conditions of the study participation have often been misunderstood: research subjects do not realize that not all striking findings which occur in a MRI are reported (participants are actually only notified of results which require further clarification). There is also further evidence indicating that the conditions of participation are misunderstood: The possibility of false-positive or false-negative findings is often misjudged (cf. Fig. 12.1).

The participants are also informed, in the context of a conversation before the MRI examination, that it may take six to eight weeks before they get a written notification. Although this consultation is done directly before the examination and the participants are asked about the assumed time span until the notification of an incidental finding immediately after the examination, more than 50% of the participants assume a shorter time span. The average expected time span is 4.18 weeks. In total, only 29% of the participants knew the correct time span. This result suggests that even simple, non-medical contents of information can only be accurately remembered by some participants.

Respondents refer to the topic of the written information concerning the MRI scan in interviews only when they are asked directly. This is certainly due to

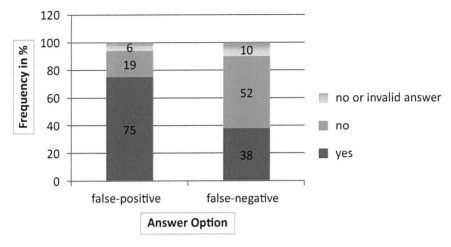

Fig. 12.1 Awareness of false-positive and false-negative findings

different reasons which can only be speculated upon here: It could be the case that participants do not want to admit that they either did not or only superficially read the documents or that they did not understand them fully. One reason for the lack of interest of some participants is that they have had a MRI examination before and therefore think that they are already aware of the relevant information. However, they didn't understand that this time they are research subjects and the MRI examination in the context of SHIP is different from a medical check-up, not only regarding the length of the examination, but also that different information is necessary. The research character of the examination and the differences resulting from it compared to usual MRI examinations in the health care context are not recognized. To the participants the sense of being given information is (a) to clarify if they meet the prerequisites for participation and (b) the preparation for what could be unpleasant during the examination (e.g. noise or confinement).

Nevertheless the information is evaluated as sufficient and comprehensible by most of the participants. They do not have the feeling that information is missing and they do not demand changes in the current information procedure.

12.5.2   Dealing with Incidental Findings

Regardless of the dealing with incidental findings of the MRI examination within SHIP the participants were asked in the presented study what kind of handling *they* would prefer if they had an influence over it. As shown before no consensus regarding these questions (Disclosure/Non-Disclosure, what is relevant to be reported etc.) could be found in the codices and the research-ethical discourse. Therefore it seemed interesting to ask the persons actually affected.

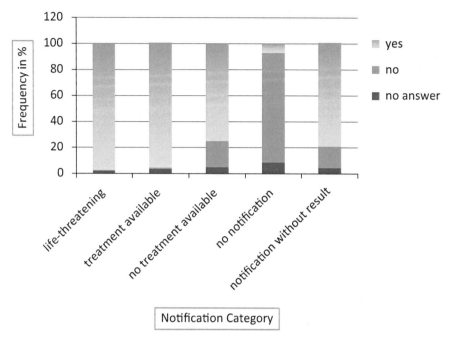

Fig. 12.2 What kind of results the participants wish to receive, *n*=409

Regarding the question of reporting incidental findings, the opinions of the participants are clear: the procedure of not notifying them about the results of the examinations is considered as withholding information and is thus disfavored. Especially when the results suggest health risks for those who participated voluntarily in the study, they should be reported to the people concerned, according to the interviewees. Those who made themselves available as participants for a study should benefit from it, is the unanimous opinion of the interviewees. Dealing with incidental findings in a way that participants are not notified about health relevant information—regardless of the question of whether it would be better for the results of the study or not—is even considered as cynical and irresponsible. The willingness to participate in an examination also depends on whether a notification of results is promised or denied to the participants; altruistic motives, like willingness to make oneself available for "science" are only effective when the basic condition of notification of results is fulfilled.

Concerning the question of whether participants should also be notified in the case of findings which do not need further clarification, no consistent evaluation is given. There are persons who want a notification in any case because they fear that a letter might get lost. On the other hand there are those who do not want a notification when no relevant result was found.

Moreover, the participants were asked which kind of results should be reported (cf. Fig. 12.2).

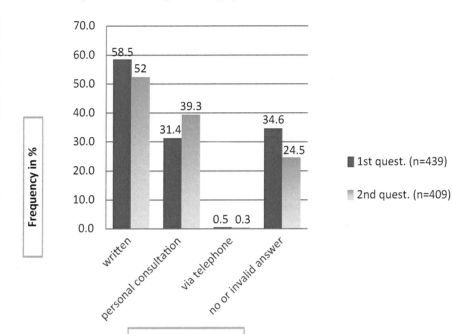

Fig. 12.3 Preferred mode of notification

Surprisingly 26 participants state that they do not want to be notified about the results. This choice, which is open to all participants, is hardly ever made in the informed consent for the SHIP MRI.

The data from the interviews confirms the results from the evaluation of the questionnaires: almost all respondents want to receive as much information as possible about the findings from the MRI examination, regardless of the question of whether the results are clinically relevant or not. Even in the view of the possibility that a reported finding could later prove to be false-positive—if this option is even discussed—the participants prefer that an unclear result should be checked again, regardless of the fears that may be caused, rather than a relevant result being incorrectly classified as not relevant for notification.

The answers to the question of how the results should be reported is demonstrated below in Fig. 12.3:

Some participants change their mind about the mode of notification in the process of filling out the two questionnaires towards the preference of a personal consultation instead of a written notification. It is striking that several participants state (although multiple answers are not allowed for this question) that they would like a written notification and a personal consultation. Most of the respondents, however, prefer a written notification although this can be a source of stress. It also does not meet the norm (cf. e.g. Maguire 2005; Fallowfield and Jenkins 2004), at least not in a clinical context. It turned out that many participants prefer a written notification because they have to decide in the questionnaire between a written and an oral notification and a combination of both is not an option. In the interview however the respondents stated, that they comprehensibly fear that choosing a personal consultation implies that they will not receive anything in written form and in such case, where they have to choose between those two alternatives, they prefer a written notification.

12.5.3 Balancing Risks and Benefits

For this study the effects resulting from the *written* form of notification in SHIP are crucial. Against this background this section first examines the risks, especially with regard to stress and possible disadvantages of participation and secondly presents the results concerning the benefits.

The results of the study concerning "stress" show that almost one quarter (23.7 %) of the participants, who received a written notification about incidental findings, were as a result moderately or heavily concerned about their health, almost one fifth (19.9 %) had moderate or heavy feelings of insecurity and less than one sixth (14.5 %) suffered from moderate or heavy insomnia. A deterioration of the psychological situation in the sense of depression symptoms as a result of the notification about incidental findings was not examined. Also the influence on daily activities and contact with family and friends is only a minor factor. However, there are factors which can modify the experienced stress in a specific way. Thus, the time after the notification is considered as a time of waiting which is experienced as stressful if the participants think (which is in some cases due to comments of medical personnel in the final consultation directly after the MRI) that something relevant, meaning something severe, was found and they will therefore receive a written notification. Moreover, under these circumstances the waiting time is experienced as stressful if no notification is received. Another factor which also induced stress is when the notification is received in an awkward moment, e.g. before the weekend when there is no chance of initial communication or clarification. Depending on the suspicion (and prior knowledge of the result), the time until the clarification might be stressful, especially if participants only get an appointment at a specialist weeks after the notification. Last but not least, prior experiences from family or friends with severe illnesses reinforce the anxiety about their health. Information about the course of disease of another person leads people to conclusions about the potential course of their own disease.

On the whole, the stress is, at least with regard to the number of the research subjects concerned, lower than originally expected. However, the participants who complain about moderate or heavy stress say in the interviews that they had to suffer from great anxiety about their health and also assumed that they would possibly die from the disease which was found. Furthermore, there was a time of insecurity until the clarification of the result, a time of anxiety, which lasted in some cases for a few weeks or even months. Besides the content of the result, the written mode of

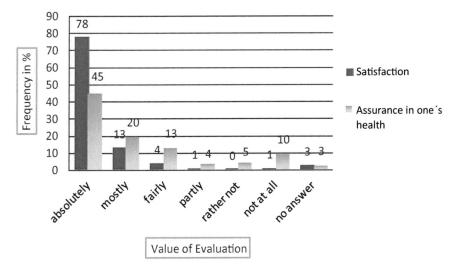

Fig. 12.4 Satisfaction of the participants

notification and also the time until the result is clarified can be identified as the main cause for the stress, according to the interviewees.

The actual benefit, i.e. the benefit of participation for the research subjects, cannot be grasped methodologically. Although it can be measured how many participants received a notification about incidental findings, this number does not indicate anything about the real benefit of participation. Firstly, this is due to the very difficult problem of false-positive and false-negative findings, i.e. neither every result nor every finding process, which did not result in a written notification, did create any beneficial effect for the participants.

There is yet another aspect which makes the evaluation more difficult, namely the fact that the participants' and the radiologists' evaluation of the relevance of the results are not consistent. In particular, two cases have to be seen critically: Firstly, where the degree of danger is estimated higher by the people concerned than by the radiologists. This is the case with 12 participants who think that the suspicion they received concerns a life-threatening disease which needs immediate treatment. The second case of misinterpretation of the relevance of the result appears when participants say that they received a finding where treatment is optional, although only findings are reported which need further clarification. Out of 131 participants who received notifications, 42 stated that the treatment is optional. Furthermore not every participant initiated the clarification of the reported results. This is probably partly linked to this misinterpretation and partly to the fact that they were already aware of the findings.

Regardless of the methodological difficulties of an objective determination of benefits, it can be said that the satisfaction of the research subjects with the examination is very high, only nine participants stated that they are "partly", "rather not" or "not at all" satisfied (cf. Fig. 12.4). Concerning the assurance in one's health, the number who stated "absolutely" (45 %) is not as high as the number of people "absolutely" satisfied with the examination (78 %).

12.6   Discussion

The following discussion will include two topics: Firstly, some methodological considerations about the carrying out of an empirical study and secondly, an ethical reflection of the results. The implicit backgrounds for this ethical reflection are the research-ethical principles and requirements of fairness from chapter 10, such as transparency and minimizing anticipated stress. Although they arose in the context of this study they are presented in Langanke et al. (chapter 10).

12.6.1   Methodological Considerations

As the questionnaires for the quantitative part of the study were developed, the challenge to find an instrument which was suitable to detect highly complex issues arose (e.g. the understanding of information concerning the possibility of false-positive and false-negative findings, cf. Raab-Steiner and Benesch 2008, pp. 50–52). This led, as indicated by the evaluation of the questionnaires, to questions which were difficult for the respondents to answer. The weaknesses of the questionnaires were compensated by a mixed-methods approach and the chance to ask deeper questions in face-to-face interviews. Restrictions concerning the validity of the interviews could possibly arise from the fact that due to delayed diagnostics, some interviews were already done a few weeks after the participation in the SHIP MRI but others almost one year after the examination. Surely, this influences which events and feelings will be remembered by the participants.

12.6.2   Ethical Reflection

In the previous sections, the necessary material for a systematic analysis was provided by analyzing the ethical regulation status and the empirical results of the survey of the research subjects. Thus, the ethical reflection can now be carried out. The advantage of considering empirical data becomes clear, as theoretically and *a priori* anticipated problems are complemented by the perspective of the participants. In this way it can be shown where codices and guidelines contain instructions which are suitable to avert potential damage from the research subjects without compromising the academic ambitions of the study. Moreover, they can indicate where action is needed with regard to solving ethical conflicts, which have not been anticipated efficiently, or to ethically analyze relevant options and their consequences.

Informed consent: Informed consent is within the three problem areas the one with explicit rules provided by the codices: The potential research subject has to be informed about the conditions of participation with care. This information has (a) to include potential risks of participation, (b) it has to be assured that all relevant information is included, (c) the form of information has to be chosen according to the understanding of the research subject, (d) it has to be assured that the participants of the study have enough time and the chance to ask further questions, (e) potential participants should have the chance to consult with a person of their choice before. On the whole (f) the consent to participate has to be given voluntarily, i.e. it must not be forced or manipulated by improper incentives. Lastly (g) it has to be possible for the participants to revoke their consent to the participation in the study.

The results of the survey presented in this chapter demonstrate that relevant conditions of participation are understood insufficiently on the part of the participants to the extent that they overestimate the benefit of the participation in the SHIP MRI, and moreover think that the MRI examination is some kind of health screening. Thus, they think that the examination will serve as clarification of medical conditions and thus makes other medical check-ups unnecessary (cf. also Schmidt et al. 2013, p. 1348). This result may be surprising at first because information in the context of the SHIP MRI is given with great care and effort. However, this result is not so surprising anymore if one considers that the phenomenon of therapeutic and diagnostic misconception within research studies is quite common and empirically documented.

Some problems which influence the understanding of IC documents negatively were identified with the help of theoretical consideration in a statement by the Swiss Academy of Medical Sciences—Positionspapier der Schweizerischen Akademie der Medizinischen Wissenschaften (SAMW) und der Arbeitsgemeinschaft der Ethikkommissionen (AGEK): Schriftliche Aufklärung im Zusammenhang mit Forschungsprojekten (2012):

- Lack of intelligibility of the written documents,
- Combination of juristic issues with the purpose of information in the IC documents,
- Carrying out the examination in a clinical context,
- Transfer of criteria of the doctor-patient relationship to the doctor-participant relationship by research subjects.

Especially in the case of the MRI examination, the setting of the SHIP study could also be an additional negative factor with regard to the validity of the information: The conditions of participation in the basic and special examinations differ from that of the MRI examination. The diagnostic validity of the examinations from the basic and special program is less debatable than for the whole-body MRI examination (SHIP 2009, p. 12).

Furthermore the participants always receive the results of the basic and special examinations, not only when something has been detected which requires notification. The validity of the results of the basic and special examinations corresponds to the best diagnosis known to us at this time so that the risk of false-positive and false-negative findings should be lower than for the MRI results. The information about the conditions of participation in the MRI examination which differ from the other SHIP examinations in the points mentioned above, are not dealt with in a separate document but are also found in the SHIP brochure (from page 9). This could lead to misinterpretations. It cannot and will not be answered here which of the presented factors causes the participants' insufficient understanding of the information documents. It will also not be considered whether the detected deficits in understanding are linked to the fact that participants are often not willing to analyze the documents because they are familiar with the MRI within a health care context and have not understood that the SHIP MRI is within a study context. This would require another study. However, the consequences of the detected deficits in understanding are quite clear: most importantly, the misinterpretation of the significance and seriousness of the findings (Schmidt et al. 2013, p. 1348), which—as the results indicate—lead to participants either not clarifying the results or not going to medical check-ups, at least for a while.

The difficulties to fulfill the implementation of the norms of informed consent cannot be linked to the regulation deficits, but are rather due to different circumstances, which could be partly avoided through improving the information procedure. There is the problem that especially the written information documents, the IC documents in particular, are often not only the basis for the information but are also used for avoiding risks of liability, which has a negative effect on the intelligibly (Thiele 2011, p. 122 ff.).

Dealing With Incidental Findings There are nearly no binding guidelines for dealing with incidental findings. It is difficult to draw guidance from the research-ethical discourse as it consists of many different arguments and partly contradicting recommendations.

For example, there are several reasons for and against the notification of incidental findings, but there is no regulating consensus besides the obvious ethical principles, of granting the right not to know and the consideration of the duty to render assistance in an emergency. More interestingly, the interviewed participants of the SHIP MRI examination had a clear opinion on the question of whether (relevant) incidental findings should be reported. The data suggests that many subjects would not consider participating if the algorithms were adjusted in the sense of a non-disclosure strategy.

The fact that the SHIP study decided against a non-disclosure strategy concerning the notification in the case of certain, and in particular in the case of severe MRI results, firstly meets the expectations of the participants and secondly helps to minimize the risk of failure to render assistance in an emergency. Thirdly, the examination-ethical perspective is taken into account.

There is at the most indirect guidance in the codices regarding the question of whether there should also be a notification if no relevant results were found, where it is pointed out that the danger of receiving no notification could be interpreted as a "clean bill of health".

In the questionnaire of the presented study, the majority (79%) chose the option to be notified about the fact that no suspicious results were found. The interviews, however, do not show such homogeneous opinion about this. Apparently, some participants think that after a certain amount of time where they have not received a notification that nothing (or nothing relevant) was found while others wish to be notified in any case in order (a) to be certain that no notification got lost and (b) to be able to close the chapter "MRI examination". However, the algorithms within the SHIP MRI examination do not provide such notification in order to avoid misinterpretations. Rather, in the context of the informed consent, it is warned against hasty interpretations of the results.

Within the field of research ethics it is intensely debated who should evaluate study results and how it should be done because this involves serious implications for the validity and the number of incidental findings (Kumra et al. 2006, p. 1001). Yet, a homogeneous estimation of this problem on the level of regulation (codices) is not recognizable in the ethical discussion. An evaluation of the participants' opinion regarding this question within this survey seemed neither useful nor necessary because the chosen procedure for the MRI examination meets the highest standards.

Concerning the question of which results should be reported, the majority of the participants answered in the post-questionnaire that they want to get as much information as possible. At least 307 (= 75 %) of the participants wanted a notification of results even when a treatment of the disease is possible but not necessary. This tendency was confirmed in the interviews. The participants want to be notified about the results, regardless of the relevance for their health.

That the validity of the whole-body MRI remains unclear to the extent that for some findings the specificity and the sensitivity are unknown is not reflected critically by the participants.

The codices are silent about the topic of which results should be reported, except the statement of the German Society of Human Genetic (Stellungnahme der Deutschen Gesellschaft für Humangenetik zu genetischen Zusatzbefunden in Diagnostik und Forschung 2013, pp. 2–3). On the level of the research-ethical discourse rival positions are taken, especially along the boundaries between the different disciplines (genetics, radiology and neuropsychology). Two issues emerge regarding the question about the types of findings relevant for notification. Firstly, there are opinions saying that the paternalistic withholding of relevant information is allowed (e.g. in the case of diseases which cannot be treated). Secondly, there is an emphasis on the impossibility of adequate information and notification in the case of many potential findings and the unnecessary disturbance of participants through the notification of less valid results.

The procedure chosen for SHIP is a compromise. Although certain results are not reported, e.g. those with an unclear predictive value or findings concerning the spine (SHIP 2009, p. 13), all other results which need further clarification are reported in written form, except in the case, that clarification was already initiated, based on an ad-hoc diagnosis.

Thus, the wish of the research subjects "to find out everything" is not fulfilled. Thereby negative consequences of less valid results, which are regularly underestimated by the participants, are avoided (Borra and Sorensen 2011, p. 850).

Lastly, the question remains of how incidental findings should be reported, which has been little recognized in the research-ethical discourse and is only explicitly accounted for in the statement of the German Society of Human Genetics (Stellung-nahme der Deutschen Gesellschaft für Humangenetik zu genetischen Zusatzbefun-den in Diagnostik und Forschung 2013).

In case of the notification of severe findings in a clinical context, especially in the case of oncological findings, oral information is the "golden standard" because the needs and reactions of the patients concerned can only be integrated through a face-to-face conversation (cf. e.g. Fallowfield et al. 2002; Baile et al. 2000). This standard has not been enforced yet for the notification of incidental findings in the research context. Thus, results of the MRI examination within SHIP are reported in written form. The majority of the respondents also state in the questionnaires that they want a written notification (59.1% pre-questionnaire; 52.5% post-ques-tionnaire); however, this result is modified by the interviews because here the par-ticipants state that they would like both a written and an oral notification, because otherwise they think that they will otherwise not get any written document which they can show to their practitioners.

The question of whether participants should receive a notification in cases where no relevant results are found can be answered with the help of the demand to mini-mize anticipated stress (cf. Langanke et al., chapter 10) because—when the rel-evant documents are formulated appropriately—no disadvantages will arise from this for the research subjects or the study. But such a notification can help to mini-mize stress through eliminating the useless waiting for the notification. It could also be formulated in a way that a misinterpretation as "clean bill of health" is prevented (Langanke and Erdmann 2011, p. 236). It should be decisively pointed out, for ex-ample, that in cases of physical discomfort a doctor should be consulted and that the examination does not at all compensate for a medical check-up.

The fairness demands presented by Langanke et al., chap. 10 are also useful for clarifying the question of who should evaluate the study results. First, one has to take into account that a restricted validity of results increases the possibility of false-positive, but also false-negative findings. At the same time false-positive findings are connected with severe and unnecessary stress for the people concerned, whereas false-negative ones could create a false sense of security for the participants. There-fore the guarantee of the greatest possible validity of the results is a demand con-cerning the minimizing of anticipated stress. This demand is fulfilled if radiologists (in the case of imaging) or geneticists (in the case of genetic examinations) carry out the interpretation, eventually with the help of specialists from other disciplines. For the problem of false-positive results in the SHIP MRI examination cf. Hegenscheid et al. (2009, p. 757).

Furthermore the demand for an empathic way of reporting the results with the help of breaking serious news algorithms, which were developed by psychologists and proven in fields such as oncology, can be justified by referring to the fact that stress can be minimized.

With regard to dealing with incidental findings the following two questions are particularly complex. Should incidental findings be reported at all? How can it be decided what kind of findings are relevant for notification?

There are prerequisites under which it seems acceptable from the perspective of a moderate research-ethical contractualism (cf. Langanke et al., chap. 10) to not report potential health-relevant information to the participants of non-interventional studies. Those prerequisites are fulfilled when the following features are met (Puls et al. 2010, p. 469):

1. The participants are informed that they will not get any results and they have consented to participate in the study knowing this.
2. The setting of the study is designed so that either no interpretation of the results takes place or the interpretation is only done when it can be ruled out that any information could be relevant for the health of the participants. The second prerequisite is met when the examiners (a) do not have the chance to medically evaluate the generated information, neither during nor after the examination and (b) the untreated data is stored under quarantine until it is certain that the results will not be relevant anymore for the persons examined. Only under this condition does the duty to render assistance in an emergency no longer appear relevant.

To what extent such a procedure, e.g. in the context of carrying out imaging studies, is feasible at all, is an open question, especially because it has to be feared that the number of people willing to participate in such a study, which has purely third-party interest, is presumably so small that the validity of the study results, i.e. the scientific quality, is compromised.

But also the decision against a strict non-disclosure strategy and for the notification, at least of "life-threatening" or "notification-relevant" results does not solve all the ethical problems, especially not if one operates with so-called "positive lists". Some problems arise because no objective criteria exist concerning the question of when a finding is worthy of notification:

1. In order to avoid conflicts with the legal duty to render assistance in an emergency, an intersubjectively communicable definition of "life-threatening" has to be formulated. Is the suspicion of a tumor a "life-threatening" finding, for instance?
2. Who decides which possible findings are on the "positive list" of the reported findings and thus defines the relevance for the people concerned?
3. How should untreatable (but life-threatening) findings be dealt with? Proposals for solutions regarding this problem, as provided by the Stellungnahme der Deutschen Gesellschaft für Humangenetik zu genetischen Zusatzbefunden in Diagnostik und Forschung (2013) or in Rudnik-Schöneborn (2014) bear the danger of paternalism which robs participants of the chance to spend their remaining lifetime more consciously (cf. Schmücker 2012, p. 15).
4. How will ethical problems be solved which arise when something is detected, which would be relevant in the eyes of the researcher or interpreting doctor, but which is not on the positive list?

If the decision is made against the use of a positive list and a mode is chosen so that the notification-relevance of findings is bound to their belonging to an open category of notification-relevant findings, there is the challenge to make transparent which findings belong to this category. This can only be done with the help of examples and counter-examples. Furthermore, risk of therapeutic and diagnostic misconception increases because such a mode can cause participants to misinterpret the study as a health screening. Moreover, there is the problem, at least in the case of genetic examinations, that (a) information cannot be provided due to the great number of possible findings and that (b) research subjects could be over-challenged by the notification of many findings.

From the notification of incidental findings, in the case of longitudinal observation studies, there can also be the risk of a systematic bias because the participants usually consult a doctor to clarify the incidental findings. How great the risk is that the validity of longitudinal observation studies will be compromised through such self-induced interventions, is indeed currently examined for the SHIP MRI in the context of a German Research Foundation (DFG) study "Incidental findings in a population cohort: A risk for valid prospective analyses of health relevant outcomes".

In the view of the author an answer to the question of which incidental findings of a study should be reported, can only lie in an approach which comes close to the chosen procedure for the MRI examination within SHIP. The following aspects, however, should be improved:

- Before the subjects make their decision to participate it must be certain that they do not consider the SHIP MRI as a health screening.
- In the course of the notification of the results there has to be an understanding regarding their restricted validity.

On the whole, the decision to link the notification relevance also to the validity of the results and thus to neither accept a paternalistic withholding of results nor an unnecessary disturbance through notification of the most likely irrelevant results, has to be supported from an ethical perspective.

12.7 Deductions for GANI_MED

The results from the empirical-ethical study regarding the effects of incidental findings from the SHIP MRI could be used for the GANI_MED project in Greifswald.

The process of informed consent within GANI_MED was adjusted to the extent that a maximum transparency could be achieved, e.g. through implementing a checklist for interviewers. Moreover, there are trainings on a regular basis for the interviewers. The focus here is on avoiding signals which could support a therapeutic and diagnostic misconception (Erdmann et al. 2013).

Regarding the notification of incidental findings which are possibly generated in the MRI examination of the head in the Paro cohort (for this special case see Langanke et al., chap. 10), the following procedure was chosen: Firstly, all participants in general are going to receive a notification, also when nothing was found. Secondly participants with a relevant diagnosis are invited to a consultation with the senior physician of the Clinic and Polyclinic for Neurology so that the results and their consequences cannot only be discussed in person but so that the participants also have a direct contact person who will advise them regarding the results and further clarification (Langanke and Erdmann 2011).

Acknowledgments For important suggestions regarding content and method I want to thank Prof. Dr. Thomas Kohlmann, Prof. Dr. Heinrich Assel, Prof. Dr. Henry Völzke and PD Dr. Carsten-Oliver Schmidt. I would also like to thank Wenke Liedtke for her work on the illustrations and Antje Holtmann for the translation into English.

References

Additional Protocol to the Convention on Human Rights and Biomedicine Concerning Biomedical Research (2005) http://www.coe.int/t/dg3/healthbioethic/Activities/02_Biomedical_research_en/195%20Protocole%20recherche%20biomedicale%20e.pdf. Accessed 03 Jul 2014

Appelbaum PS, Roth LH, Lidz CW et al (1987) False hope and best data: consent to research and the therapeutic misconception. Hastings Cent Rep 17(2):20–24

Baile WF, Buckmann R, Lenzi R et al (2000) SPIKES-a six-step protocol for delivering BAD news: application to the patient with cancer. The Oncologist 5:302–311

Borra RJ, Sorensen AG (2011) Incidental findings in brain MRI research: what do we owe our subjects? J Am Coll Radiol 8:848–852

Convention for the protection of human rights and dignity of the human being with regard to the application of biology and medicine: Convention on Human Rights and Biomedicine. Council of Europe (1997) http://conventions.coe.int/treaty/en/Treaties/Html/164.htm. Accessed 03 Jul 2014

Erdmann P (2014) Zufallsbefunde aus bildgebenden Verfahren in populationsbasierter Forschung: Eine empirisch ethische Untersuchung. Mentis. (forthcoming)

Erdmann P, Langanke M, Assel H (2013) Zufallsbefunde-Risikobewusstsein von Probanden und forschungsethische Konsequenzen. In: Steger F (ed) Medizin und Technik. Risiken und Folgen technologischen Fortschritts. Mentis, Münster, pp. 15–47

Fallowfield L, Jenkins V (2004) Communicating sad, bad, and difficult news in medicine. Lancet 363:312–319

Fallowfield L, Jenkins V, Farewell V et al (2002) Efficacy of a Cancer Research UK communication skills training model for oncologists: a randomized controlled trial. Lancet 359:650–656

Grabe HJ, Assel H, Bahls T et al (2014) Cohort profile: Greifswald Approach to Individualized Medicine (GANI_MED). J Transl Med 12(144). doi:10.1186/1479-5876-12-144

Hegenscheid K, Kühn JP, Völzke H et al (2009) Whole-body magnetic resonance imaging of healthy volunteers: pilot study results from the population-based SHIP study. Fortschr Röntgenstr 181:748–759

International Guidelines for Ethical Review of Epidemiological Studies (1991) http://www.cioms.ch/images/stories/CIOMS/guidelines/1991_texts_of_guidelines.htm. Accessed 03 Jul 2014

International Ethical Guidelines for Biomedical Research Involving Human Subjects (2002) Prepared by the Council for International Organizations of Medical Sciences (CIOMS) in collaboration with the World Health Organization (WHO). http://www.cioms.ch/publications/guidelines/guidelines_nov_2002_blurb.htm. Accessed 03 Jul 2014

International Ethical Guidelines for Epidemiological Studies (2009) Prepared by the Council for International Organizations of Medical Sciences (CIOMS) in collaboration with the World Health Organization (WHO)

Kumra S, Ashtari M, Anderson B et al (2006) Ethical and Practical Considerations in the Management of Incidental Findings in Pediatric MRI Studies. J Am Acad Child Adolesc Psychiatry 45:1000–1006

Lamnek S (2010) Qualitative Sozialforschung. Beltz, Weinheim

Langanke M, Erdmann P (2011) Die MRT als wissenschaftliche Studienuntersuchung und das Problem der Mitteilung von Zufallsbefunden: Probandenethische Herausforderungen. In: Theissen H, Langanke M (eds) Tragfähige Rede von Gott. Festgabe für Heinrich Assel zum 50. Geburtstag 9. Februar 2011. Verlag Dr. Kovač, Hamburg, pp. 198–240

Leitlinien der Bundesärztekammer zur Qualitätssicherung der Magnet-Resonanz-Tomographie. Gemäß Beschluss des Vorstandes der Bundesärztekammer (2000) Deutsches Ärzteblatt 97(39):A2557–A2568

Maguire P (2005) Breaking bad news: talking about death and dying. Medicine 33:29–31

(Muster-)Berufsordnung für die in Deutschland tätigen Ärztinnen und Ärzte-MBO-1997-in der Fassung der Beschlüsse des 114. Deutschen Ärztetages 2011 in Kiel (2011) http://www.bundesaerztekammer.de/page.asp?his=1.100.1143. Accessed 3 Jul 2014

Positionspapier der Schweizerischen Akademie der Medizinischen Wissenschaften (SAMW) und der Arbeitsgemeinschaft der Ethikkommissionen (AGEK). (2012) Schriftliche Aufklärung im Zusammenhang mit Forschungsprojekten. Schweizerische Ärztezeitung 93:1299–1301

Plano Clark VL, Creswell JW (2008) A sequential explanatory mixed methods design to explain findings, editors' introduction. In: Plano Clark VL Creswell JW (eds) The mixed methods reader. Sage, Nebrasca, pp. 466–467

Puls R, Hamm B, Hosten, N (2010) MRT ohne Radiologen-ethische Aspekte bei bevölkerungsbasierten Studien mit MR-Untersuchung. Fortschr Röntgenstr 182:469–471

Raab-Steiner E, Benesch M (2008) Der Fragebogen: Von der Forschungsidee zur SPSS-Auswertung. Facultas, Wien

Raspe H, Hüppe A (2012) Research Ethics Committees: identifying and weighing potential benefit and harm from clinical studies. WMJ 58(1):2–6

Rid A, Emanuel EJ, Wendler D (2010) Evaluating the risks of clinical research. JAMA 304(13):1472–1479

Rudnik-Schöneborn S, Langanke M, Erdmann P et al (2014) Ethische und rechtliche Aspekte im Umgang mit genetischen Zufallsbefunden: Herausforderungen und Lösungsansätze. Ethik in der Medizin 26(2):105–119

Schmidt CO, Hegenscheid K, Erdmann P et al (2013) Psychosocial consequences and severity of disclosed incidental findings from whole-body MRI in a general population study. Eur Radiol 23:1343–1351

Schmücker R (2012) Zufallsbefunde-was gebietet die Menschenwürde? Preprints and Working Papers of the Centre for Advanced Study in Bioethics Münster 2012/36. https://www.uni-muenster.de/imperia/md/content/kfg-normenbegruendung/intern/publikationen/schmuecker/36_-_schm__cker_-_zufallsbefunde_und_menschenw__rde.pdf. Accessed 20 Jun 2014

SHIP (2009) Leben und Gesundheit in Vorpommern. Teilnehmerinformation. Version S2-V 3.0.

Stellungnahme der Deutschen Gesellschaft für Humangenetik zu genetischen Zusatzbefunden in Diagnostik und Forschung (2013) http://www.gfhev.de/de/leitlinien/LL_und_Stellungnahmen/2013_05_28_Stellungnahme_zu_genetischen_Zufallsbefunden.pdf. Accessed 03 Jul 2014

Thiele F (2011) Autonomie und Einwilligung in der Medizin. Eine moralphilosophische Rekonstruktion. Mentis, Paderborn

Völzke H, Alte D, Schmidt CO et al (2011) Cohort profile: The study of health in Pomerania. Int J Epidemiol 40:294–307

WMA Declaration of Helsinki-Ethical Principles for Medical Research Involving Human Subjects (2013) Adopted by the 18th WMA General Assembly, Helsinki, Finland, June 1964 and amended by the: 29th WMA General Assembly, Tokyo, Japan, October 1975, 35th WMA General Assembly, Venice, Italy, October 1983, 41st WMA General Assembly, Hong Kong, September 1989, 48th WMA General Assembly, Somerset West, Republic of South Africa, October 1996, 52nd WMA General Assembly, Edinburgh, Scotland, October 2000, 53rd WMA General Assembly, Washington DC, USA, October 2002 (Note of Clarification added), 55th WMA General Assembly, Tokyo, Japan, October 2004 (Note of Clarification added), 59th WMA General Assembly, Seoul, Republic of Korea, October 2008, 64th WMA General Assembly, Fortaleza, Brazil, October 2013. http://www.wma.net/en/30publications/10policies/b3/. Accessed 22 May 2014

Part VI
Health Economic Assessment
of Individualized Medicine

Chapter 13
Individualized Medicine: From Potential to Macro-Innovation

Steffen Flessa and Paul Marschall

Abstract Individualized Medicine has the potential to change the relationship of doctors and patients, to alter the rules, institutions and regulations of the health care sector and even to influence societal values. However, there are major barriers preventing the key stakeholders to adopt this new approach to medicine. The aims of this contribution are: First, to analyze these barriers. Second, to investigate, whether Individualized Medicine has the potential to become a macro-innovation. It concludes that Individualized Medicine is still in an early stage of the development and adoption process so that it has to mature in a niche before it can become the new standard solution for the health care system for allocating scarce resources.

Keywords Micro-, meso- and macro-innovation · Innovation · Barriers to innovation · Individualized medicine · Stakeholders

13.1 Introduction

The single-cause-single-effect model (Schwartz et al. 2003) is a basic paradigm of medicine. Accordingly *one* (biological) cause exists for getting a specific disease. Having it removed makes it possible to return into a healthy state. This paradigm turned out to be appropriate with respect to many infectious diseases for which a non-ambiguous pathogen is exclusively responsible. Accordingly its elimination (e.g. via antibiotics) is sufficient for a complete cure. However, chronic degenerative diseases require a multi-cause-multiple-effect model due to two reasons: First, they are the consequence of many factors like genetics, metabolism, behavior or

S. Flessa (✉) · P. Marschall
Rechts- und Staatswissenschaftliche Fakultät, Lehrstuhl für Allgemeine Betriebswirtschaftslehre und Gesundheitsmanagement, Ernst-Moritz-Arndt-Universität Greifswald, Friedrich-Loeffler-Str. 70, 17487 Greifswald, Germany
e-mail: steffen.flessa@uni-greifswald.de

P. Marschall
e-mail: paul.marschall@uni-greifswald.de

© Springer International Publishing Switzerland 2015
T. Fischer et al. (eds.), *Individualized Medicine,* Advances in Predictive, Preventive and Personalised Medicine 7, DOI 10.1007/978-3-319-11719-5_13

environmental conditions. Second, there is no deterministic or non-linear relation between the occurrence of risk factors and the outbreak of disease. Consequently an aging society needs a paradigm shift regarding model conceptions. This has often been asked for but implemented inadequately so far (Bryan 2009).

The change in epidemiologic thinking additionally requires a change of the health care provision paradigm. According to the single-cause-single-effect model, all patients suffering from a specific disease may be treated as a homogenous group. However, according to a multi-cause-multi-effect model numerous constellations of individual biology (e.g. genetics, metabolism), environmental factors (e.g. exposition, living space, and environment) and of behavior (e.g. risk exposure) have to be distinguished. Whereas a "one size fits all" approach was sufficient during the first and second phase of epidemiological transition and during the dominance of infectious diseases it is no longer adequate with regard to the complexity of chronic-degenerative diseases. Thus, prevention, diagnostics and treatment tailored to individuals are necessary. This calls for Individualized Medicine (IM) (Cortese 2007; Hoffmann et al. 2011).

This demand is criticized by the argument that IM is only a buzzword for adjusting medicine to the individual patient who has always been the core of medical ethos (Gadebusch Bondio 2011). According to this view IM cannot be considered as an innovation but as a re-launch at most. Other critics claim that IM is a great promise which is neither tenable nor fundable (Der Spiegel 2011). It is rather a marketing instrument of the pharmaceutical industry and could thus never become a standard paradigm of medicine (Bartens 2011).

It is the aim of this chapter to discuss the potential of Individualized Medicine—within the area of the conflict of these hypotheses—as a macro-innovation for the health care system in Germany. Additionally, possible barriers will be presented which could prevent the acceptance of this innovation as a standard solution. Therefore the question will be analyzed of whether Individualized Medicine only marks a niche phenomenon for some selected diseases or whether it will lead to a comprehensive shift in paradigm in health care. To this end, the concept of Individualized Medicine is defined in the second section while separating itself from other expressions which are often mentioned in the same context. The third section refers to IM as an innovation and presents conditions which IM has to comply to in order to have the potential to be a macro-innovation. This could change thinking patterns within medicine fundamentally. For this purpose, in the fourth section selected stakeholders of health care will be examined with regard to the challenges IM poses for them as well as conditions under which they could become promoters of IM. The chapter is then concluded with a short outlook.

13.2   Definitions

The term 'Individualized Medicine' has not yet been universally defined (Sohn et al. 2010). Enumerative definitions predominate in literature (Hüsing et al. 2008) due to the fact that demarcation from classical differential diagnostics is not entirely possible. Therefore, Individualized Medicine in the strict (s.s.) and in the wide sense (s.l.) will be distinguished. IM s.s. refers to so-called Advanced Therapy Medical Products (ATMP). Such medical products are manufactured for an individual and are used exclusively for this person. These include, for instance, gene therapy (Leyen et al. 2005) and tissue engineering (Lanza et al. 2007), as well as regenerative technologies (Kube 2010). While prostheses and implants are already made to measure, the individualization is usually limited to their external shape. Those products could in principle be used for other patients as well. Unique products, however, correspond to the genotype of the patient in a way that does not allow any transmission to other patients. An example would be the removal, culturing and reimplantation of a patient's own cartilage (Regenerative Medicine) (Meisel 2005). In this way complete organs will be manufactured individually in the future. In any case, ATMP are limited to a single individual. So far these technologies are available in exceptional cases only.

IM s.l. also represents the turning away from the orientation of prevention, diagnosis and treatment for the total population. However, this concept involves categorizing each patient into groups. This sense considers IM as a stratified medicine (Hüsing et al. 2008). In this regard, the prototype—sometimes even used synonymously (Boston Consulting Group 2011)—is the so-called pharmacogenomics (respectively pharmacogenetics) (Sohn et al. 2010; Vegter 2008b), i.e. the choice of drug as well as dosage and administration based on the patient's individual information. In addition to the classical information on phenotype (e.g. age, height, weight), biomarkers as the result of molecular medicine (genomics, metabolomics, proteomics), as well as imaging techniques (e.g. MRT), are mainly to be used for this purpose. These markers of stratification determine the classification of patients into subgroups associated with a specific pharmacotherapy.

The paradigm of patient group stratification can be transferred to other therapeutic, diagnostic and preventive measures (Boston Consulting Group 2011). The development of a tailored program for diabetes prevention, for instance, could be depending e.g. on the existence of the HbA1c biomarker. That is to say people willing to exercise prevention measures could be stratified in various groups based on their biological disposition. There is a fundamental difference from the classical differential diagnostics, whose primary goal was the exact determination of a disease in order to assign the patient to a standardized path of treatment for that disease. However, it will be referred to Individualized Medicine, if based on differential diagnostics, additional biomarkers (along with imaging) are used in order to identify specific prognoses on the efficacy as well as the risk of certain preventive, diagnostic and therapeutic measures for individual subtypes. In this way physicians are able to choose a path of treatment corresponding to the specific stratum (see Fig. 13.1).

The indistinct definition in literature complicates demarcation. Thus, the term 'personalized medicine' (e.g. Boston Consulting Group 2011) is often used synonymously for IM although the terms 'person' and 'individual' are to be distinguished sociologically (Kamps 2010). While the term 'individual' puts emphasis on the words unified, systemic and unique, the expression 'person' implies a certain role

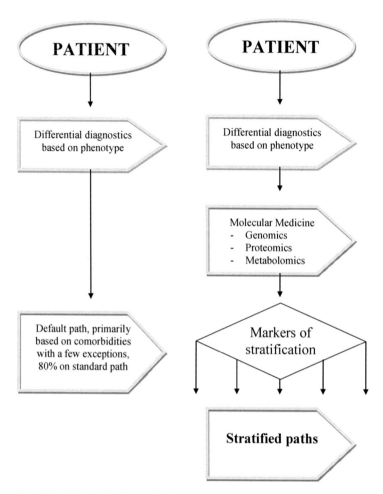

Fig. 13.1 Differential diagnostics (*left*) and stratification (*right*). (Source: own figure)

the individual has to fill. From this point of view it is more likely to speak of Individualized instead of Personalized Medicine due to the emphasis on the uniqueness of a human being. However there is often no distinction made in practice.

Finally, IM has to be put into perspective with Evidence-Based Medicine (EBM). EBM assumes all relevant decisions on prevention, diagnosis and therapy being based on proven efficacy. There is no contradiction to IM, which also calls for an evidence base for each individual unique product as well as every stratified path. Problems occur only due to the fact, that stratification implies a significantly smaller group size per stratum, which complicates the base of evidence (Henderson and Schumacher 2011). This poses a methodological difficulty, not a contradiction.

13.3   Innovation Model

Something is referred to as an innovation if a good, a process, an idea or a paradigm is novel. Further, it has to be used by at least one person as a solution to a problem. There are academics who deny Individualized Medicine's attribute as a novelty (Gadebusch Bondio and Michl 2010; Gadebusch Bondio 2011; Felzmann 2011), thus understand it merely as a pseudo-innovation. According to other academics IM can only be regarded as an innovation in the very early stage of a process of developing, adopting and adapting. Thus, a statement on whether it represents an incremental improvement in medicine or a quantum leap cannot be made at this point in time (Hüsing et al. 2008). Therefore it is sensible to ask, what kind of innovation IM is and about its potential.

In general, there is a distinction between micro-, meso- and macro-innovations (Ritter 1991). A micro-innovation exists, if the innovation changes only the lowest system level and does not require the adjustment of any other element of that system. In health care a micro-innovation can be identified, if only the physician-patient-relationship or direct prevention, diagnosis or treatment is affected and neither further stakeholders are added nor norms or structures of the health care system have to be adjusted. However, if this is the case one can speak of a meso-innovation. An implant (e.g. in endoprosthetics) for example, is an innovation which usually only affects surgical processes. The structure and further processes of the hospital, health insurance, rehabilitation and the outpatient sector, remain mostly unaffected. The introduction of minimally invasive surgery, however, is an innovation, which required complete adjustment of bed capacity, processes, remuneration structures, rehabilitation, etc. Micro-innovations always lead to a meso innovation, if the scope of change is either very large or the micro level is instable. In these cases, the innovation is not able to compensate fluctuations through adjustment of its own regulators and has to pass them on to the meso level.

A macro-innovation refers to a situation, where the whole social system with its essential moral values, regulations, processes and structures is affected. Macro-innovations arise, if the novelty cannot be absorbed by the meso structure any more, in which case the fundamental system elements of society have to adjust. The introduction of mobile phones for example, does not only imply further development of technological infrastructure (Heidorn 2011). The paradigm of permanent accessibility and informability also bares far reaching consequences on the entire social system and shapes our patterns of thinking and behavior far beyond the mobile phone per se (Höhn et al. 2006). Accordingly, in health care the term macro-innovation should be used, if essential stakeholders, structures, processes and paradigms of society are changed, for example if the general understanding of health and disease with its complex consequences on the social value system is affected. Consequently the question remains, if Individualized Medicine can be interpreted as micro-, meso- or macro-innovation. Figure 13.2 gives an outline of these relations.

In general micro-innovations do have a much higher chance of getting accepted in comparison to meso- or even macro-innovations (Rogers 1983). However, the

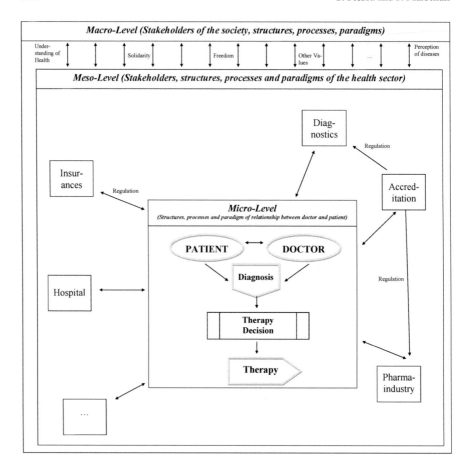

Fig. 13.2 Micro-, meso- and macro-innovation. (Source: adapted from Flessa and Marschall (2012))

willingness to accept increases in case of an expanding innovation, i.e. if based on this innovation, problems which were unsolvable so far, can now be overcome. Alternatively, crowding out innovations require the former system solution being considerably worse than the new one and transformation costs to be low in comparison to the advantage of the improvement. For micro-innovations this proves to be the case quite frequently. Crowding out macro-innovations on the other hand only has a chance in situations of radical social change (Ritter 1991).

The so far known procedures of IM do not only crowd out conventional procedures of medicine. They also generate novel structural and procedural solutions by applying individual incentives of prevention (e.g. by determining the personal risk of developing diabetes), opening up new opportunities for curing diseases which so far were not or not easily curable (e.g. cultivation of individual organs for transplantation) or by reducing severe side-effects of former interventions (e.g. adverse

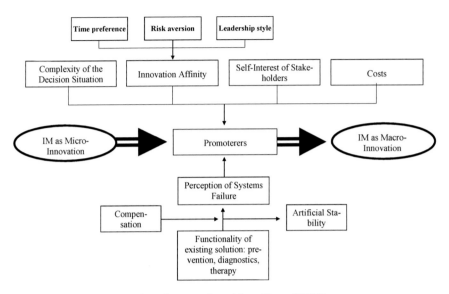

Fig. 13.3 Barriers to innovation. (Source: adapted from Flessa (2002))

drug reactions). It can be stated, that the already existing IM procedures, especially in the field of cancer therapy, are not pseudo-innovations but a huge step into a new dimension of therapy.

Nevertheless, it might be expected that barriers (see Fig. 13.3), which conflict with innovation adoption, do exist for individual stakeholders of the health care system. Such barriers include the costs and the risks of innovation adoption, bureaucratic obstacles, resistance to change, as well as the individual self-interest and fears of the stakeholders. Typically they implicate that an innovation is not accepted as a new system solution, but rather continues to be a niche. From there it matures to become a potential (Dopfer 1990) or an innovation embryo (Ritter 1991), whose adoption as a basic paradigm of health care is being oppressed. For a certain timeframe the existing system solution will be stabilized artificially (metastability) (Ritter 1991), until the pressure becomes too strong for the old system regime to be operational. As illustrated in Fig. 13.4, an evolutionary jump occurs, i.e. the innovation technology which matured within the niche, will now—for many by surprise—be adopted as the standard solution. At the point of bifurcation, the existence of these potentials determines the direction of further development.

Therefore the question of which barriers could prevent IM being adopted as the standard technology of medicine should be analyzed. Thus, in the following the most important stakeholders will be studied in regard to the consequences IM has for them. Based on that, it will be analyzed, whether IM in the sense of a micro-innovation concerns only physicians and patients or if health care per se and the whole society are affected as well.

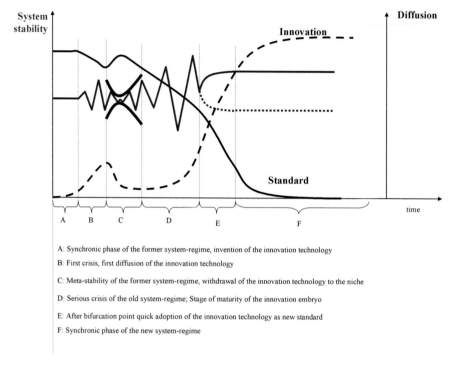

A: Synchronic phase of the former system-regime, invention of the innovation technology

B: First crisis, first diffusion of the innovation technology

C: Meta-stability of the former system-regime, withdrawal of the innovation technology to the niche

D: Serious crisis of the old system-regime; Stage of maturity of the innovation embryo

E: After bifurcation point quick adoption of the innovation technology as new standard

F: Synchronic phase of the new system-regime

Fig. 13.4 Change of the system regimes. (Source: following on Ritter (1991))

13.4 Selected Stakeholders and Their Incentive Structure

13.4.1 Patients

A medical micro-innovation has the best chances of being adopted. However, it only has implications for patients and physicians and their relation to each other. Normally there should not be any barriers to innovation against Individualized Medicine from the perspective of a patient, who is acting completely rationally. This is due to an additional benefit without an additional cost for him in the case of health insurances bearing the costs. In case of bounded rationality there may exist some personal emotional barriers, which might imply objections against this new technology. However, Individualized Medicine can make a contribution to prevent diseases effectively, to better diagnose illnesses, to provide safer and more effective therapies and to prevent the recurrence of diseases with higher probability. An individualized prevention, which creates a risk profile for the individual patient based on his or her genetic disposition, his or her individual metabolism etc. and presents detailed prevention measures (e.g. with regard to nutrition), is in general useful for the patient—regardless of the patient's adherence. Additionally, the procedures of IM already in existence, enable a partial or complete cure of diseases, which were so far either not treatable or only with significant risks. For example, immunoadsorption with subsequent IgG substitution in the case of heart failure (Felix et al. 2008) represents a burdensome procedure for patients, that only permits the partial or complete cure of dilative cardiomyopathy (DCM) for a comparably low share of patients. With the help of extensive procedures of molecular medicine, it will almost certainly be possible to determine whether a specific patient will gain from immunoadsorption or not (Ameling et al. 2012). It will be a great advantage for the patient, even if he is excluded from immunoadsorption therapy. In that case he does not have to pass through the repeating procedures, which last for hours—without success.

Parts of the methods of Individualized Medicine are currently provided without the funding of health insurance companies. The total sequencing of human genome (e.g. Biotech 2011), for example, could be relevant regarding the prevention of diseases. However, this is not covered by the health insurance companies. The costs are, at least for the time being, still high at several thousand Euros and it appears that the service will be of no relevance for most of the patients in Germany for at least a few years.

Pharmacogenomics is the best documented field of Individualized Medicine in literature, which is related to health economics (Sohn et al. 2010). Most articles, however, implicitly or explicitly take the perspective of health insurance, while the patient's perspective is typically (exception: Bala and Zarkin 2004) only reported in terms of quality of life. The reviews consistently show the positive contribution pharmacogenomics has on a patient's quality of life (Dervieux and Bala 2006; Flowers and Veenstra 2004; Phillips and Van Bebber 2004; Vegter et al. 2008).

13.4.2 Physicians

Usually, the medical doctor in charge has the aim to cure his patients to the maximum extent possible using the scientific knowledge that is available to him. Other things being equal, he will therefore apply IM innovations eagerly, if he is of the opinion that this will lead to a better outcome in treatment. This statement is not ambiguous with respect to individualized prevention as well as pharmacogenomics, since improved patient benefit will regularly be achieved by higher safety. However concerning procedures of Individualized Medicine s.s., the risks are currently unforeseeable (e.g. durability of unique items). Consequentially, affinity towards innovation and risk amongst physicians is of relevance for the adoption of innovation.

The positive effects of IM s.l. are accompanied by a certain tendency of persistence, respectively a status-quo bias (Samuelson and Zeckhauser 1988) among physicians. This refers to a tendency of keeping existing procedures and of accepting new knowledge only if it is deemed necessary. Max Planck criticized the indolence of new perceptions not being enforced by their empirical cogency but only through the death of old experts of a discipline (Leggewie and Welzer 2009), which proves to be quite rational. This is due to the fact that innovation adoption requires an investment in learning, which is connected to direct and indirect costs. Therefore, physicians will also tend to rather solve a problem by optimization of an established method within an existing paradigm than by confronting it with a comprehensive innovation of thinking patterns. However, in the medium-term most of the physicians will be open to IM. Considering the inadequacy of treatments of many chronic-degenerative diseases they have no other choice in order to follow the principles of their professional ethics.

On the micro level a certain willingness to implement IM can thus be expected. This will be the case not only in areas where IM represents an all new optional therapy but also in the case of diseases for which IM implies a replacement of previous standard solutions.

13.4.3 Institutional Health Care Providers

Certainly, it has to be noted that physicians themselves are integrated in health care institutions either as an employee (e.g. in a hospital) or as a self-employed person (e.g. as a panel doctor). In this function they have to consider objectives, which exceed the physician-patient-relationship and thus affect the meso-level.

Foremost it has to be considered that, other things being equal, an individualized path of treatment (Oberender 2005) is more cost-intensive than the standardized path. This is the consequence of stratification, which requires detailed medical history and laboratory diagnosis. Due to lump sum remuneration (remuneration according to the DRG system as well as the quarterly lump sum according to the general assessment standard for outpatient services) IM can appear to be disadvantageous, if through the process of stratification, costs cannot be cut. Currently, in the field of pharmacogenomics only, a few studies on potential savings to service providers have been published (You et al. 2009; Pichereau et al. 2011). Consequently, for a hospital or a medical practice there is a high level of uncertainty connected to a decision for or against the adoption of IM.

This uncertainty is amplified by the fact that stratified patient paths imply a higher risk for institutional providers. Figure 13.5 illustrates examples for a default path as well as an individualized path. The variation of costs of the standardized path is relatively low since they can only occur due to the difference in time consumption of fixed tasks. The individualized treatment path, however, leads to a division of patients to numerous parallel paths running different interventions. In this way the variation of costs per case is spread significantly. Even in case of a constant expected value of costs Individualized Medicine implies a higher risk of financial loss. This being said, the variation of costs per case is higher, the smaller the number of cases per path.

Alternative treatment plans with the same underlying diseases did exist prior to the introduction of IM. The rule of thumb is that 80 % of cases, as well as 80 % of the total financial cost, should be connected to the standardized path (Dykes and Wheeler 2002), in order to achieve a relatively high homogeneity of cost. This is

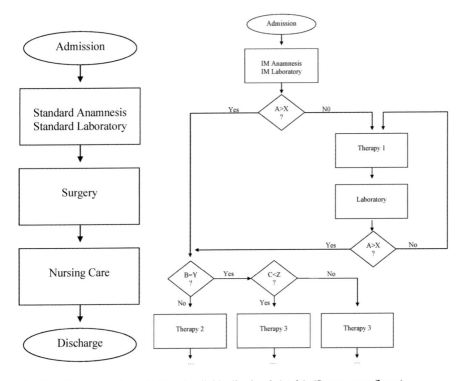

Fig. 13.5 Standardized path (*left*) and Individualized path (*right*). (Source: own figure)

particularly relevant with regard to the DRG system since a DRG is based on a group as economically homogenous as possible. This means, cases within a DRG either follow the same default path or the number of deviations from that default path is limited. A significant deviation from the standardized path such as those resulting from a serious secondary diagnosis leads to categorization into another DRG (Flessa 2013).

In contrast to the logic of DRG, which is based on standardization, the basic principle of IM presents a highly-structured path of treatment within a DRG. The patient's genotype as well as phenotype influences the process of treatment individually, which is not reflected in a different remuneration under the current DRG system. Therefore, IM and DRGs seem to be in contradiction with each other. DRGs demand the largest possible homogenous groups, while IM implies stratification up to the treatment of individual cases. Similar relations are expected for lump sums according to the general assessment standards for outpatient services.

High variation in costs per case with a fixed payment from health insurance companies per case implies a potential loss for service providers. This would also be the case, if the expected value of cost corresponds with the health insurance payments for, for example, a case group under the DRG system. Consequently, smaller providers in particular will be critical of IM innovation. This barrier to innovation can be overcome by alternative financial instruments. This includes the significant disassociation from DRGs in order to map the individualized and thus cost inhomogeneous patient paths, each within one DRG. This could be a useful way to assure the funding of IM services for several years. However, it remains unclear, whether it is possible to administer thousands of DRGs, each having only a small number of cases. This also applies to possible additional charges (*Zusatzentgelt* in Germany), as well as newer examination and treatment methods. As long as IM remains a niche phenomenon, minor adjustments to the financing system (e.g. a few additional charges and the levels of severity of existing DRGs) can overcome the barriers for service providers. If IM, however, turns into a new paradigm and thus into the standard solution of patient care, different insurance payments depending on the individual, as in the scale of charges for physicians (e.g. GOÄ in Germany) will be inevitable. Clearly, IM proves to be a meso-innovation, which could question the entire financial system.

13.4.4 Pharmaceutical and Biotechnical Industry

The research and development of new drugs is very expensive and requires extensive economic evaluation, as referred to in additional literature (Schöffski et al. 2008). Individualized Medicine and pharmacogenomics in particular provide a number of challenges to the industry, which have not yet fully been recognized in literature.

Firstly, the development of drugs is extended by the simultaneous exploration of corresponding biomarkers. Innovative medicine is usually accompanied by a specific (gene) test, the so-called Companion Diagnostics (Boston Consulting Group 2011). This leads to a much closer integration of diagnostics, medical technology and pharmaceuticals than was previously necessary. These industries have to be seen as part of an overall package, due to the sole customer's interest in the entire product consisting of biomarkers, implementation of medical technology and intervention (e.g. drugs). The customer is no longer interested in only one of these components.

Secondly, this leads to the development of a new dynamic of coordination and cooperation of the companies involved. For example, pharmaceutical companies need to decide whether it is more productive to purchase innovative biotechnology companies or to simply cooperate. Numerous mergers in the recent past (e.g. Novartis Molecular Diagnostics) show the dynamics of this industry in terms of M&A.

The third challenge which IM presents is the rational development of drugs. Target and control groups of drug development have to be sufficiently large, since regulatory authorities require proof of significant safety and efficacy. However, IM implies small groups ('orphan drugs') (Henderson and Schumacher 2011). In the development of companion drugs this problem is exacerbated due to the necessity of, in many cases, several biomarkers. Consequently, only multivariate analyzes with high numbers of cases, is likely to provide appropriate evidence. Within this discussion, the approval of IM based diagnosis and treatment meets its limits. On the one hand, large groups are required for case studies. On the other hand, only stratified small groups are feasible in both the phase of research as well as in the phase of implementation.

Fourthly, a solution in the intensive cooperation of companies, research institutions and government agencies is necessary. IM requires access to large biological databases which can no longer be kept in a single location. Large and international studies, which can find the proverbial 'needle in a haystack' based on their large data stock, are in great demand. Therefore, IM has the potential to increase the willingness of businesses to cooperate among each other as well as with research institutes.

Finally, the ideas are transferable to unique therapeutic measures. In terms of therapy specifically tailored to an individual patient, it can be expected that as the quantity of this type of therapy is increased, the unit cost per therapy will significantly decline. However, the number of implantations will increase disproportionately. Therefore, the health care system will be confronted with costs whose extent can hardly be foreseen. This development extends far beyond the health system. A social consensus on the value of life as well as the willingness to pay for health is required. IM could potentially inspire and strengthen this current discussion.

In order to analyze the willingness of the pharmaceutical and biotechnological industry to innovate, alternative investment assets need to be taken into consideration. After initial hesitation pharmaceutical companies are now fully adopting IM innovation, according to statements of representatives of that industry made at the PerMediCon 2011 (Pfundner 2011). Some academics ask, whether the interest in IM is not simply the result of a lack of alternatives. Declining numbers of authorization of real innovations in the pharmaceutical sector support this proposition (Kearney 2010; Rajan 2009). The majority of products having been approved in recent years are so-called Me-Too-Products. The companies distinguish themselves from competitors mainly be offering expanded services such as molecular diagnostics (Hochwimmer 2010). IM is likely to be of particular interest for companies striving for leadership in innovation and thus for leadership in quality. Companies setting the market price (especially those in the field of generics) are unlikely to be promoters of IM. Accordingly, the number of diagnostic tests being developed for drugs beyond the patent term is very low. IM drugs are mainly focused on oncology or treatment of HIV infection, i.e. on innovative high priced products (Verband der forschenden Arzneimittelhersteller 2011). However, niche providers offering highly specialized and very expensive drugs, are unlikely to obtain authorization in the future without offering companion diagnostics. They could potentially develop to become a so-called 'Niche Buster', since they mainly offer high priced drugs. These are of great economic importance to a pharmaceutical company (Greiner and Knittel 2011).

This proves IM to be an innovation, which has the potential to change the existing structures and processes in the pharmaceutical and biotechnological industry. However, it should primarily be of more interest to the industry to develop drugs for large groups of the population with the exception of niche suppliers.

13.4.5   Health Insurance Companies

Most studies of Individualized Medicine are written from the point of view of the health insurance companies. To them, a comparison of the expected values of cost with and without innovative technology is primarily relevant in the decision making process, since an insurance company should have a large risk pool. In fact, there are examples (Vegter et al. 2008a) of innovative technologies of Individualized Medicine being dominant, that is leading to absolute cost savings over the standard technology. In this case, a promotion of IM would be beneficial to health insurance companies in the long run. However, two other cases could potentially arise. First, the costs per case can be higher than those using the standard technology. In this case, the insurance company has to consider, whether or not the higher costs are justified by other effects (e.g. higher quality of life, satisfaction with the insurance company). An example is the above-mentioned immunoadsorption, which leads to substantial additional costs per case. Second, the costs per case of IM may be lower than the current standard of care. However, the process of IM could contribute to a significant increase in volume. Serious adverse reactions, for example, could be avoided through prior risk selection. Therefore methods, which have until now only been used in a few exceptional cases, could be adopted for larger groups of patients.

A health insurance company's willingness to innovate depends on its aims and its position in the market, in addition to individual factors such as risk and the time preference of decision makers. In this context, commercial private insurance companies, social insurance companies in competition with others as well as social insurance companies acting as a monopolist have to be distinguished.

A private insurance company acting commercially will determine its program of services as well as the scope of services in a way to maximize the difference of premiums and reimbursements in the long term. Assuming competition, an insurance company can only influence the price in the case of no market transparency, or if it differs from other insurance companies by offering a unique service component. Therefore private health insurance companies have to strive to attain a comparative advantage over their competitors through offering innovative services. Therefore, adding modern technologies to their portfolio is beneficial. The better the information available to the insured person is, the higher the pressure to innovate. Since commercial insurance companies tend to target educated people with higher income they are likely to adopt methods of IM to their range of services earlier and to a larger extent. The higher inclination to innovate shown by commercial insurance companies is documented in literature (Wild 2008). A commercial private insurance company's tendency to innovate will only be lower if that company enjoys a monopolistic status.

A social health insurance company, which is in competition with other health insurance companies, will seek to acquire as many customers as possible. Other things being equal, an increase in membership implies both, lower administrative costs per member as well as a lower risk for the company. Thus, an insurance company has an incentive to offer IM services to its members, thereby differentiating itself from its competitors. However, in Germany these opportunities are severely limited. Only within the limits of special contracts (e.g. pilot projects), are social health insurance companies able to go beyond the legally prescribed performance program.

If a social insurance company has a monopolistic status (as is the case in England), there is little need to drive innovation in the short term. Since customers have no choice, they cannot directly respond to an unsatisfactory performance policy by changing their behavior in terms of demand. Their only way to respond is the time consuming activity of going through the political institutions.

To summarize, one can say that health insurance companies will promote innovation only when it promises profits, cost savings, acquisition of new customers and lower risks. It would be unrealistic to consider health insurance companies as acting completely selflessly solely for the benefit of people. Even non-profit organizations have to at least cover their expenses. Health insurance companies will therefore hardly promote risky and expensive technologies of Individualized Medicine, if the technologies do not promise a competitive advantage.

13.5   Individualized Medicine: A Macro-Innovation?

The analysis of selected stakeholders in health care shows, that IM is an innovation, which not only affects the individual physician-patient-relationship, but it has the potential to influence various players, structures, processes and paradigms of health care. The effect on financing mechanisms, intersectional patient flow paths, cooperation as well as coordination of stakeholders and the marketing of health insurance companies will be substantial. Therefore, IM can at least be considered as a meso-innovation. At this point in time it is not yet clear whether only individual methods of the IM market can create a breakthrough (e.g. pharmacogenomics and tissue engineering) or whether Individualized Medicine as a whole will become the new standard. As shown, the stratification of prevention, diagnostics and treatment focusing on individuals is able to change the entire health care system, especially its underlying pattern of thought. This scenario could happen if a multi-cause multi-effect paradigm based on so-called -omics sciences takes the lead. Consequently, the complete network consisting of genetics, metabolism, living space and behavior are considered overall as a stochastic risk system, whose results consist of various states of health and disease. However, this would only be the case if individual innovational ideas are not developed alone, but instead the pattern of thought of IM as a whole is implemented in the health care system. So far, the penetration of the market was significantly slower than originally expected (Sohn et al. 2010). The mentioned IM-induced change of paradigm in health care has not yet taken place and cannot yet be guaranteed in the future.

Furthermore, the question arises, whether or not IM is an innovation, which can change the society as a whole including its system of values as well as key assumptions, in such a way that one could speak of a macro-innovation. A total social phenomenon is evident, if for example fundamental social concepts and arrangements are changed. IM has the potential to alter the social conception of health and disease. Mostly, people refer to someone as being sick, who is showing acute symptoms of a certain disease. This is associated with a binary way of thinking, which clearly distinguishes healthy from ill. The existence of risk factors alone is not proof enough, to be assigned to the state 'sick' (cf. Põder and Assel chap. 9, pp. 165–180). IM in contrast challenges this thinking by creating a definition of disease, which removes itself from clinical symptoms. Genetic testing (e.g. for individualized prevention) implies the creation of the stage 'not-yet-sick'. This stage would describe a person, who has the genetic predisposition to a disease that is not yet developed. The perception of what illness is about is complicated by the fact that a disposition to disease is usually only represented by a probability of occurrence. In most cases this probability is different from one. Consequently, Individualized Medicine has a predictive value (Evans et al. 2001), so the understanding of "health and disease of society" may change (Gadebusch Bondio and Michl 2010).

Such a redefinition of a fundamental concept of society could have far reaching consequences on social values such as freedom and solidarity. The current consensus, with a few exceptions, is that a lack of prevention does not lead to a loss of entitlement to medical services. Dentures represent the exception to the rule due to, among other things, a distinct correlation between prevention (regular visits to a dentist) and curative services (dentures) exists. In contrast, numerous influencing factors (multi-cause) exist for most other diseases without possibilities to specify a more precise assignment of these factors to a particular outcome (multi-effect). IM drives to change this. By using risk scores, for example, one can determine relatively accurately, what kind of prevention promises the best possible outcome for a patient with a certain risk profile (Parikh et al. 2008).

With the new disease concept of 'not-yet-sick', the risk of developing a disease, as well as the success of prevention, can be individually predicted. Therefore, assigning a personal liability to patients is easier, when he or she did not exercise prevention despite known risks. The social value of solidarity—in this case between the healthy and the sick—is much more strongly challenged, because, in the extreme case, the solidarity between healthy and sick could be a required solidarity between people complying with prevention and people objecting to prevention. The exclusion from solidarity funding could also be promoted, if it turns out that the development of certain unique therapeutics or companion drugs is profitable only for population groups with a certain genetic disposition.

In addition, the freedom of individuals to know or ignore his or her own risk of developing a disease is considerably called into question. IM is potentially questioning the freedom that people have to behave in a healthy or unhealthy way without them suffering any negative consequences in terms of their insurance premiums or their rights to services.

In conclusion, IM has the potential to alter the conception of human beings. IM represents the application of molecular medicine outside the niche of rare inheritable disorders. In this sense, IM can be seen for most people as the first contact with these modern scientific methods. Are human beings predetermined at birth by their genetic code or are they shapeable? Is their behavior predetermined by their metabolism or are they able to choose? To what degree is the human decision a free one (Libet et al. 1983; Roth 2005; Roth 2008)? So far, these issues have been discussed mainly in the narrow circle of experts or the few people affected, such as in the context of genetic diagnostics of inheritable diseases (e.g. Schöffski 2000). In case of a broad use of IM, these issues relate to chronically ill people, i.e. to the majority of the population. This process of thought could be strengthened by the achievements of regenerative medicine, which ultimately aims to generate unique therapeutics as "spare parts" for human beings. This idea has so far only been scrutinized in the narrow circles of ethicists. However, in the future it could be of more relevance to large proportions of society.

While this is not directly an effect of IM, but of genetics, IM could introduce the scientific way of thinking to most people on all levels of society for the first time. Consistent individual prevention, diagnosis and treatment will undoubtedly have a social impact. To date their range has hardly been accessible to the public (Gadebusch Bondio 2011).

In a nutshell: it can therefore be stated, that Individualized Medicine is an innovation, which could potentially change the health care system fundamentally. It also has the potential to shape society as a macro-innovation. Currently a fundamental shift of paradigm is unlikely, due to numerous barriers which still exist. We consider IM to still be in the stage of development. Therefore, it should be maintained in niches in order to overcome initial difficulties and to mature into an innovation. If the continuing aging of the population and the increasing proportion of people suffering from chronic degenerative diseases leads to a situation, where mono causal diagnostics and therapeutic methods are no longer satisfactory, IM is an alternative ready to initiate a paradigm shift. From this perspective, the fact that many projects of IM are publicly funded in Germany in order to provide the necessary niche for further development can be considered positively.

References

Ameling S, Herda LR, Hammer E, Steil L, Teumer A, Trimpert C, Dorr M, Kroemer HK, Klingel K, Kandolf R, Volker U, Felix SB (2012) Myocardial gene expression profiles and cardiodepressant autoantibodies predict response of patients with dilated cardiomyopathy to immunoadsorption therapy. Eur Heart J 34(9):666–675. doi:10.1093/eurheartj/ehs330

Bala MV, Zarkin GA (2004) Pharmacogenomics and the evolution of healthcare: is it time for cost-effectiveness analysis at the individual level? Pharmacoeconomics 22(8):495–498

Bartens W (2011) Jedem seine Pille. Süddeutsche Zeitung. 18 March 2011

Biotech G (2011) DNA-Sequenzierung vom europäischen Marktführer. GATC. http://www.gatc-biotech.com/de/lp/genom-sequenzierung.html?gclid=CMuIvIa92KoCFcG9zAodSSk5-w. Accessed 18 Oct 2011

Boston Consulting Group (2011) Die Personalisierte Medizin. In: Boston Consulting Group (ed) Medizinische Biotechnologie in Deutschland 2011. Boston Consutling Group, München, pp 20–46

Bryan A (2009) Public health theories. In: Wilson F, Mabhala M (eds) Key concepts in public health, Sage, London, pp 21–25

Cortese D (2007) A vision of individualized medicine in the context of global health. Clin Pharmacol Therapeut 82(5):491–493

Der Spiegel (2011) Das große Verspechen. Der Spiegel, issue 32/2011

Dervieux T, Bala MV (2006) Overview of the pharmacoeconomics of pharmacogenetics. Pharmacogenomics 7(8):1175–1184

Dopfer K (1990) Elemente einer Evolutionsökonomik: Prozeß, Struktur und Phasenübergang. In: Witt U (ed) Studien zur evolutorischen Ökonomik I. Duncker & Humblot, Berlin, pp 19–47

Dykes PC, Wheeler K (2002) Critical Pathways—interdisziplinäre Versorgungspfade DRG-Management-Instrumente, vol 3. Huber, Bern

Evans JP, Skrzynia C, Burke W (2001) The complexities of predictive genetic testing. BMJ 322(7293):1052

Felix SB, Dörr M, Herda LR, Beug D, Staudt A (2008) Immunadsorption als Therapieverfahren der dilatativen Kardiomyopathie. Der Internist 49(1):51–56. doi:10.1007/s00108-007-1991-x

Felzmann T (2011) Personalisierte Medizin: Revolution in der Behandlung oder „neuer Wein in alten Schläuchen". doi:http://haematologie-onkologie.universimed.com/artikel/personalisierte-medizin-revolution-der-behandlung-oder-%E2 %80%9Eneuer-wein-a

Flessa S (2002) Gesundheitsreformen in Entwicklungsländern. Lembeck, Frankfurt a.M.

Flessa S, Marschall P (2012) Individualisierte Medizin: vom Innovationskeimling zur Makroinnovation. Pharmaco Economics German Res Art 10(2):53–67

Flessa S (2013) Grundzüge der Krankenhausbetriebslehre, Bd 1. 3, aktualisierte Auflage Oldenbourg , München

Flowers CR, Veenstra D (2004) The role of cost-effectiveness analysis in the era of pharmacogenomics. Pharmacoeconomics 22(8):481–493

Gadebusch Bondio M (2011) Personalisierte Medizin: Kritisches Bild eines komplexen Phänomens. Dtsch Arztebl 108(4): A –173

Gadebusch Bondio M, Michl S (2010) Individualisierte Medizin: Die neue Medizin und ihre Versprechen. Dtsch Arztebl 107:1062–1064

Greiner W, Knittel M (2011) Wirtschaftliche Potenziale der Individualisierten Medizin. Pharmaco-Economics—German Res Art 9(1):45–54

Heidorn S (2011) Das Handy—Eine Innovation für die Zukunft? Mr.-SMS.de. http://www.mr-sms.de/handy-zukunft.html. Accessed 18 Oct 2011

Henderson R, Schumacher M (2011) Clinical epidemiology and individualized medicine. Biom J 53(2):167–169

Hochwimmer G (2010) Big Pharma vor einer neuen Ära. Going Public—Das Kapitalmarktmagazin 44–47

Hoffmann W, Krafczyk-Korth J, Voolzke H, Fendrich K, Kroemer HK (2011) Towards a unified concept of individualized medicine. Personalized Medicine 8(2):111–113

Höhn R, Pongratz S, Tobias M (2006) Innovative Informations-und Kommunikationstechnik. In: Pfriem R et al (eds) Innovationen für eine nachhaltige Entwicklung. Deutscher Universitätsverlag, Wiesbaden, pp 79–94

Hüsing B, Hartig J, Bührlen B, Reiß T, Gaiser S (2008) Individualisierte Medizin und das Gesundheitssystem. Arbeitsbericht Nr. 126. Forschungszentrum Karlsruhe, Karlsruhe

Kamps H (2010) Massgeschneiderte Medizin: Der wichtige Unterschied zwischen Individuum und Person. Dtsch Arztebl 107(50):A-2490/B-2164/C-2120

Kearney AT (2010) Innovative Pharmaindustrie als Chance für den Wirtschaftsstandort Deutschland. Eine Studie im Auftrag von PhRMA (Pharmaceutical Research and Manufactourers of America), dem Branchenverband der forschenden Pharmaindustrie in den USA, und der deutschen LAWG (Local American Working Gorup). Accessed 21 Oct 2014

Kube P (2010) Prospektive Entwicklung der Regenerativen Medizin: Eine empirische Analyse der Wettbewerbsfähigkeit Deutschlands. Kovac, Hamburg

Lanza R, Langer R, Vacanti J (2007) Principles of tissue engineering (Tissue engineering intelligence unit) (Gebundene Ausgabe). Academic, New York

Leggewie C, Welzer H (2009) Das Ende der Welt, wie wir sie kannten. S. Fischer, Frankfurt a. M.

Leyen HEvd, Wendt C, Dieterich HA (2005) Gentherapie und Biotechnologie. Ansätze zu neuen Therapieformen in der Medizin. Wissenschaftliche Verlagsgesellschaft, Stuttgart

Libet B, Gleason CA, Wright EW, Pearl DK (1983) Time of conscious intention to act in relation to onset of cerebral activity (readiness-potential). The unconscious initiation of a freely voluntary act. Brain 106(Pt 3):623–642

Meisel HJ (2005) Bandschreibenregeneration durch körpereigene Zelltherapie. Pharma News—Zelltherapie 1. Quartal 2005

Oberender P (2005) Clinical Pathways–Facetten eines neuen Versorgungsmodells. Kohlhammer, Stuttgart

Parikh NI, Pencina MJ, Wang TJ, Benjamin EJ, Lanier KJ, Levy D, D'Agostino RB, Kannel WB, Vasan RS (2008) A risk score for predicting near-term incidence of hypertension: the framingham heart study. Ann Intern Med 148(2):102

Pfundner H (2011) Unternehmensstrategie Personalisierte Medizin. vol Podiumsdiskussion, 22. Juni 2011. PerMediCon, Köln

Phillips KA, Van Bebber SL (2004) A systematic review of cost-effectiveness analyses of pharmacogenomic interventions. Pharmacogenomics 5(8):1139–1149

Pichereau S, Le Louarn A, Lecomte T, Blasco H, Le Guellec C, Bourgoin H (2011) Cost-effectiveness of UGT1A1* 28 genotyping in preventing severe neutropenia following FOLFIRI therapy in colorectal cancer. J Pharm Pharm Sci 13(4):615–625

Rajan KS (2009) Biokapitalismus: Werte im postgenomischen Zeitalter. Suhrkamp, Berlin

Ritter W (1991) Allgemeine Wirtschaftsgeographie. Oldenbourg, München

Rogers EM (1983) Diffusion of innovations. Free, New York

Roth G (2005) Das Gehirn und seine Wirklichkeit: kognitive Neurobiologie und ihre philosophischen Konsequenzen. 1 Aufl. Nachdr. edn. Suhrkamp, Frankfurt a. M. (First published)

Roth G (2008) Persönlichkeit, Entscheidung und Verhalten. Klett-Cotta, Stuttgart

Samuelson W, Zeckhauser RJ (1988) Status quo bias in decision making. J Risk Uncertainty 1(1):7–59

Schöffski O (2000) Gendiagnostik: Versicherung und Gesundheitswesen - Eine Analyse aus ökonomischer Sicht. Verlag Versicherungswirtschaft, Karlsruhe

Schöffski O, Fricke FU, Guminski W (2008) Pharmabetriebslehre. Springer, Heidelberg

Schwartz FW, Siegrist J, Troschke Jv, Schlaud M (2003) Gesundheit und Krankheit in der Bevölkerung. In: Schwartz FW, Badura B, Busse R et al (eds) Public Health–Gesundheit und Gesundheitswesen, vol 2. Urban & Fischer, München, pp 23–47

Sohn S, Dornstauder P, Schöffski O (2010) Die Nutzung von Geninformationen für eine personalisierte Pharmakotherapie: Stand, Zukunftspotenziale und wirtschaftliche Implikationen. PharmacoEconomics—German Res Art 8(2):109–118

Vegter S, Boersma C, Rozenbaum M, Wilffert B, Navis G, Postma M (2008a) Pharmacoeconomic evaluations of pharamacogenetic and genomic screening programmes. Pharmacoeconomics 26(7):569–587

Vegter S et al (2008b) Pharmacoeconomic evaluations of pharamacogenetic and genomic screening programmes. Pharmacoeconomics 26(7):569–587

Verband der forschenden Arzneimittelhersteller (2011) In Deutschland zugelassene Arzneimittel für die personalisierte Medizin. http://www.vfa.de/de/arzneimittel-forschung/datenbanken-zu-arzneimitteln/individualisierte-medizin.html. Accessed 17 Oct 2011

Wild F (2008) Die Verordnungen von neuen Arzneimitteln bei Privatversicherten im Vergleich zu GKV-Versicherten. Gesundh ökon Qual manag 13(1):15–18

You J, Tsui K, Wong R, Cheng G (2009) Potential clinical and economic outcomes of CYP2C9 and VKORC1 genotype-guided dosing in patients starting warfarin therapy. Clin Pharmacol Therapeut 86(5):540–547

Chapter 14
Assessing Individualized Medicine— The Example of Immunoadsorption

Paul Marschall, Timm Laslo, Wolfgang Hoffmann, Kerstin Weitmann and Steffen Flessa

Abstract Biomarkers can be used for the prediction of treatment response within the context of widespread diseases. The aim of this contribution is to address this issue from the economic perspective. Based on time studies and investigations of used resources at the University Medicine Greifswald some preliminary results for the costs of the Immunoadsorption therapy with subsequent IgG (IA/IgG) substitution and the corresponding gene expression analysis are provided. Under the current setting the latter can be regarded as diagnostics for deciding, if IA/IgG is appropriate. Currently, both parts of the IM tandem are not implemented in combination in clinical routine. It is argued that the reimbursement system has a critical role for providing incentives for health care providers to translate research into routine. In addition some preliminary results of outcome evaluation based on disease-related quality of life are provided.

Keywords Health economic evaluation · Individualized Medicine · Costs · Quality of life · Time study · Heart failure · Dilated cardiomyopathy · Immunoadsorption with subsequent IgG substitution · Gene expression analysis

P. Marschall (✉) · T. Laslo · S. Flessa
Rechts- und Staatswissenschaftliche Fakultät, Lehrstuhl für Allgemeine Betriebswirtschaftslehre und Gesundheitsmanagement, Ernst-Moritz-Arndt-Universität Greifswald, Friedrich-Loeffler-Str. 70, 17487 Greifswald, Germany
e-mail: paul.marschall@uni-greifswald.de

T. Laslo
e-mail: laslo@laslo.de

S. Flessa
e-mail: steffen.flessa@uni-greifswald.de

W. Hoffmann · K. Weitmann
Institut für Community Medicine Abt. VC, Universitätsmedizin Greifswald, Ellernholzstr. 1–2, 17489 Greifswald, Germany

W. Hoffmann
e-mail: wolfgang.hoffmann@uni-greifswald.de

K. Weitmann
e-mail: kerstin.weitmann@uni-greifswald.de

© Springer International Publishing Switzerland 2015
T. Fischer et al. (eds.), *Individualized Medicine,* Advances in Predictive, Preventive and Personalised Medicine 7, DOI 10.1007/978-3-319-11719-5_14

14.1 Introduction

The identification of adequate biomarkers is an important characteristic of research on Individualized Medicine (IM), with applications including diagnosis, selection of targeted therapies and determination of disease prognosis (Ziegler et al. 2012). In the medical section of this book it is pointed out that the concept of biomarker can also be adopted to widespread diseases (cf. Dörr et al. chap. 6, pp. 81–92). Thus, one of the examples outlined is Dilated Cardiomyopathy (DCM), which is one of the common causes of heart failure. Immunoadsorption therapy with subsequent IgG substitution (IA/IgG) is a novel therapeutic option that has been evaluated during the last years in DCM patients. However, response rates to this therapeutic intervention are characterized by a wide inter-individual variability and until recently individual response was not predictable (Staudt et al. 2010). By analyzing the patient's genes, it is now possible to predict, if a patient who suffers from DCM, benefits from IA with respect to improvement of cardiac function (Ameling et al. 2012). Thus, this can be regarded as the translation of the general principle of pharmacogenomics into the field of common diseases.

In general, the clinical effectiveness of a therapy could potentially be assessed by clinical parameters. But the medical perspective is only one viewpoint. As we are living in a world of scarce resources an economic assessment has to be implemented as a prerequisite for the successful translation of an innovative therapy into clinical practice. Thus it has to be investigated whether the advantages (i.e. benefits) resulting from the implementation of this therapeutic intervention improve upon the corresponding disadvantages (i.e. costs). Innovative treatments are beneficial from the societal perspective, if its cost-effectiveness ratio is more advantageous compared to an existing alternative. In particular direct medical costs are shouldered by health insurers. Their costs represent hospital revenues. Thus, it is an implicit key condition that hospitals have the chance to invoice such treatments on a regular basis. If not, this can be interpreted as a barrier against IM (cf. Flessa and Marschall chap. 13, pp. 253–271). Furthermore, the patient's subjective perspective also has to be considered: What are the consequences of a performed therapy on his or her quality of life (QoL)?

It is the aim of this chapter to use established health economic methods to analyze for the first time the economic potential of immunoadsorption: First, what are the costs of IA/IgG diagnostics and therapy from the hospital perspective? As a precondition the procedural limits of IA/IgG must be defined. Second, what are the corresponding consequences from the patients' viewpoint? Therefore, this chapter is divided into six sections. Following this introduction, Section 2 provides some essential background information about IA/IgG. Section 3 presents the methods used. Information about the study site, the data and its analyses are presented. Within that context we provide basics of health economic assessment. This is followed by Section 4 which shows the results based on pilot studies. Section 5 provides a discussion of the findings. Finally, Section 6 concludes this chapter.

14.2 Immunoadsorption: Diagnostics and Therapy

Chronic heart failure (CHF) is one of the most common syndromes worldwide (Bleumink et al. 2004). In Germany it was listed within the three leading causes of death in 2011 (Statistisches Bundesamt 2012). In face of the strong age-dependence of CHF the demographic change will lead even to a further increase of prevalence and incidence rates (Kühn et al. 2014). There are many different definitions of heart failure. It can be defined as an abnormality of cardiac structure or function leading to failure of the heart to deliver oxygen at a rate commensurate with the requirements of the metabolizing tissues, despite normal filling pressures (Dickstein et al. 2008). Clinically, heart failure can be understood as a syndrome in which patients have symptoms (e.g. breathlessness, ankle swelling, and fatigue) and signs (e.g. elevated jugular venous pressure, pulmonary crackles, and displaced apex beat) resulting from an abnormality of cardiac function (McMurray et al. 2012). CHF is usually classified based on symptoms. The most commonly used classification system was established by the New York Heart Association (NYHA) and places patients in one of four categories based on their limitation during physical activity (Hunt et al. 2001).

Dilated cardiomyopathy (DCM) is one the most important causes of both CHF and the need for heart transplantation. It is characterized by ventricular enlargement and impaired myocardial function. In adults, DCM arises more commonly in men than in women. A wide range of underlying causes may lead to the development of DCM (Felker et al. 2000). Besides rather rarely occurring gene mutations, viral infections and myocardial inflammation are increasingly recognized as common among patients with DCM. Furthermore, autoimmune disorders have been implicated in the development of DCM. Particularly, cardiac-specific antibodies have been reported in DCM patients (Staudt et al. 2002) and their pathogenic role has been proven in animal models (Okazaki et al. 2003). This observation is the basis for the application of IA/IgG to treat DCM. By this therapeutic procedure immunoglobulins are extracted from the plasma of DCM patients, leading to an improvement of hemodynamic variables in the majority of affected patients, according to several clinical trials (Felix et al. 2002).

IA/IgG diagnostics can be conceived as a process following *CHF basic diagnostics*, which is not specific for immunoadsorption (Fig. 14.1). Important basic diagnostic interventions are anamnesis, laboratory tests, echocardiography, X-ray-imaging of the chest (thoracic radiograph), cardiopulmonary exercise testing (spiroergometry) or cardiac catheterization. They are applied to determine whether a patient suffers from heart failure; to confirm the diagnosis; determine the cause of symptoms; and to evaluate the degree of underlying cardiac pathology (McMurray et al. 2012). Thus, they cannot be understood as individualized treatment in the sense of IM, but are important components of the diagnostic basic workup.

Endomyocardial biopsy (EBM) can be regarded as a particular diagnostic tool, in the case of patients who were identified as being in *general* suited for IA: First, it is an integral part of the diagnostic workflow to clarify the concrete diagnosis of heart failure based on histological and immunohistological analyses. Second, a small piece of tissue from the heart muscle can also be used to obtain information that can be used to predict the response to IA/IgG (Ameling et al. 2012). Within the context of its second meaning, EBM can potentially be the starting point for *IA/IgG-diagnostics* (Fig. 14.1), which is for the most part, the application of gene expression analysis. This methodology was applied by Ameling et al. (2012) to identify responders to IA/IgG therapy, who can be characterized by specific myocardial gene expression patterns. Gene expression analysis is a multi-phase procedure, which consists of homogenization, RNA isolation, array analysis and a final evaluation. (Newgard et al. 2010; Yee and Ramaswamy 2010).

Diagnostic tools such as those implemented in CHF basic diagnostics or IA/IgG diagnostics are necessary for choosing the appropriate treatment strategy. There is in general a distinction between symptomatic and causal DCM treatment strategies. The former aims at improving symptoms and prognosis as well as at increasing quality of life, while the latter intends to reduce or stop the progression of the disease. Symptomatic standard treatments consist of medical treatment (e.g. with angiotensin-converting enzyme (ACE) inhibitors, β-blockers, or diuretics), and non-pharmacologic treatment (e.g. exercise, dietary changes, or surgery). IA therapy is counted among the causal treatment approaches—beside antiviral and immunosuppressive therapy, respectively.

IA/IgG therapy can in a simplified way be described as an extracorporeal blood purification procedure, by which circulating autoantibodies are unspecifically removed by extracting selected immunoglobulins from the patient's plasma (Dörr 2002). In general, patients with DCM, in whom an autoimmune and/or inflammatory origin may be postulated, are eligible for this therapeutic option. Possible exclusion criteria for this procedure include, for example, acute infectious diseases, chronic alcoholism, postpartum cardiomyopathy, or heart failure due to other known origins (Ameling et al. 2012). Different schemes for conducting an IA/IgG have been allied for DCM patients.

At the University Medicine Greifswald, Clinic for Internal Medicine B, IA/IgG is currently performed in five sessions, on five consecutive days. Individuals, who undergo an IA/IgG treatment, have to be admitted as an inpatient at the intermediate care unit. On the day of admission, the basic examinations are performed in order to determinate the current functional status of the disease (e.g. analyzes of general blood parameters, echocardiography and cardiopulmonary exercise testing). This is followed by the insertion of the central venous catheter (CVC) which is necessary for the proper conduction of IA/IgG. The position of the CVC is then checked by X-ray in order to guarantee a continuous running blood flow. Afterwards, the first course of IA is started. Since a specific removal of cardiac autoantibodies is not feasible, an unspecific extraction of immunoglobulins has to be performed, of which all antibodies are a subfraction. For that purpose a machine is used to separate a patient's blood into blood cells and plasma. The plasma containing the immunoglobulins (including the antibodies) is then passed through two immunoadsorption columns, alternating between the columns for each pass. The columns contain a special ligand (Protein A or special peptides) that binds to some subtypes of

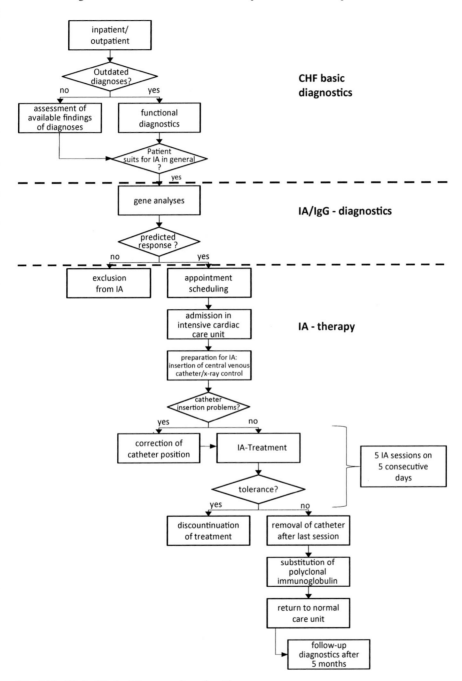

Fig. 14.1 Clinical Path of Immunoadsorption Therapy

immunoglobulins (Immunosorba®, Fresenius, Bad Homburg, Germany). While the first column is loaded with immunoglobulins, the second is rinsed of the antibodies in a process known as regeneration to prepare for another cycle. After the removal of immunoglobulins/antibodies from the plasma, it rejoins the blood cells and is given back to the patient. Each IA session is continued until 2–3 fold of plasma volume has passed the IA columns, resulting in a treatment time of approximately 4–6 h per session. The therapy is repeated on the second to fifth day. In case of a critical or important medical condition such as signs of a developing infection or cardiovascular problems, the therapy can be stopped temporally. Whether IA is continued, for example, on the next day, or canceled definitely depends on some critical factors like blood values or the cardiovascular status of the patient. On the last day of therapy, 4 h after finishing the course of IA, the dialysis catheter is removed. After IA and subsequent IgG depletion, standard immunoglobulins are substituted for safety reasons in order to restore the IgG plasma levels and thereby reduce the risk of infection. The concentration of standard immunoglobulins (e.g. Privigen®) given is calculated based on the body weight of each patient (0.5 g/kg). Regularly, patients are given appointments for follow-up diagnostics after five months.

14.3 Methods

14.3.1 Study Site and Data

The study was carried out at the University Medicine Greifswald, a maximum care clinic with around 900 beds during the study period. Two complementary projects were implemented, on the one hand to analyze the costs of IA from the hospital perspective, and on the other hand to analyze the consequences from the patients' point of view.

So far, aiming to predict a response to the procedure of IA, as a tandem solution, consisting of therapy and diagnostics, has not yet been implemented into clinical practice. Therefore, in order to acquire a deeper understanding of the IA/IgG therapy and the gene expression analysis, which is necessary for deciding, whether this therapy is adequate, the *cost project* was implemented by two sub-studies. Both were based on the time study approach (Fehrle et al. 2013).

Between April and June, 2011, the clinical pathway of immunoadsorption therapy was embraced based on observations of nine DCM patients at the Clinic for Internal Medicine B. Before patients were included in the study they were informed about its aims and procedure and they had to sign an agreement. Having identified the staff activities, the recorded time was tracked in standardized data-entry forms. In addition, the resources used such as the materials and overheads were documented. On grounds of quality assurance the study began with a test phase, in which the starting and end points of each activity were defined. Additionally, the hospital administration provided costing data for the calculation of actual average treatment costs, according to the principles of the German dual financing.

At the Interfaculty Institute for Genetics and Functional Genomics data about the gene expression analysis where collected between May and July, 2012. During the period under review, the corresponding activities and use of resources (materials, buildings, equipment and overheads) which would be ideally necessary under normal clinical IA/IgG operation were identified, measured and assessed. In this context the homogenization of 12 samples and the subsequent RNA-isolation of four samples were recorded making use of existing protocols. Additionally, the analyses of 32 tissue-samples during four subsequent days took place. Collaborators of the institute were also consulted in cases of missing information or data.

To investigate the effects of IA therapy, data provided by the Transregional Collaborative Research Center SFB Transregio 19 (SFB/TR 19) were used. Under this framework some university hospitals, as well as university-based and non-university-based research institutions, have joined forces to focus on DCM to identify structural determinants of myocardial inflammation and/or viral infection, the most frequent cause of the disorder (Angelow et al. 2008). The sample consisted of all DCM patients covered by the SFB/TR 19 at the University Medicine Greifswald, Germany, from 2005 to 2012, with at least one follow-up, as well as complete information about their quality of life assessment, where available. Data from 253 patients (= 504 cases) were considered for further analyses.

14.3.2 Costing

The perspective of the study determines the entire process of cost estimation. For assessing IA, the hospital perspective was chosen. Costs, i.e. resource use, must be distinguished according to its dependence on the number of goods or services produced. In hospitals fixed costs do not change with the number of IA/IgG patients, or the samples which were studied by gene expression analysis, e.g. building depreciation. In contrast, variable or marginal costs rise with every additional IA/IgG patient, e.g. drug cost. Consequently, the average fixed cost per patient decreases with an increasing number of services provided (decline of marginal costs). The costs at the Clinic for Internal Medicine were calculated mainly as average costs (excluding depreciation of buildings and equipment); and at the Institute for Genetics and Functional Genomics as average costs, considering depreciation. These assumptions were chosen against the background of the German system of hospital financing. In the case of IA therapy, costs are (partially) refunded by the DRG system, which aims at refunding operating costs. These do not cover, for example, the depreciation of buildings or device. However, there is currently no (partial) equivalent in the current DRG system for explicitly implementing the costs of gene analysis, in the case of IA. Therefore total costs were calculated, which provide average costs per sample.

Staff costs were calculated by multiplying the recorded time with specific calculation rates for each staff group (nursing, medical technical assistant, assistant and senior physician). The rates, which are defined per minute, take into account the employer's contributions (social insurance and retirement provisions) and delayed earnings (e.g. on-call duties and night shift premiums). The rates were provided by the Department of Controlling. The costs of the expendables were calculated based on information provided by the purchasing department, the hospital pharmacy and the manufacturer of the used immunoadsorption system. Average prices were multiplied with the documented quantities. In the gene expression analysis, study protocols and information about the workflow, which were provided by the suppliers of test kits, were used additionally.

Costs for department overheads (cleaning, nutrition, and transport) were allocated by keys. Further overhead costs referred to administrative services for some departments (legal and personnel affairs, finance and purchasing), delivery and pick-up service, disposal of waste, technical department, and the pharmacy were also considered. All costs for external services and overheads were divided by the total inpatient days of the Clinic of Internal Medicine B and subsequently multiplied by the identified treatment days to get the proportional costs for external services and overheads per patient for an IA treatment. Total costs per sample for conducting gene analysis were calculated by full cost pricing.

14.3.3   Quality of Life Effects

For assessing the patient's perspective of the therapy outcome the Minnesota Living with Heart Failure Questionnaire (MLHFQ) was used. The MLHFQ is a disease related quality of life (QoL) instrument, which does not cover the overall health related wellbeing (Rector and Cohn 1992). It is widely used in heart failure studies. The MLHFQ is regularly given to the SFB/TR 19 patients for self-assessment. It contains 21 questions which aim to determine how heart failure affects the physical, psychological and socio-economic condition of patients. The questions relate to the signs and symptoms of heart failure, social relationships, physical and sexual activity, work and emotions. The answer is chosen from a scale from 0 (none) to 5 (very much). This implies, that greater score values are synonymous with a deteriorating QoL.

Effects of IA/IgG on QoL were investigated by applying the regression technique. The classical linear regression model takes the form

$$Y = \beta_0 + \beta_1 X_1 + \varepsilon, \tag{14.1}$$

where Y, a dependent variable, is explained by a vector of independent variables X_i. The β_i are unknown regression coefficients, β_0 represents a constant and ε is the error term reflected in the residuals. Regression equation (14.1) can be estimated by Ordinary Least Squares (OLS), which minimizes the sum of squared distances between the observed responses in a set of data, and the fitted responses from the regression model. OLS yields best linear unbiased estimates (BLUE) on the assumption of independent identically distributed (iid) observations with constant mean and variance, if several important assumptions about the way in which the observations are generated are not violated. If for example the relationship between dependent and independent variables is not linear or the dependent variable is limited in some way, OLS estimates are biased, even asymptotically. In this case diagnostic analyses or misspecification tests are needed for finding out more adequate formulations (Greene 2008).

To analyze the influence of IA/IgG on the patients' quality of life, the MLHFQ-Score was used as the dependent variable. The corresponding values consist of positive natural numbers and are highly skewed, as in general many patients have an overall good level of QoL, which corresponds to a low MLHFQ score. Thus the assumptions of linear regression are not adequate. In this case Generalized Linear Models (GLM), which can be considered as a flexible generalization of linear regression, can be regarded as a solution. GLM is made up of a linear predictor

$$\eta_i = \beta_0 + \beta_1 X_{1i} + \dots \beta_1 X_{1i}, \tag{14.2}$$

and two functions: First, a link function that describes how the mean, $E(Y_i)=\mu_i$, depends on the linear predictor $g(\mu_i)=\eta_i$ Second, a variance function that describes how the variance, $var(Y_i)$ depends on the mean $var(Y_i)=\varnothing V(\mu)$, where the dispersion parameter $\varnothing$ is a constant. The link function provides the relationship between the linear predictor and the mean of the distribution function. According to a X^2 goodness-of-fit test, negative binomial distribution has the best fit regarding the MLHFQ values. Thus, a GLM model with log-link and negative binomial distribution was chosen as the regression approach.

To allow for nonlinear associations between response and independent variables, fractional polynomials (FP) were considered. This is an important and powerful extension to the polynomial transformation of the dependent variable, which uses the combination of polynomial and logarithmic functions (Royston and Sauerbrei 2008). The general m-degree fractional polynomial function of a continuous variable x is given by

$$\beta_0 + \beta_1 x^{(p_1)} + \beta_2 x^{(p_2)} + \dots + \beta_m x^{(p_m)}, \tag{14.3}$$

where $p_1, \dots, p_m$ are fractional powers and the round bracket notation denotes the Box-Tidwell transformation (Box and Tidwell 1962)—that is, $x^{(p)}=x^p$ if $p \neq 0$ and $x^{(p)}=\ln(x)$ if $p=0$. It differs from the conventional polynomial in that the power p can be a noninteger number. An integer suffix indicates the degree of FP. For assessing the model fit the Akaike information criterion (AIC) and Bayesian information criterion (BIC) were checked. Smaller values of AIC or BIC indicate a better-fitting model.

The variables listed in Table 14.1 were included in the correlation and regression analyses:

Table 14.1 Description of variables

Variable	Description
MLHFQ	MLHFQ (Minnesota Living with Heart Failure Questionnaire) Score (0=highest quality of life; 105=worst quality of life)
IA	1, if immunoadsorption therapy with subsequent IgG substitution
NYHA	Severity of heart failure, I-IV
Sex	1, if man
Age	age at time of IA, in years
Diff	Number of days between IA and QoL assessment
Responder	1, if identified as a responder by Ameling et al. (2012)

14.4 Results

14.4.1 Costs

During the study period only one EMB was performed. This intervention took around 2 h in total. Three medical technical assistants, one physician and one laboratory staff member were involved. The corresponding staff costs for EBM were 162.02 €. 88% of total time which was necessary for performing IA/IgG therapy was provided by the nursing staff. On average, the nursing costs per treatment were around 246.51 €, including documentation. Physicians spent around 2 h in total providing services for the average patient, equivalent to costs of 69.73 €.

Table 14.2 presents an overview of the main cost components of IA/IgG over time on average. The total costs of a 5-day IA/IgG therapy add up to about 19,351 €. Around 95% of the costs are for materials. Less than 2% are staff costs. It is also evident, that the costs depend on the corresponding day of treatment. The first day was especially expensive due to implementing the filter systems, which cost 11,378 €. There is only a small fluctuation concerning the costs for days 2–4. On day 5 immunoglobulins (Privigen®) involve costs of 1,491.96 €. It is striking, that around 95% of the existing costs can be attributed to expendables. Personnel costs are only accountable for 1.63% of total costs.

In addition overheads are quite low. Costs are also sensitive to the specific case. During the study period two different IA systems were used, with different cost consequences. The new one induced about 6% higher costs for expendables than the old one.

In general, the total process of gene analysis took 52 h, 55 min and 48 s, on average. Within this timeframe one staff member was on average engaged for 8 h, 9 min and 52 s. In particular, array analysis was responsible for around 61% of the required total time. Table 14.3 presents the costs of gene analysis in terms of cost types. The stated cost figures represent different sample sizes. In 2011, 18 patients were treated using IA therapy. It is remarkable, that the total costs per sample decrease enormously with an increasing number of samples. This is due to fixed costs, such as depreciation for buildings and equipment.

Table 14.2 Average costs of immunoadsorption therapy by cost type, in €

	Day 1	Day 2	Day 3	Day 4	Day 5	Total
Personnel						
Staff costs (nursing)	56.20	56.20	45.30	42.58	46.23	*246.50*
Staff costs (MD)	42.03	0.83	1.29	1.54	1.33	*47.02*
Ward round	4.53	4.53	4.53	4.53	4.53	*22.66*
Materials						
General materials	349.33	349.33	349.33	349.33	349.33	*1,746.65*
Filter system	11,377.78	0	0	0	0	*11,377.78*
IA-Kit	530.89	530.89	530.89	530.89	530.89	*2,654.45*
Laboratory	217.48	216.98	216.98	216.98	216.98	*1,085.38*
Drugs	8.45	8.66	8.59	8.03	6.17	*39.90*
Immunoglobulins	0	0	0	0	1,491.96	*1,491.96*
Overheads						
Admin. allocation	105.45	105.45	105.45	105.45	105.45	*527.25*
Department overheads	22.21	22.21	22.21	22.21	22.21	*111.06*
						19,350.61

Table 14.3 Total costs of gene analysis by cost type, in €

Cost type	*n*=18	*n*=50	*n*=100	*n*=250
Materials (non-durables)	431.79	431.79	431.79	431.79
Materials (durables)	32.10	11.55	5.78	2.31
Equipment	3,304.81	1,189.73	596.38	237.95
Maintenance	9.44	3.40	1.70	0.68
Building depreciation	103.15	37.13	18.57	7.43
Cleaning	9.11	3.28	1.64	0.66
Operating costs (district heating)	17.02	6.13	3.06	1.23
Waste disposal	0.20	0.20	0.20	0.20
Staff costs	154.81	154.81	154.81	154.81
Total costs per sample	*4,067.49*	*1,843.09*	*1,218.99*	*842.12*

14.4.2   Effects

The study population consisted of 253 individuals (206 [81.4%] men and 47 [18.6%] women). 79 patients (31.2%) were treated using IA/IgG. The database included information for 504 cases in total, 103 (20.4%) of them with an IA/IgG. For each person up to eight observations were available. Table 14.4 presents descriptive statistics by cases.

Table 14.4 Descriptive statistics of included DCM patients (n=504 cases)

	Min	Max	Mean	Std. Dev.
Age (years) at examination	19.88	80.05	11.25	52.54
MLHFQ score	0	88	27.16	22.36
NYHA stage	I	IV	1.811	0.69

The following Table 14.5 summarizes the relationship between NYHA stages (I-IV; rows) and existing IA therapy (yes-no; columns) by percent.

Table 14.5 Relation between NYHA stages and IA therapy, by % (n=504 cases)

		IA therapy	
		No	Yes
NYHA stage	I	28.41	17.90
	II	50.59	49.34
	III	18.56	31.87
	IV	2.44	0.19

Thus patients with IA/IgG can be associated with a higher NYHA severity stage.

The effect of IA therapy on QoL (=independent variable, expressed by MLHFQ) was investigated by GLM regression with negative binomial distribution. The results for the full sample (N=504) were:

$$\text{MLHFQ} = \underset{(260.12)^{***}}{3.208} + \underset{(16.31)^{***}}{0.318\ \text{IA}} + \underset{(9.73)}{0.200\ \text{Sex}}$$

$$- \underset{(-4.99)^{***}}{0.021\ \text{Age}\,(p1)} + \underset{(4.99)^{***}}{0.345\ \text{Age}\,(p2)} + \hat{u}, \tag{14.4}$$

$$\text{with Age}\,(p1) = \left(\frac{10,000,000}{\text{Age}}\right) - 43.8798186$$

$$\text{Age}\,(p2) = \sqrt{\frac{10,000,000}{\text{Age}}} - 6.624184372$$

t statistics are in parentheses. With the exception of sex all coefficients are significant at the 1% level. The information criteria for that estimation were: AIC: 23.06497 and BIC: 6188.204. Accordingly, DCM patients with IA therapy do have a lower QoL at the time of filling out the MLHFQ questionnaire, compared to people without it. The results were quite similar to an alternative approach with a regression model based on the Poisson distribution. The latter provided a slightly less significant effect of IA on QoL. However, the information criteria suggested a slightly better model fit. The regression was re-run, considering an interaction effect between NYHA stage and IA located by correlation analysis with the following results:

$$\text{MLHFQ} = \underset{(211.83)^{***}}{3.341} - \underset{(-5.04)^{**}}{0.296} \text{ IA} + \underset{(11.17)^{**}}{0.283} \text{ IA} \times \text{NYHA } (p1) + \underset{(7.69)}{0.200} \text{ Sex}$$

$$+ \underset{(3.33)^{***}}{0.160} \text{ Age } (p1) - \underset{(-4.39)^{***}}{0.0699} \text{ Age } (p2) + \hat{u}, \qquad (14.5)$$

with

$$\text{IA} \times \text{NYHA } (p1) = \text{IA} \times \text{NYHA} - 4.112149533$$

$$\text{Age } (p1) = \left(\frac{Age}{10,000}\right)^2 - 3.769180927$$

$$\text{Age } (p2) = \left(\frac{Age}{10,000}\right)^2 - 7.317630517$$

indicating that the association between patients with IA/IgG therapy and worse ML-HFS (high values), which is expressed by the positive sign in (14.5), is due to more severe types of heart failure. This model provides a better fit than estimation (14.4), with AIC: 21.91117 and BIC: 4822.7. However, the significance of the estimated IA coefficient declined to the 5% level (**). Subsequently, the effect of IA/IgG over time for the subsample of patients with IA therapy ($n=82$) was analyzed. The results for the estimated model based on the negative binomial distribution considering fractional polynomials are:

$$\text{MLHFQ} = \underset{(3.31)^{***}}{3.557} - \underset{(3.31)^{***}}{0.00138} \text{ Diff } (p1) - \underset{(-5.181)^{***}}{0.056555} \text{ Diff } (p2) + \hat{u}, \qquad (14.6)$$

with

$$\text{Diff } (p1) = \left(\frac{\text{Diff} + 1}{100}\right)^{-1} - 1.11686190$$

$$\text{Diff } (p2) = \left(\frac{\text{Diff} + 1}{100}\right)^{3} - 0.7177969082$$

again, t statistics are in parentheses and the stars denote the statistical significance. It is striking that the time difference from IA therapy does have a nonlinear effect on the perceived level of QoL as illustrated in Fig. 14.2. The coefficients are quite small, but the effects are highly significant. Compared to the models in which IA was included as dummy variable [(14.4), (14.5)], the values of the information criteria are larger, indicating a worse fit: AIC: 1755.524 and BIC: 1762.44.

Finally, the effects of IA therapy on QoL for the patients included in the Ameling et al. (2012) paper were studied. After the exclusion of cases, for which an MLHFQ score was only available for dates IA ex ante, 35 observations remained, with 20 non-responders and 15 responders. 20 out of 35 MLHFQ questionnaires were

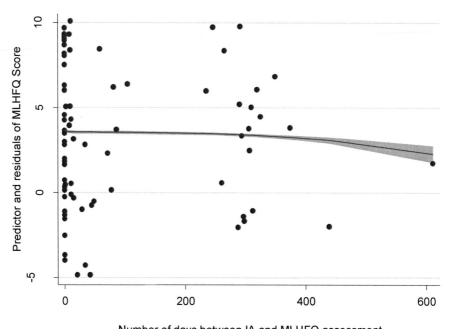

Fig. 14.2 Consequences of IA therapy on quality of life over time, $n=82$

completed during IA sessions. The results of the model with the corresponding negative binomial regression considering fractional polynomials are:

$$\text{MLHFQ} = \underset{(3.633)\text{\tiny{\textbf{*\kern-1pt*\kern-1pt*}}}}{3.633} + \underset{(2.22)*}{0.198} \text{Diff}\,(p1) - \underset{(-2.51)*}{0.755} \text{Diff}\,(p2)$$
$$+ \underset{(1.36)}{0.0795}\,\text{Responder} + \hat{u}, \tag{14.7}$$

with
$$\text{Diff}\,(p1) = \left(\frac{\text{Diff}+2}{100}\right)^2 - 0.5028835918$$

$$\text{Diff}\,(p2) = \left(\frac{\text{Diff}+2}{100}\right)^3 - 0.3566163071$$

t statistics are again in parentheses. The superscripts * and *** denote significance at the 10% and 1% levels. The corresponding results which describe the fit of the model are: AIC: 14.40528 and BIC: 202.2439. In comparison to the larger sample, the fit of the model improved. In particular, based on the small sample size it was not possible to identify, whether predicted responders do benefit more from IA with respect to QoL.

14.5 Discussion

This chapter examined the costs and consequences of IA for patients suffering from DCM. The results from the cost studies demonstrated the high costs of IA/IgG diagnostics and therapy. Under the current G-DRG reimbursement scheme the hospital only generates revenues by IA/IgG therapy. Additional charges (*Zusatzentgelt*) which satisfy staff costs and the costs for expendables can be earned by arrangement between the specific hospital and the health insurance companies. These additional charges only refer to the therapy activity in the strict sense, i.e. the process step starting with the connection to the column and finishing with the completion of the hemodialysis. Further activities must be covered by the specific DRG. Additional charges for IA therapy do not contain immunoglobulins. As these additional charges are fixed for specific hospitals, there is currently no solution for the whole of Germany. In the study year 2011, the University Medicine Greifswald received per IA/IgG case 13,000.00 €, which was similar to the revenues of other German hospitals, which also provide IA/IgG therapy. For example the Passau Hospital received 12,872.20 € (Klinikum der Stadt Passau 2011), the University Hospital of Bonn 11,091.00 € (Universitätsklinikum Bonn 2011), and the University Medicine Göttingen 13,500.00 € (Universitätsmedizin Göttingen 2011). According to our findings, there was a difference between actual costs and revenues of 6350.61 € in the case of Greifswald. By now additional costs have to be invoiced, for example by Patient Clinical Complexity Level (PCCL), or operation/procedure-keys (OPS) within the billing of the complete case. There is no incentive for implementing gene expression analysis. This could be avoided if extra additional charges would be introduced or the treatment of DCM would be better depicted within the DRG frame. According to the hopes of IM the number of promising candidates for being implemented into clinical routine will rise in the near future. If every individualized therapy would be described by a DRG, the system would no longer be manageable. Thus, there is no real incentive for a hospital to introduce the complete diagnostics and therapy tandem. For a successful implementation of IA/IgG, including diagnostics, it must also be guaranteed that costs for IA systems and further equipment are funded for, according to the principle of dual financing by the state. Currently, such costs are still cross-financed by research projects for example. Thus, from an economic point of view the current setting is not necessarily beneficial, as patients undergo a painful IA therapy who probably do not profit from it.

At the moment EMBs are performed at the University Medicine Greifswald for two reasons: First, as a standard tool within the diagnostic workup for left ventricular dysfunction. Second, tissue samples are the base material for gene expression analysis. As this intervention is not without procedural risks for the patient, there are clear recommendations, under which circumstances EMB should be performed, and when it should not be considered (Cooper et al. 2007). There are currently ambitious efforts being made to develop diagnostic tools for the prediction of response to IA/IgG therapy that may be performed without the need to obtain material by EBM. If it would be possible to switch to blood plasma, the patient's burden would be reduced enormously. In case of success, this would probably, other things being equal, result in decreasing costs. The essential cost structure would not be changed. Our findings from the gene analysis sub-study show, that the total costs per sample are highly sensitive to sample size. This is the consequence of the importance of fixed costs, which are independent from the number of analyzed samples. If it would be possible to use the existing equipment for more analyses, the corresponding costs per sample would decrease: the costs per additional sample (marginal costs) are thus diminishing. This could be interesting in case of one institute covering the demand of many hospitals which perform IA/IgG. However, the potential for decreasing marginal costs concerning IA/IgG therapy is rather limited.

The use of QoL instruments is well established in several disciplines, which study treatment effects. The findings from our research on IA/IgG outcomes clearly show that there is an obvious link between IA/IgG and well-being. In comparison to patients without IA/IgG, their perceived QoL is lower, but this is due to a higher severity level. Furthermore, age matters. This meets expectations and can be interpreted as the consequence of multimorbidity and other signs of aging. There is a non-linear relationship of the time span between IA implementation and the date of the reported QoL report, and the perceived QoL. This calls for further research. In particular the long-term benefits of immunoadsorption have to be evaluated beyond the medical literature (Dandel et al. 2012). Unfortunately, the existing information about the QoL assessment of the patients in the Ameling et al. (2012) paper is rather small. To compare the differences between identified IA responders and non-responders, as well as patients from the control group over time, several data points are necessary. There are currently available only two or more QoL statements from two responders, and two or more QoL statements from two non-responders. This is insufficient. However, lack of reliable data is common among IM research projects. In addition, it could be helpful to compare our results based on MLHFQ with other established QoL instruments, which are used for assessing heart failure effects, e.g. the Quality of Life in Severe Heart Failure Questionnaire (Wiklund et al. 1987) or the Kansas City Cardiomyopathy Questionnaire (Green et al. 2000). Furthermore, generic QoL instruments, like EQ-5D or SF 36 could be implemented. This would be an important extension to disease-specific QoL, as an improvement in clinical parameters, with respect to heart failure, does not necessarily go hand in hand with the improvement of well-being (Tepper 2009). There are some further options to picture outcomes, controlling the effects for patients, identified in advance as responders, and patients, eligible for IA, respectively, which have not been implemented. For example data about important clinical parameters could be used to calculate cardiologic risk scores about predicted mortality. Models such as the Heart Failure Survival Score, the Seattle Heart Failure Model (Levy et al. 2006), the PACE risk score or the SHOCKED predictors (Alba et al. 2013) could be used.

The results of this study have to be considered in the light of the following limitations. First, the complete IA/IgG package, i.e. companion diagnostics and therapy, are not yet installed in tandem. Only the latter is so far regularly implemented in practice. Consequently, the conducted cost studies can rather be understood as exploratory analyses and should not be confused with reality. The cost figures presented in this chapter represent a snapshot in time. However, the data highlights some important problems of IA/IgG and its translation into clinical practice. Thus, the historical data was not updated. Second, time studies are time intensive. Therefore, it was not possible to observe a large number of cases. This has the consequence, that the calculated costs might be biased and that the dispersion of the measured values is too small. Third, it was not possible to identify each activity and all used resources. In the time study at the Interfaculty Institute for Genetics and Functional Genomics the analyses of negative notropic activity of antibodies was not considered. Fourth and last, the assessment of IM is still in its infancy. Therefore it was not possible to undertake a comprehensive evaluation of this example of IM. Because IM is not yet installed as a tandem solution in medical practice, the corresponding findings reproduce the costs of research rather than of medical practice. Assessing the relevance of the full costs of the immunoadsorption does also mean that these cost figures have to be compared with the costs of cardiac transplantation and its effectiveness. The available information about the effects of IA therapy is still rather limited.

14.6   Conclusions

This chapter examined an application of IM/IgG within the field of common diseases from an economic perspective. The example of immunoadsorption shows, that the successive implementation of the diagnostics and therapy tandem into clinical practice is not easy. Economic barriers still exist and the current reimbursement system in Germany does not provide enough incentives for translating this type of IM into clinical routine, as costs are not fully covered.

There is still a need for further research. Costs are changing and this has consequences on the incentive structure of stakeholders and the corresponding path of implementation. At the moment, it is not easy to assess the outcome effects of IM over time, which is important for evaluating IM.

Acknowledgments This work is part of the research project Greifswald Approach to Individualized Medicine (GANI_MED). It was also supported by the German Research Foundation (DFG) via a grant to the SFBTRR 19.

The authors would like to thank S. Ameling from University of Greifswald, Interfaculty Institute for Genetics and Functional Genomics; and N. Bading, S. Blaut, M. Fritsche, S. Gailute, A. Heidmann, C. Heinrigs, M. Hiltscher, K. Hopert, S. Hübner, S. Ringhand, A. Sachse, M. Schubert, S. Sebeke, K. Sienknecht, S. Simonsen, T. Strödel, T. Suchsland and J. Zwingmann for their participation in the time studies.

References

Alba AC, Agoritsas T, Jankowski M, Courvoisier D, Walter SD, Guyatt GH, Ross HJ (2013) Risk prediction models for mortality in ambulatory patients with heart failure a systematic review. Circ Heart Fail 6(5):881–889

Ameling S, Herda LR, Hammer E, Steil L, Teumer A, Trimpert C, Dorr M, Kroemer HK, Klingel K, Kandolf R, Volker U, Felix SB (2012) Myocardial gene expression profiles and cardiodepressant autoantibodies predict response of patients with dilated cardiomyopathy to immunoadsorption therapy. Eur Heart J 34(9):666–675. doi:10.1093/eurheartj/ehs330

Angelow A, Schmidt M, Weitmann K, Schwedler S, Vogt H, Havemann C, Hoffmann W (2008) Methods and implementation of a central biosample and data management in a three-centre clinical study. Comput Methods Programs Biomed 91(1):82–90. doi:10.1016/j.cmpb.2008.02.002

Bleumink GS, Knetsch AM, Sturkenboom MC, Straus SM, Hofman A, Deckers JW, Witteman JC, Stricker BH (2004) Quantifying the heart failure epidemic: prevalence, incidence rate, lifetime risk and prognosis of heart failure The Rotterdam Study. Eur Heart J 25(18):1614–1619. doi:10.1016/j.ehj.2004.06.038

Box GE, Tidwell PW (1962) Transformation of the independent variables. Technometrics 4(4):531–550

Cooper LT, Baughman KL, Feldman AM, Frustaci A, Jessup M, Kuhl U, Levine GN, Narula J, Starling RC, Towbin J (2007) The role of endomyocardial biopsy in the management of cardiovascular disease A Scientific Statement from the American heart association, the American college of cardiology, and the European society of cardiology endorsed by the heart failure society of America and the heart failure association of the European society of cardiology. Eur Heart J 28(24):3076–3093

Dandel M, Wallukat G, Englert A, Lehmkuhl HB, Knosalla C, Hetzer R (2012) Long-term benefits of immunoadsorption in β1-adrenoceptor autoantibody-positive transplant candidates with dilated cardiomyopathy. Eur J Heart Fail 14(12):1374–1388

Dickstein K, Cohen-Solal A, Filippatos G, McMurray JJ, Ponikowski P, Poole-Wilson PA, Strömberg A, van Veldhuisen DJ, Atar D, Hoes AW, Karen A., Mebazza A, van Feldhuisen DJ, Atar D, Hoes AW, Karen A., Mebazza A, Nieminen M, Priori S, Swedberg K (2008) ESC Guidelines for the diagnosis and treatment of acute and chronic heart failure 2008 the task force for the diagnosis and treatment of acute and chronic heart failure 2008 of the European society of cardiology. Developed in collaboration with the Heart Failure Association of the ESC (HFA) and endorsed by the European Society of Intensive Care Medicine (ESICM). Eur J Heart Fail 10(10):933–989. doi:10.1016/j.ejheart.2008.08.005

Dörr M (2002) Hämodynamische Effekte einer Immunadsorption mit nachfrolgender Immunglobulin-G-Substitution bei Patienten mit dilatativer Kardiomyopathie—Ergebnisse einer prospektiven und randomisierten Studie. Humboldt-Universität zu Berlin, Berlin

Fehrle M, Michl S, Alte D, Götz O, Flessa S (2013) Zeitmessstudien im Krankenhaus. Gesundheitsökonomie Qualitätsmanagement 18(01):23–30

Felix SB, Staudt A, Landsberger M, Grosse Y, Stangl V, Spielhagen T, Wallukat G, Wernecke KD, Baumann G, Stangl K (2002) Removal of cardiodepressant antibodies in dilated cardiomyopathy by immunoadsorption. J Am Coll Cardiol 39(4):646–652

Felker GM, Thompson RE, Hare JM, Hruban RH, Clemetson DE, Howard DL, Baughman KL, Kasper EK (2000) Underlying causes and long-term survival in patients with initially unexplained cardiomyopathy. N Engl J Med 342(15):1077–1084

Green CP, Porter CB, Bresnahan DR, Spertus JA (2000) Development and evaluation of the Kansas City Cardiomyopathy Questionnaire: a new health status measure for heart failure. J Am Coll Cardiol 35(5):1245–1255

Greene WH (2008) Econometric analysis, 6 edn. Prentice Hall, New Jersey

Hunt SA, Baker DW, Chin MH, Cinquegrani MP, Feldman AM, Francis GS, Ganiats TG, Goldstein S, Gregoratos G, Jessup ML (2001) ACC/AHA guidelines for the evaluation and management of chronic heart failure in the adult: executive summary a report of the American college of cardiology/American heart association task force on practice guidelines (committee to revise the 1995 guidelines for the evaluation and management of heart failure) developed in collaboration with the international society for heart and lung transplantation endorsed by the heart failure society of America. J Am Coll Cardiol 38(7):2101–2113

Klinikum der Stadt Passau (2011) DRG-Entgelttarif für das Klinikum Passau im Anwendungsbereich des KHEntgG und Unterrichtung des Patienten gemäß § 8 KHEntgG

Kühn K, Marschall P, Dörr M, Flessa, S (2014), Ein Markov-Modell zur Kostenprognose der Herzinsuffizienz in Mecklenburg-Vorpommern, Gesundh ökonomie Qual manag 19, S. 168–176.

Levy WC, Mozaffarian D, Linker DT, Sutradhar SC, Anker SD, Cropp AB, Anand I, Maggioni A, Burton P, Sullivan MD (2006) The seattle heart failure model prediction of survival in heart failure. Circulation 113(11):1424–1433

McMurray JJ, Adamopoulos S, Anker SD, Auricchio A, Böhm M, Dickstein K, Falk V, Filippatos G, Fonseca C, Gomez-Sanchez MA (2012) ESC Guidelines for the diagnosis and treatment of acute and chronic heart failure 2012 the task force for the diagnosis and treatment of acute and chronic heart failure 2012 of the European society of cardiology. Developed in collaboration with the Heart Failure Association (HFA) of the ESC. Eur Heart J 33(14):1787–1847. doi:10.1093/eurheartj/ehs104

Newgard CB, Stevens RD, Wenner BR, Burgess SC, Ilkayeva O, Muehlbuaer MJ, Sherry AD, Bain JR (2010) Comprehensive metabololic analysis for understanding of disease. In: Ginsburg GS, Willard HF (eds) Essentials of genomic and personalized medicine. Elsevier Inc., San Diego, pp 97–107

Okazaki T, Tanaka Y, Nishio R, Mitsuiye T, Mizoguchi A, Wang J, Ishida M, Hiai H, Matsumori A, Minato N (2003) Autoantibodies against cardiac troponin I are responsible for dilated cardiomyopathy in PD-1-deficient mice. Nat Med 9(12):1477–1483

Rector TS, Cohn JN (1992) Assessment of patient outcome with the Minnesota Living with Heart Failure questionnaire: reliability and validity during a randomized, double-blind, placebo-controlled trial of pimobendan. Am Heart J 124(4):1017–1025

Royston P, Sauerbrei W (2008) Multivariable model-building: a pragmatic approach to regression anaylsis based on fractional polynomials for modelling continuous variables, vol 777. Wiley, New Jersey

Statistisches Bundesam (2012) Todesursachen in Deutschland 2011. Fachserie 12, Reihe 4–2, Wiesbaden.

Staudt A, Böhm M, Knebel F, Grosse Y, Bischoff C, Hummel A, Dahm JB, Borges A, Jochmann N, Wernecke KD (2002) Potential role of autoantibodies belonging to the immunoglobulin G-3 subclass in cardiac dysfunction among patients with dilated cardiomyopathy. Circulation 106(19):2448–2453

Staudt A, Herda L, Trimpert C, Lubenow L, Landsberger M, Dörr M, Hummel A, Eckerle L, Beug D, Müller C (2010) Fcγ-receptor IIa polymorphism and the role of immunoadsorption in cardiac dysfunction in patients with dilated cardiomyopathy. Clin Pharmacol Ther 87(4):452–458

Tepper M (2009) Immunadsorptionstherapie bei Patienten mit dilatativer Kardiomyopathie: Einfluss auf Morbidität und Lebensqualität. Freie Universität Berlin, Berlin

Universitätsklinikum Bonn (2011) Behandlungskostentarif, Unterrichtung des Patienten gemäß § 8 KHEntgG, Anlage 1 zu den Allgemeinen Vertragsbestimmungen

Universitätsmedizin Göttingen (2011) Krankenhausentgelttarif für stationäre und teilstationäre Behandlung in der Universitätsmedizin Göttingen

Wiklund I, Lindvall K, Swedberg K, Zupkis RV (1987) Self-assessment of quality of life in severe heart failure: an instrument for clinical use. Scand J Psychol 28(3):220–225

Yee AJ, Ramaswamy S (2010) DNA microarrays in biological discovery and patient care. In: Ginsburg GS, Willard HF (eds) Essentials of genomic and personalized medicine. Elsevier, San Diego, pp 73–88

Ziegler A, Koch A, Krockenberger K, Großhennig A (2012) Personalized medicine using DNA biomarkers: a review. Hum Genet 131(10):1627–1638

Chapter 15
How Individualized is Medicine Today?

The Case of Heart Failure in the G-DRG System

Timm Laslo, Paul Marschall and Steffen Flessa

Abstract Heart failure is currently one of the most cost-intensive diseases in Germany. It also represents one of the most common reasons for hospitalization. Currently the remuneration of hospitals in the German DRG system (G-DRG) is carried out by using a lump sum for each case of inpatient treatment. The system must be able to map complex cases. This contribution aims to analyze how a pathway for clinical care is apparent for heart failure using the example of the base DRG F62. Therefore, a comprehensive data set from the University Medicine Greifswald is used. By analyzing cost homogeneity of the base DRG F62 it becomes clear, that there are high variations within case groups. Heart failure is treated in a very individualized way today. It is concluded, that highly complex diseases such as heart failure can challenge the G-DRG-system to its limits. Therefore, an adequate adjustment in order to reflect the risk for health care providers in an acceptable way will be required. Implementing methods for risk adjustment could be one possible solution.

Keywords Heart failure · G-DRG system · Highly complex diseases · Cost homogeneity · Patient pathway · Future of DRG systems

T. Laslo (✉) · P. Marschall · S. Flessa
Rechts- und Staatswissenschaftliche Fakultät, Lehrstuhl für Allgemeine Betriebswirtschaftslehre und Gesundheitsmanagement, Ernst-Moritz-Arndt-Universität Greifswald, Friedrich-Loeffler-Str. 70, 17487 Greifswald, Germany
e-mail: laslo@laslo.de

P. Marschall
e-mail: paul.marschall@uni-greifswald.de

S. Flessa
e-mail: steffen.flessa@uni-greifswald.de

© Springer International Publishing Switzerland 2015
T. Fischer et al. (eds.), *Individualized Medicine,* Advances in Predictive,
Preventive and Personalised Medicine 7, DOI 10.1007/978-3-319-11719-5_15

15.1   Introduction

Heart failure is one of the most frequent and cost intensive diseases in the German health care system. Additionally, this diagnosis is one of the most common causes of hospitalization. In 2008, 2.26% of inpatient treatment costs were caused by heart failure, whereas in 2004, it was only 1.65% (Neumann et al. 2009). Due to demographic change a further increase in prevalence and incidence can be expected (Kühn et al. 2014). Thus, the significance of heart failure in medical care will continue to increase. Currently the remuneration of hospitals in the G-DRG system is carried out by using a lump sum payment for each case of inpatient treatment. Consequently, the remuneration system needs to be able to map complex cases, such as cases of heart failure, with a focus on performance for those providing services. The aim of this chapter is, firstly, to analyze how a pathway for clinical care is apparent for heart failure. Secondly, to determine criteria for the clinical course of symptoms of heart failure. Thus the cost of treatment will be identified. Information on the medical background as well as an introduction to the German DRG system is given in the second section of this chapter. Section 3 illustrates the methodology followed by presenting the findings in Section 4. A discussion of these findings is given in Section 5. The chapter closes, stating that the German DRG system will face a major challenge in mapping complex syndromes in the near future.

15.2   Theoretical Background

15.2.1   Medical Background

The clinical picture of heart failure has been defined by the World Health Organization in 1995 as the inability of the heart to supply the organs of peripheral regions with a sufficient amount of oxygen-enriched blood (Richardson et al. 1996). Existing definitions in literature do not understand heart failure as an unrelated disease. They much rather consider it as a comprehensive complex of diseases as well as various symptoms, whose causes and effects not only affect the cardiovascular system but also additional control systems of the human body. From a medical point of view, heart failure is not considered a disease per se. It is much rather understood as a clinical syndrome whose spectrum ranges from a mild asymptomatic course of disease to ailments suffered by the patient even in states of rest (Böcker et al. 2004). A syndrome is considered as the simultaneous occurrence of various symptoms with regard to one disease (Siems et al. 2009). Michels and Schneider (2010), as well as Loscalzo and Möckel (2011), share this view of heart failure. Jackson et al. (2000) describe heart failure as a multisystemic disorder. According to that definition, this is characterized by a reduction in function of the heart, the skeletal muscles and the kidneys as well as in the stimulation of the sympathetic nervous system in combination with a pattern of neurohormonal changes.

The prevalence of heart failure in the Western population is about two percent. A prevelance higher than ten percent is observed in individuals older than 75 years (McMurray et al. 2012). This increase can be, among other factors, traced back to higher survival rates in cases of acute coronary syndrome (ACS), in particular heart attack. Causes of heart failure are diverse (Steffel and Lüscher 2011; Neuhold and Hülsmann 2008). A large number of cases are associated to coronary heart disease and arterial hypertension (Furger 2011; Neuhold and Hülsmann 2008). According to Longo et al. (2011) about 60–75 % of patients affected by either one of these pre-existing conditions will eventually suffer from heart failure. The localization of the dysfunction generally represents an important criterion for etiological classification. This results in the classification of right-sided, left-sided or global heart failure (Classen et al. 2009). Right-sided heart failure occurs when only the right ventricle of the heart is affected. Accordingly, left-sided heart failure occurs in the event of a malfunction located in the left ventricle. In case of a global heart failure both ventricles experience functional impairment (Renz-Polster and Krautzig 2012; Lederhuber and Lange 2010)

The medical severity of heart failure, the so-called NYHA (New York Heart Association) classification can be divided into four levels. According to this classification system, level 'I' signifies the lowest and level 'IV' the highest medical severity.

15.2.2   G-DRG System

The overall objective of DRG systems is to define homogeneous case groups with respect to the underlying resource intensity. Consequently, the term DRG is used for cost homogeneous groups of cases based on related diagnosis and procedures. Depending on the current system, these groups demonstrate a more or less pronounced homogeneity (Metzger 2004). The DRG system represents a patient classification system, which classifies all of the inpatient treatment cases of a hospital into economically sensible groups of cases. The primary goal in the definition and formation of case groups is to minimize the spread of cost of one group of cases through the use of variance reduction (InEK 2004; Metzger 2004). Therefore, the DRG system usually assigns each case of treatment to one case group associated with a specific remuneration fee, regardless of the frequency of internal transfers (Metzger 2004).

The homogeneity coefficient (HC), a measure of the uniformity within one class, is used to verify this attribute. This is defined as the following (InEK 2004):

$$ HC_i = \frac{1}{1 + \dfrac{\sigma_i}{\mu_i}} \qquad (15.1)$$

with

HC$_i$	Homogeneity coefficient of class i
σ$_i$	Standard deviation within class i
μ$_i$	Arithmetic mean within class i

The homogeneity coefficient provides information on the diversification within a class (InEK 2004). In relation to case costs, it serves, among other things, as a measure for assessing the quality of a DRG system (InEK 2011). If the ratio of standard deviation and mean value is less than or equal to one, a HC higher than or equal to 0.5 is obtained. In this situation, the standard deviation has to be less than the arithmetic mean. That is, the lower the ratio of standard deviation and mean value, the lower is the diversification within the data and the higher the homogeneity within the class (InEK 2004). The Institute for the Hospital Remuneration System in Germany (InEK) considers the proportion of cost homogenous case groups starting at a homogeneity coefficient of 60% (InEK 2011). The rule of thumb in literature reckons a "sufficiently high level of homogeneity" as being reached at about 60% (Flessa 2010). In addition to factors such as the degree of patient classification, the cost intensity of hospitals as well as the properly case related allocation of cost, especially the quality and the integrity of medical case coding, influences the degree of homogeneity (Klauber et al. 2004).

By using a grouping algorithm, it is possible to assign a treatment case to the corresponding DRG or case group based on the available data (Rochell and Roeder 2003). The definition manuals for the German DRG system provide the basis for the entire grouping process (Rapp 2010). The grouping algorithms are performed by specific software, so-called groupers. Such a grouping is begun through the analysis of a patient-specific data set consisting of information on age, gender, reason for admission, holidays, admission weight for newborns, length of stay, status of stay, ventilation time and the reason for discharge. An assessment of plausibility is additionally performed. This ensures the compatibility of the documented code for surgery and procedures as well as secondary diagnoses with age, gender and the primary diagnosis (InEK 2012a). The German modification of the tenth revision of the International Statistical Classification of Diseases and Related Health Problems (ICD-10-GM) forms the current base for coding, i.e. encrypting and the subsequent documentation of the diagnoses and procedures. The main part of this directory is divided into chapters specific to organs and their localization (DIMDI 2010). Within the ICD-10-GM the diagnosis of heart failure is classified under the denotation I50.

It is economically relevant, in terms of the use of resources, that medical procedures are coded based on the catalogue of codes for surgeries and medical procedures (*Operationen- und Prozedurenschlüssel, OPS*) (DIMDI 2010). The occurrence of complications, as well as comorbidities, can exacerbate the treatment of a disease additionally and therefore increase the cost. By means of the Clinical Complexity Level (CCL) matrix, the secondary diagnoses impacting the level of severity are assessed. They usually lead to a significantly higher consumption of resources. Consequently, this crucially affects the grouping and thus the remuneration of inpatient treatment cases (InEK 2012b). The level severity of complications or comorbidities is a measure of seriousness assigned to all secondary diagnoses. Its value varies between 0 and 4 for surgical and neonatal episodes of treatment. The value for medical episodes of treatment varies between 0 and 3. A combination of medical evaluation and statistical analyses determines its value (InEK 2012b).

The final decision of which CCL value is actually assigned to a diagnosis depends on whether the code is a valid comorbidity or complication (CC) or whether it is classified as mild, moderate, severe or extremely severe CC in relation to the base-DRG of the dataset (InEK 2012b). CCLs are combined to a Patient Clinical Complexity Level (PCCL) using a complex process. The PCCL is calculated for each case of treatment by means of a smoothing formula and indicates the overall case severity (Flessa 2010):

$$PCCL = Min\left(a; round\left\{\begin{array}{l} \dfrac{\ln\left\{1+\sum_{i=k}^{n} CCL_i \times e^{-\alpha \times (i-k)}\right\}}{\ln\left(\dfrac{3}{\alpha}\right)/4} \end{array}\right\} \begin{array}{l} \text{w/o secondary diagnosis} \\ \text{otherwise} \end{array}\right)$$

(15.2)

with

α	Constant (α=4)
k	= 1 in case MDC 15 otherwise 2
n	Number of secondary diagnoses (max. 14)
a	Maximum number of PCCL (a=4)

The PCCL is integer and is measured in a five-tier system depending on the severity. It takes a value between 0 and 4, with zero representing no comorbidity and/or complication, one mild, two moderate, three severe and four equaling an extremely severe comorbidity and/or complication (InEK 2012b). The assignment of the base-DRG to an economic level of severity, respectively a billable DRG, concludes the process (Rochell and Roeder 2003). The distinct assignment of the case of treatment to one of the cost homogeneous case groups of the German DRG system is the result of this grouping algorithm.

Since the beginning of 2014 a modified version of formula (15.2) for calculating PCCL has been used. In detail, number 1 in the numerator of formula (15.2) was dropped. Accordingly, in case of weak CCL manifestations high PCCL fluctuations do not occur so pronounced, due to the fixed value of the denominator (InEK 2014).

The distinction is based on a typical notation for DRGs using a four-digit alphanumeric sequence of characters, such as F62A. The first letter indicates the corresponding Major Diagnostic Category (MDC). The two-digit number refers to the associated partition. In combination they show the base DRG. The last letter subdivides the base DRG into case classes which reflect the economic severity. The related consumption of resources indicated by the letter A is the highest. It

Table 15.1 Subdivision of the base-DRG F62. (Source: BDA 2010)

DRG	Description
F62A	Heart failure and shock in combination with extremely severe CC, dialysis or resuscitation or complicated diagnosis
F62B	Heart failure and shock in combination with extremely severe CC, without dialysis, without resuscitation, without complicated diagnosis
F62C	Heart failure and shock without extremely severe CC

decreases with the continuing alphabet. Letter Z indicates no further subdivision (InEK 2012a).

At the time of data collection in the years 2008 to 2011, there were three economic levels of severity for the G-DRG "heart failure and shock", represented by the G-DRGs F62A, F62B and F62C (Table 15.1).

15.3   Methods

The analyzed data, which was collected between 2008 and 2011, was taken from the financial controlling department of a maximum care hospital. All cases were indicated by the main diagnosis of heart failure (ICD I.50*) and were assigned the base DRG F62 (heart failure and shock). A descriptive analysis of the data was generated in order to identify the pattern of treatment.

To this end, a total of 712 records were identified within the base DRG F62. 59 cases of which were allotted to DRG F62A, 167 cases to DRG F62B and 486 cases to DRG F62C.

The following variables were used for the purpose of this study (Table 15.2):

In order to record physician services in monetary terms performed procedures according to the code for surgeries and medical procedures have been transformed to GoÄ-points according to the German scale of fees for physicians (*Gebührenordnung für Ärzte*). Homogeneity analyses in relation to physician services, nursing services and length of stay were performed to allow an assessment of the complexity of comorbidities in the context of the DRG system. In the next step multivariate regression analyses were conducted in order to determine the causal background. It should therefore be ascertained which factors influence the intensity of inpatient treatment in heart failure.

As a rationale for multivariate analyses, the hospital stay is divided into three core areas. This is carried out in order to assess which factors influence the intensity of treatment, respectively the patient pathway. The three core areas are provision of physician services, provision of nursing services and provision of services of accommodation. For each of these three areas the influencing variables are investigated.

According to the "Rotterdam-study" of 2004, age as well as the patient's gender does influence the probability of suffering from heart failure (Bleumink et al. 2004).

Table 15.2 Study variables

Variable	Description
Sex	1, if man
Age	Age in years
ND	Secondary diagnoses according to ICD 10
OPS	Procedures according to OPS
PCCL	Overall case severity (PCCL)
Length of stay	Overall length of stay in days
Intensive days	Length of stay in ICU
Ventilation time	Ventilation time in hours
Death	Death during hospitalization
DRG	Invoiced economic severity
NYHA	Severity of heart failure, 1–4
Nursing health care	Nursing care services in PPR-minutes (incl. Intensive care)
Physician health care	Physician services in GoÄ-points

In addition to the patient's age a study by He et al. (2001) indicate diseases such as diabetes mellitus, obesity, hypertension, vascular and coronary heart disease as independent factors influencing the development of heart failure (He et al. 2001). Thus, comorbidities play an essential role as well. They are taken into account using the variable PCCL in this model and reflect the patient related overall severity. The medical severity of heart failure is characterized by the NYHA class. This value fluctuates in an integer range of [1;4] just like the PCCL.

Furthermore, it is assumed that the length of stay in an intensive care unit can economically be considered as a cost driver. Ventilation time (measured in hours) is another cost driver since the ventilation of a patient represents an additional economic burden. Performing a hemodialysis influences costs as well due to its relevance to the remuneration of the base DRG F62. If a patient in this base DRG is treated with hemodialysis, the case is automatically assigned the highest economic severity.

In addition, it is assumed that the economic severity within the base DRG affects the utilization of physician and nursing resources as well as the length of stay. A high number of required resources of a single case should also be mapped by a high economic severity. The death of a patient during hospitalization may have an impact on the use of resources as well. The variable "death" indicates whether a patient died during the period of inpatient treatment.

It is understood that each of the given variables affect the utilization of physician resources and the need for nursing care resources. In addition, it is presumed that the use of physician resources has an influence on the utilization of nursing resources. In estimating the length of stay, the use of physician as well as nursing resources are assumed to be the most important factors.

The standard model Fig. 15.1 describes the dependencies of the variables among each other, taking into account the outlined relationships.

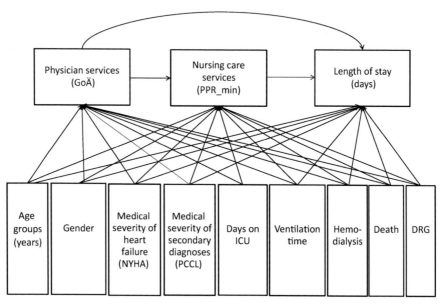

Fig. 15.1 Interdependences of variables

The next step consisted of performing a systematic data cleansing to identify disturbances and to exclude them. This is necessary in order to ensure the application of appropriate econometric techniques of analysis. Part of the data cleaning process was to exclude all cases of treatment not within the average length of stay (corresponding to outliers as well as extreme values) from further analysis.

Accordingly, the original data set of 712 cases was reduced by 171 cases of short-stay patients and 30 cases of longer-stay patients to a number of 511 cases. Cases missing some data were removed from the data set in a second step. A total of 48 cases with missing information on either documentation of the code for surgeries and procedures, on determination of the patient's NYHA class or on coded minutes for nursing care were removed. Hence, a final sample size of 463 cases for the base DRG F62 was achieved for the following multivariate analysis.

To explain the demand for physician directed health care, nursing care and the period of hospitalization, a regression analysis was conducted. The general linear regression model takes the form

$$Y_i = \beta_0 + \beta_1 X_{1i} + \ldots + \beta_1 X_{1i} + \varepsilon, \qquad (15.3)$$

with Y_i, a dependent variable or response, being explained by a linear function of independent variables. X_i and an error term ε reflect the residuals. β_i stand for unknown regression coefficients; β_0 represents a constant. This regression equation can be estimated by Ordinary Least Squares (OLS). This method minimizes the sum of squared distances between the observed responses in a set of data and the fitted responses from the regression model. Based on diagnostic analyses or misspecification tests, it is possible to identify estimation techniques and models providing a best fit to the data (Greene 2008).

15.4 Results

15.4.1 Descriptive Analyses

The descriptive analyses showed that within the 712 cases being examined a variety of 787 secondary diagnoses were coded. The maximum number of secondary diagnoses in one case alone within the data set is 28. Differentiated according to economic severity, F62A shows a maximum of 28, F62B a maximum of 26 and F62C a maximum of 21 secondary diagnoses per case. On average a patient being hospitalized for heart failure has about eleven secondary diagnoses. This mean value derived from the overall number of secondary diagnoses (7620) divided by the number of examined cases (712). Those 7620 secondary diagnoses include double counting, i.e. secondary diagnoses that occurred in more than one patient. The exclusion of duplications results in 1307 different secondary diagnoses over all 712 cases. Overall, 89.6% of the examined cases showed more than five secondary diagnoses.

In relation to the base DRG a total of 2660 procedures were coded according to the OPS catalogue of codes for surgeries and medical procedures. 497 procedures are attributable to DRG F62A, 734 to DRG F62B and 1519 procedures to DRG F62C. This corresponds to an average of about 6.89 procedures per case in DRG F62A, an average of about 4.39 procedures per case in DRG F62B and an average of about 3.12 procedures per case in DRG F62C. Thus, the average number of procedures per case shows an increase with higher levels of economic severity. A total of 189 different OPS codes were generated. On average, 3.7 procedures per case were coded according to OPS. In the DRG system the overall medical severity of each case is reflected by the PCCL. As explained, the PCCL has a value in the range of "0" to "4". 65.9% of cases examined have a PCCL equaling or higher than "3".

Considering the medical severity of heart failure, also according to the New York Heart Association (NYHA), it can be observed that an increase of medical severity coincides with an increase of economic severity. A total of 72.1% of cases have a NYHA class of "4" (Fig. 15.2).

The results of the descriptive analysis suggest heart failure as being a highly complex disease. The impact of these results on the homogeneity will be examined.

Referring to the homogeneity coefficients in terms of physician and nursing services as well as the length of stay as shown in Table 15.3, the assumption of heart failure being a very complex disease is confirmed. The InEK only takes cases into account that are considered inliers, e.g. with a length of stay between the lower

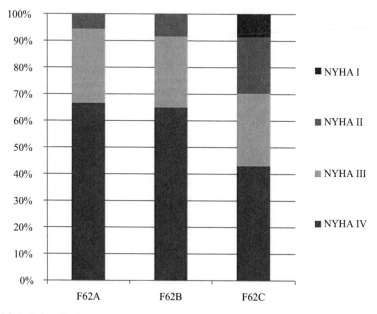

Fig. 15.2 Relative distribution of NYHA-stages to each DRG

Table 15.3 Coefficients of homogeneity

	F62A	F62B	F62C
PS_Inlier/Outlier	0.52	0.41	0.50
PS_Inlier	0.54	0.42	0.49
NS_Inlier/Outlier	0.51	0.57	0.47
NS_Inlier	0.73	0.57	0.53
LoS_Inlier/Outlier	0.81	0.61	0.56
LoS Inlier	0.69	0.69	0.69
LoS_Inlier _InEK	0.66	0.69	0.69

Note: *PS* physician services (GoÄ), *NS* nursing services (PPRmin), *LoS* length of stay (days)

and upper limits. Cases which go below or exceed one of the two boundaries for length of stay are not taken into consideration in the DRG-browser provided by the InEK. Regarding physician services, the highest homogeneity is at 0.54. Thus, the homogeneity coefficient for physician services lies below the target specification of 0.6, given by the InEK. In principle, one speaks of homogeneity starting at a value higher than 0.5. When examining nursing services, a less problematic situation can be observed. In this case only one value was below 0.5. In terms of the length of stay no value below 0.5 was found. With the exception of one value in DRG F62C, each homogeneity coefficient was higher than 0.6 (Table 15.3).

15.4.2   Multivariate Analysis

15.4.2.1   Physician Service

Some statements can be made for a level of significance of less than or equal to 0.05. Cases with a NYHA class of IV use significantly fewer physician resources than cases with a NYHA class of III. Furthermore, a significant increase in the demand for medical services coincides with an increase in days spent in intensive care. This also applies to performing a hemodialysis with a level of significance of 0.1. It can be noted that the demand for physician services expanded significantly with an increase of the economic level of severity. The quality of the estimation can be evaluated with the help of R^2. This value provides the explained variance's proportion to the total variance. It is very low in this case, with a value of 0.107. Taking the number of predictors into account it is even less (adj. R^2). This mainly equals a very high diversification. This again can be interpreted as an indication of a highly individualized medical treatment. In order to detect collinearity, the Variance Inflation Factor (vif) and the Tolerance were calculated. Neither of these two indicators provides evidence of collinearity in this calculation (Table 15.4).

Table 15.4   Estimates of physician resources measured in GoÄ points

	Model		Colinearity diagnostics	
	Coefficient	Standard error	Tolerance	vif
Constant	7.387	0.184		
Sex	−0.080	0.091	0.917	1.090
Age 60–69	0.114	0.127	0.869	1.151
Age 70–79	0.136	0.100	0.879	1.138
Age 80–99	−0.017	0.100	0.945	1.059
PCCL	−0.045	0.053	0.477	2.097
NYHA 1	0.236	0.329	0.906	1.104
NYHA 2	0.026	0.164	0.791	1.264
NYHA 4	−0.227**	0.102	0.761	1.315
Intensive days	0.066**	0.021	0.709	1.410
Ventilation time	0.020	0.018	0.762	1.313
Hemodialysis	0.545*	0.282	0.804	1.243
Death	−0.2410	0.239	0.927	1.079
DRG F62A	0.398**	0.199	0.596	1.677
DRG F62 B	0.265*	0.136	0.491	2.035
R^2	0.107			
Adj. R^2	0.079			
SE	0.931			

Note: The table presents estimates using OLS. The dependent variable is "Utilization of Physician Resources"

* and ** denote significance at the 10 and 5% level, respectively

15.4.2.2   Nursing Services

Based on a level of significance of less than or equal to 0.05 the following statements can be made: In analogy to the estimates of physician services no statistical significance on an appropriate level can be observed regarding age and sex
(Table 15.5). However, a significantly higher demand for nursing resources goes
along with increasing medical severity. This also applies to the severity of heart failure. Again, the higher the severity of a disease, the higher the demand for nursing
resources. An increased need for physician services leads to a significantly higher
demand for nursing services as well. Surprisingly, patients on hemodialysis show a
significantly reduced need for nursing services. The presence in ICU increased the
demand for nursing resources at a significance level of 0.01. The demand for nursing services becomes greater with increasing economic severity on the same level
of significance. The death of a patient with heart failure has no significant influence
on the demand for nursing services during the hospitalization.

The estimates of nursing services show a value of 0.622 for R^2. In comparison
to the estimates of physician services this value is much better. The indicators vif
and Tolerance also do not provide evidence for collinearity for the presented values.

15.4.2.3   Length of Stay

According to the conducted estimates, cases of a patient age 60–69 show a significantly higher length of stay than cases of patients of younger ages (Table 15.6). Cases with NYHA class I and II exhibit a significantly shorter length of stay than cases
with NYHA class III (comparison category). The higher the case severity the significantly longer the length of stay. According to the estimates, an increase in ventilation time, hemodialysis as well as death influence the length of stay negatively at
a significance level of 0.05. An increased utilization of physician as well as nursing
resources enhances the length of stay at a significance level of 0.01. Likewise, an
increase in length of stay affects the economic severity in a positive way.

The estimates for length of stay show a value of 0.426 for R^2. Thus, it is above
the value of R^2 of the estimates for the utilization of physician services and below
the value for utilization of nursing resources. No signs of collinearity can be identified by the parameters vif and Tolerance.

15.5   Discussion

Some years ago McMurray et al. (1998) presented the idea that "heart failure is
common, costly, disabling and deadly". As demonstrated, the interdependencies
in production, i.e. those in the treatment of heart failure are highly complex. The

Table 15.5 Estimates of nursing resources measured in PPR minutes

	Model		Colinearity diagnostics	
	Coefficient	Standard error	Tolerance	vif
Constant	6.741	0.096		
Sex	0.004	0.046	0.917	1.090
Age 60–69	0.009	0.065	0.868	1.152
Age 70–79	−0.46	0.051	0.871	1.148
Age 80–99	−0.006	0.051	0.945	1.059
PCCL	0.058**	0.027	0.477	2.097
NYHA 1	−0.548**	0.169	0.905	1.105
NYHA 2	−0.189**	0.084	0.791	1.265
NYHA 4	0.110**	0.053	0.753	1.329
Intensive days	0.213***	0.011	0.681	1.469
Ventilation time	−0.008	0.009	0.760	1.316
Hemodialysis	−0.349**	0.145	0.800	1.249
Death	0.164	0.123	0.924	1.082
DRG F62A	0.367***	0.103	0.592	1.688
DRG F62B	0.288***	0.070	0.486	2.058
Physician health care	0.000*	0.000	0.867	1.153
R^2	0.622			
Adj. R^2	0.609			
SE	0.477			

Note: The table presents estimates using OLS. The dependent variable is "Utilization of Nursing Resources"

*, ** and *** denote significance at the 10, 5, and 1% level, respectively

studied cases show a high rate of randomization (scattering), in particular with regard to the utilization of physician services. The DRG system, however, follows the logic of standardization. Also, this is the basic requirement for the application of a clinical pathway. According to Dykes and Wheeler (2002), as well as Flessa and Marschall (2012), 80% of cases and total cost should be on one standard path in order to achieve cost homogeneity. However, the obtained results do not indicate a standardization of treatment. Instead a high level of individualized treatment is evident.

High spreads of case costs in lump sum based reimbursements provide a risk of loss for service providers, even if the expected value of cost would correspond to the remuneration of a case class. Maximum care providers are likely to face a higher risk in the treatment of highly complex diseases, such as heart failure, than providers of lower levels of care. These providers could refer complex cases to a hospital of a higher level of care. Maximum care providers do not have this option and, therefore, are forced to treat the patient.

Table 15.6 Estimates of length of stay in days

	Model		Colinearity Diagnostics	
	Coefficient	Standard error	Tolerance	vif
Constant	1.458	0.082		
Sex	0.006	0.039	0.920	1.087
Age 60–69	0.125**	0.055	0.868	1.152
Age 70–79	0.000	0.043	0.872	1.147
Age 80–99	0.019	0.043	0.945	1.058
PCCL	0.045*	0.023	0.476	2.103
NYHA 1	−0.298**	0.142	0.901	1.109
NYHA 2	−0.123*	0.071	0.789	1.267
NYHA 4	0.058	0.044	0.755	1.325
Ventilation time	−0.016**	0.008	0.782	1.279
Hemodialysis	−0.273**	0.121	0.809	1.235
Death	−0.215**	0.103	0.924	1.082
DRG F62A	0.384***	0.088	0.566	1.766
DRG F62B	0.295***	0.059	0.486	2.056
Physician health care	0.000***	0.000	0.847	1.180
Nursing health care	0.000***	0.000	0.623	1.605
R^2	0.426			
Adj. R^2	0.407			
SE	0.400			

Note: The table presents estimates using OLS. The dependent variable is "Length of Stay"
*, ** and *** denote significance at the 10, 5, and 1% level, respectively

Generating cost homogenous groups of medical related cases based on actual cost and compensating those in a homogenous way is the goal of the DRG system. In the treatment of complex cases, a lump sum based remuneration requires an exact calculation of costs as well as a high quality in coding. Cost calculation is influenced significantly by the hospitals supplying data to the InEK. This delivery of data takes place on a voluntary basis.

Economically, efficient providers, i.e. those able to exactly determine their cost, have no incentive to supply data to the InEK, and the majority of data supplying hospitals are not able to exactly determine their cost. In reality, many hospitals have not implemented a full cost unit accounting system. However, a functioning cost unit accounting system is necessary in order to provide valid data to the InEK.

Although the introduction of so-called NUBs (new methods for examination and treatment) as well as additional remuneration fees could reduce the individual risk for service providers, this will not change the system of standardization as such. The more additional fees introduced for the treatment of a specific disease, the more one approaches the principle of individual-service remuneration.

As shown above, the high dispersion in intensity of treatment is a critical factor which makes the treatment of heart failure appear difficult. The demonstrated high rate of randomization, especially in the utilization of physician services, makes a compensation based on a diagnosis related lump sum per case difficult. This is an indication that a highly individualized medical treatment of groups of cases is necessary. In order to still create a fair remuneration for the individual service provider, the existing DRGs could be split up. In this way the individualized and therefore cost inhomogeneous treatment paths could each be represented by one DRG. This would result in the existence of a variety of different DRGs for heart failure with a small number of cases. The difficult administrative management of DRGs with low numbers of cases is however a big disadvantage. In practice, the economic levels of severity within the base DRG F62 were reduced from three to two in 2012.

From an economic point of view, it would possibly appear reasonable that solely the most economic service providers take over the treatment of complex cases. Such a measure of risk adjustment could lead to this problem. To this end, implementing a fair comparison among all service providers needs to be realistic. The availability of corresponding data of the service providers for such calculations would be required. Currently, this is only possible for very few, highly complex and resource intensive procedures with manageable numbers of cases and few service providers, such as in the field of coronary surgery (BQS-Institut 2014). In the area of heart failure, however, such a solution will prove to be difficult, due to the high number of cases as well as service providers.

For a couple of years, the social health insurance system in Germany is pursuing the concept of risk structure compensation. This happens in order to manage the distribution of resources to individual health insurance companies and the therein insured individuals (Bundesversicherungsamt 2008). The resource management through risk structure compensation also has competitive effects. Social health insurance companies receive a higher financial support from the health fund for insured individuals showing a high individual risk of illness than for those with a lower individual risk of getting a disease.

A concept similar to the one of risk structure compensation could also go into effect for calculations of reimbursement for the treatment of patients with heart failure. Aspects such as genetic disposition, environmental factors or the patient's lifestyle could therefore be considered.

So far, even less literature on the economic analysis of inpatient treatment for heart failure from the perspective of service providers can be found. This chapter presented various alternatives and discusses options that may lead to a more equitable compensation of inpatient treatment of heart failure based on performance. The regression analysis suffers from neglecting the count data nature of the dependent variable in the estimation of the period of hospitalization. This shortcoming can be overcome by using an appropriate model which takes into account the specific nature of the corresponding data distribution.

15.6 Conclusions and Future Prospects

Highly complex diseases such as heart failure can challenge the G-DRG-system to its limits. Therefore, an adequate adjustment in order to reflect the risk for health care providers in an acceptable way will be required. The base DRG F62 is a good example for a highly complex syndrome for which no standardized patient pathway can be applied to. Additionally, generating cost homogeneous case groups for heart failure proves to be problematic. The homogeneity coefficients of the base DRG F62 primarily indicate high variations within those case groups (especially in the utilization of physician resources). As shown in this chapter the syndrome of heart failure is treated in a very individualized way today. As a consequence, the implementation within DRGs involves the illustrated problems. Considering the consequences of demographic change, it can be noted that the G-DRG system will face a major challenge in mapping complex syndromes such as heart failure. Implementing methods for risk adjustment could be one possible solution.

References

BDA (2010) Fallpauschalen-Katalog, GDRG Version 2010
Bleumink GS, Knetsch AM, Sturkenboom MC, Straus SM, Hofman A, Deckers JW, Witteman JC, Stricker BH (2004) Quantifying the heart failure epidemic: prevalence, incidence rate, lifetime risk and prognosis of heart failure The Rotterdam Study. Eur Heart J 25(18):1614–1619. doi:10.1016/j.ehj.2004.06.038
Böcker W, Denk H, Heitz P et al. (2004) Pathologie, 3rd edn. Urban & Fischer Verlag, München
BQS-Institut (2014) Risikoadjustierung: Differenzierte Modelle für die Herzchirurgie, Gefässchirurgie und Dekubiusprophylaxe. http://bqs-institut.de/innovationen/risikoadjustierungsmodelle.html. Accessed 23 March 2014
Bundesversicherungsamt (2008) Wie funktioniert der morbiditätsorientierte Risikostrukturausgleich
Classen M, Diehl V, Kochsiek K, Böhm M, Hallek M, Schmiegel W (2009) Innere Medizin, 6th edn. Urban & Fischer, München
DIMDI (2010) Basiswissen Kodieren, Eine kurze Einführung in die Anwendung von ICD-10-GM und OPS
Dykes PC, Wheeler K (2002) Critical Pathways - interdisziplinäre Versorgungspfade DRG-Management-Instrumente, vol 3. Huber, Bern
Flessa S (2010) Grundzüge der Krankenhausbetriebslehre, 2nd edn. Oldenbourg, München
Flessa S, Marschall P (2012) Individualisierte Medizin: vom Innovationskeimling zur Makroinnovation. Pharmacoecon Ger Res Artic 10(2):53–67. doi:10.1007/BF03320778
Furger P (2011) Innere Medizin quick, 3rd edn. Thieme, Stuttgart
Greene WH (2008) Econometric Analysis, 6th edn. Prentice Hall, Upper Saddle River
He J, Ogden LG, Bazzano LA, Vupputuri S, Loria C, Whelton PK (2001) Risk factors for congestive heart failure in US men and women: NHANES I epidemiologic follow-up study. Arch Intern Med 161(7):996–002
InEK (2004) Abschlussbericht, Weiterentwicklung des G-DRG-Systems für das Jahr 2005, Teil 1
InEK (2011) Abschlussbericht, Weiterentwicklung des G-DRG-Systems für das Jahr 2012, Teil 1
InEK (2012a) German Diagnosis Related Groups, Version 2012/2013 Definitionshandbuch Band 1

InEK (2012b) German Diagnosis Related Groups, Version 2012/2013, DRG-Definitionshandbuch Band 5

InEK (2014) Abschlussbericht, Weiterentwicklung des G-DRG-Systems für das Jahr 2014

Jackson G, Gibbs C, Davies M, Lip G (2000) ABC of heart failure: pathophysiology. BMJ 320(7228):167

Klauber J, Robra B-P, Schellschmidt H (2004) Krankenhausreport 2003, Schwerpunkt G-DRGs im Jahre 1. Stuttgart

Kühn K, Marschall P, Dörr M, Flessa S (2014) Ein Markov-Modell zur Kostenprognose der Herzinsuffizienz in Mecklenburg-Vorpommern. Gesundh ökonomie Qual manag 19:168–176.

Lederhuber HC, Lange V (2010) Kardiologie—Basics, 2nd edn. Urban & Fischer, München

Longo DL, Loscalzo J, Jameson JL, Hauser SL, Fauci AS, et al. (2011) Harrison's principles of internal medicine, vol 2, 18th edn. McGraw-Hill Medical, New York

Loscalzo J, Möckel M (eds) (2011) Harrisons Kardiologie. ABW Wiss.-Verlag, Berlin

McMurray J, Petrie M, Murdoch D, Davie A (1998) Clinical epidemiology of heart failure: public and private health burden. Eur Heart J 19:P9–16

McMurray JJ, Adamopoulos S, Anker SD, Auricchio A, Böhm M, Dickstein K, Falk V, Filippatos G, Fonseca C, Gomez-Sanchez MA (2012) ESC guidelines for the diagnosis and treatment of acute and chronic heart failure 2012 The task force for the diagnosis and treatment of acute and chronic heart failure 2012 of the European Society of Cardiology. Developed in collaboration with the Heart Failure Association (HFA) of the ESC. Eur Heart J 33(14):1787–1847. doi:10.1093/eurheartj/ehs104

Metzger F (2004) DRGs für Einsteiger: Lösungen für Kliniken im Wettbewerb. Wissenschaftliche Verlagsgesellschaft, Stuttgart

Michels G, Schneider T (2010) Klinikmanual Innere Medizin. Springer Medizin Verlag, Heidelberg

Neuhold S, Hülsmann M (2008) Herzinsuffizienz. Wien Klin Wochenschr 3(1):17–33

Neumann T, Biermann J, Neumann A, Wasem J, Ertl G, Dietz R, Erbel R (2009) Herzinsuffizienz: Häufigster Grund für Krankenhausaufenthalte—Medizinische und ökonomische Aspekte. Dtsch Arztebl Int 106(16):269–275. doi:10.3238/arztebl.2009.0269

Rapp B (2010) Praxiswissen DRG - Optimierung von Strukturen und Abläufen. s edn. W. Kohlhammer, Stuttgart

Renz-Polster H, Krautzig S (eds) (2012) Basislehrbuch Innere Medizin, 4th edn. Urban & Fischer, München

Richardson P, McKenna R, Bristow M, Maisch B, Mautner B, O'connell J, Olsen E, Thiene G, Goodwin J, Gyarfas I (1996) Report of the 1995 World Health Organization/International Society and Federation of Cardiology Task Force on the definition and classification of cardiomyopathies. Circulation 93(5):841–842

Rochell B, Roeder N (2003) DRG - das neue Krankenhausvergütungssystem für Deutschland, Teil 1: Einführung. Der Urologe 42(4):471–484

Siems W, Bremer A, Przyklenk J (2009) Allgemeine Krankheitslehre für Physiotherapeuten. Springer, Berlin

Steffel J, Lüscher TF (2011) Herz-Kreislauf. Springer Medizin Verlag, Heidelberg

Chapter 16
Conclusions and Recommendations

Tobias Fischer, Martin Langanke, Paul Marschall and Susanne Michl

In this conclusion we aim to sum up the most important results of the previous chapters. Thereby we focus at those findings which are lined up with and justified by the experiences of the "Greifswald Approach to Individualized Medicine" (GANI_MED).

16.1 Definition and Concept of Individualized Medicine

In 2009 the University Medicine Greifswald, Germany, launched the "Greifswald Approach to Individualized Medicine" to address some of the major challenges of "Individualized Medicine".

T. Fischer (✉)
Department für Ethik, Theorie und Geschichte der Lebenswissenschaften,
Universitätsmedizin Greifswald, Walther-Rathenau-Str. 48, 17475 Greifswald, Germany
e-mail: tobias.fischer@uni-greifswald.de

M. Langanke
Theologische Fakultät, Lehrstuhl für Systematische Theologie, Ernst-Moritz-Arndt-Universität
Greifswald, Am Rubenowplatz 2–3, 17487 Greifswald, Germany
e-mail: langanke@uni-greifswald.de

P. Marschall
Rechts- und Staatswissenschaftliche Fakultät, Lehrstuhl für Allgemeine Betriebswirtschaftslehre
und Gesundheitsmanagement, Ernst-Moritz-Arndt-Universität Greifswald,
Friedrich-Loeffler-Str. 70, 17487 Greifswald, Germany
e-mail: paul.marschall@uni-greifswald.de

S. Michl
Institut für Geschichte, Theorie und Ethik der Medizin, Universitätsmedizin der Johannes
Gutenberg-Universität Mainz, Am Pulverturm 13, 55131 Mainz, Germany
e-mail: susmichl@uni-mainz.de

© Springer International Publishing Switzerland 2015   311
T. Fischer et al. (eds.), *Individualized Medicine,* Advances in Predictive,
Preventive and Personalised Medicine 7, DOI 10.1007/978-3-319-11719-5_16

The major goals of IM require excellent clinical stratification of patients as well as the availability of genomic data and biomarkers as prerequisites for the development of novel diagnostic tools and therapeutic strategies.

Against this background we recommend the introduction of "Individualized Medicine" as a technical term. More specifically we suggest the following definition:

"Individualized Medicine" stands for current medical fields of research on the one hand, and for health care practices on the other hand. Medical fields of research can be summed up under the term Individualized Medicine if they aim at identifying, validating and integrating biomarkers into the clinical routine which allow to predict the outbreak or the course of diseases and/or the effect of therapies or their unwanted effects for certain patient groups in a better way. Preventive, therapeutic or rehabilitative health care practices, which can be included in the term of Individualized Medicine, are characterized by the fact that they use biomarkers for a systematic prediction of risks or courses of diseases and/or for the prediction of the effect of therapies or their unwanted effects.

16.2   Perspectives of Socio-Cultural and Historical Studies

Our knowledge of IM—which is often labeled as "Personalized Medicine" in the scientific and popular publications on which Part II is based—as a societal phenomenon that involves promoting, monitoring and providing stakeholders is limited. An assessment of the multiple biomedical initiatives in and outside universities and research centers, requires more than an analysis of whether such trends will live up to their clinical and economic promises and meet ethical requirements. An integration of sociological, anthropological, and historical perspectives into biomedical research enables us to consider the emergence, the present state and the future of "Personalized Medicine" as a cultural framework of visions, expectations and claims. "Personalized Medicine" as a strategy of conceptual rebranding constitutes an imaginary context for changes in medical research and practice and in society at large.

The "timescapes" of these developements in medicine follow a specific narrative according to which medical history necessarily leads to "Personalized Medicine". This narrative of an inevitable triumph is based on normative assumptions that are often implicit. The challenge for scholars in the social sciences and the humanities is to render these assumptions explicit and to reconsider them in an ethically and historically grounded discussion.

Beyond this, it is unusual for socio-cultural and historical studies to give specific recommendations. The analysis of a phenomenon, past or present, rarely leads to well-informed action-oriented advice. That said, the social sciences and the humanities allow us to understand how and why medical research centered upon the key category of "(bio-) chemical individuality" and "human variability" until the late 1950s, followed by a focus on "individualized therapies" which gave way to the concept of "Personalized Medicine" in the early twenty-first century. Against this background, current medical research and practice would be well advised to pay careful attention to the following issues:

1. Whether IM will ever be able to measure up to the expectations it gives rise to, remains to be seen. What is striking, however, is that the controversy over "Personalized Medicine" is rooted in historical opposition lines such as the alleged asymmetrical levelling of the art/science-divide. This epistemological shift in which science grows at the expense of art accompanied and drove the transformation of medical practice and research since the early nineteenth century. As a result, the idea of medical progress has been framed in terms of science and technology, rarely in terms of an art that is increasingly refined.

2. In the narrowly defined research area of pharmacogenetics, the invention of "Personalized Medicine" is a deliberate conceptual strategy. As a term, "Personalized Medicine" was coined to enable collaborations and to broaden the research scope well beyond questions of technological progress. In fact, pharmacogenetic research strategies labeled as "Personalized Medicine" suddenly appear to be part and parcel of a venerable tradition of past advances, a tradition that heralds a bright future for (Western) societies.

3. An analysis of past framings of pharmacological and genetic research centered upon "individuality" or "variability" illustrates alternate conceptualizations and past visions of the future for a better health care. It is a history of survival or oblivion of scientific byways, detours and dead-ends that draws our attention to the fact that there was not a linear technological and scientific development that led to contemporary "Personalized Medicine". Against this background, we can shed light not only on the question of what we gain but also on what we lose by framing specific research projects and traditions as "Personalized Medicine".

16.3 Medical Perspectives

The current status of medical research raises a very differentiated and, in parts decidedly, clear-cut record. The biomarker-based IM is certainly not the new paradigm of medicine, but a research approach, which may contribute to the progress in very limited areas of medical care: IM shows respectable successes especially in the field of biomarker-based stratification of oncological entities and in the field of pharmacogenetic prediction. Regardless of this success, IM exhibits a much slower progress than anticipated in the heydays of the IM hype. Especially in the field of prediction of complex diseases or courses of such diseases ('prognostic markers') great clinical success has so far failed to materialize. But, as the example of immu-noadsorption shows, there are in this area at least promising research approaches.

16.4   Concept-Based Ethical Questions

16.4.1   On a Philosophy of Individualized Medicine: Conceptual and Ethical Questions

The more medical research is directed towards IM, the more urgent becomes the recollection of medicine to its philosophical and ethical fundaments. Analysis shows that IM is based on concepts which refer to underlying philosophical and ethical problems. Conceptual implications of IM framing need close attention and should be addressed in a philosophy of IM.

Given a plurality of competing medical approaches, it would be flawed to connect IM only to a molecular and genetically based approach. IM should take a mediating role between the various approaches, such as the lifeworld-approach, the conventional approach, "alternative" healing approaches, and the molecular-genetic approach. No approach has a monopoly to truth within the realm of medicine. There could be a "spirit" of IM which combines the humanistic motives of different medical approaches.

Within the ethical dimension of IM, we hold that IM can also be connected with a eudaimonistic concept of a cheerful calmness regarding the individual human mortality, with a body-sensitive attitude of "cura sui", with a dialogical-consulting medicine, which upholds the principle of informed consent, and with a solidary, and even generous health care system. Responsibility for one's own health should not be negated in an abstract way.

After five years of GANI_MED research, our results point at the necessity to reflect the conceptual dimension of IM, both in epistemic and ethical respects. The progress of IM requires the right kinds of questions, not just the production of a mass of data. Acceleration of ongoing mass data production ("system medicine", "automatic processing") might not resolve any substantial question about IM. We proclaim that data as such are worthless if not interpreted according to innovative hypotheses.

Therefore we would like to see

1. research on the affinities between IM and "alternative" and human-ecological concepts of healing and health,
2. research of how IM ideas are embedded in the discourse about human enhancement, trans- and post-humanism,
3. research of how specific natural environments affect health, disorders, sickness, and disease.

16.4.2   The Concept of Disease in the Era of Prediction

The research and health care practice in the area of IM should be aware that increasingly precise predictive knowledge can lead to an extension of the term "disease", or more specifically a pathologization of dispositions.

It is important to explicitly address this developing trend and to identify the practical problems associated with it. The way in which IM determines its own central terms and designs as well as its own use of language is important. In the attempt to reflect and understand this trend, it is essential to philosophically think through medical terms and concepts.

It is not possible to say with certainty whether it will come to such a pathologization. However it is in any case imperative to reflect on predictive knowledge and to deal with it in a practical and sensible way.

16.5   Applied Research Ethics

16.5.1   Informed Consent in GANI_MED—A Sectional Design for Clinical Epidemiological Studies Within Individualized Medicine

An ethical analysis suggests that the consent in epidemiological and clinical epidemiological projects in the field of IM should be designed on the basis of a moderate contractual research ethical model. This model consists of three principles:

1. A contractual base norm: The consent documents form a participant′s contract. Hence the concluded agreements are to be understood as binding for both parties. The right of withdrawal of the participant is a part of this contract.
2. Transparency requirement: To create an ethically valid informed consent the participant needs to understand the conditions under which he or she takes part. Also the possible consequences of the participation which can be anticipated need to be understood. Both aspects have to be clearly distinguished: The presentation of any information—written or spoken—needs to be as transparent as possible. Furthermore, the possible consequences linked with the participation need to be addressed explicitly during the informed consent process. From a methodical point of view the requested transparency is an ideal which demands the identification and use of potentials to optimize the informed consent process and all related documents.
3. The requirement to minimize anticipated stress: This principle demands that by designing potentially stressful studies, anticipated stress is to be avoided or at least minimized. This includes the disclosure of incidental findings. This demand should not be weighed against pragmatic constraints such as limited staff and financial recourses.

16.5.2   Ethics Meets IT: Aspects and Elements of Computer-Based Informed Consent Processing

Informed consent (IC) is one of the fundamental elements of IT-based data processing within IM-related studies. Because of the mandatory requirement to process only data that have been agreed to become processed, IT must have knowledge about the respective informed consent contents as well as its link to all the data covered by it. Furthermore IT must be able to process the contents.

The quality of a consent form—and thereby its ethical validity—has significant impact on the ability to process it, which is often underestimated. Therefore, the logical structure of consent forms and the interface between them and the IT processes should be focused on more intensively than is usually the case. It is desirable both from a processing and ethical perspective to supply research with a homogeneous and uniform sectional IC architecture that facilitates to manage different study types with adequate documents which use a uniform structure, cover identical requirements with identical sections of text and specific requirements with specific sections of text.

Based on the example of GANI_MED we recommend a high-level architecture of an IT platform consisting of five building blocks:

1. data sources such as diagnostic devices, hospital-wide infrastructure systems and the IT Cohort Management,
2. an Extraction Layer working as an interface and cache between the data sources and the research database,
3. a Trusted Third Party as an organizational unit implementing measures for data privacy as well as a technical instrument to supply pseudonyms and to keep all relevant given consents of the participants available,
4. a Persistence Layer primarily storing all consented medical data together with the respective pseudonyms and
5. a Transfer Unit as an organizational unit managing the use and access regulations.

16.5.3   Handling Incidental Findings from Imaging Within IM Related Research

The taking part of MRI examinations in human subject research causes particular stress. This stress increases if the results are returned only in a written form. Another factor which causes stress is the lengthy time of uncertainty until the clarification of the shared results.

Regarding the notification of incidental findings, which are possibly generated by imaging examinations in human subject research, the following procedure is recommended:

1. All test persons in general should receive a notification, also when nothing was found.

2. Participants with a relevant diagnosis should be invited to a consultation with a physician from the study team so that the results and their consequences cannot only be discussed in person but so that the participants also have a direct contact person who will advise them regarding the results and their further clarification.

In human subject research, even when there is a thorough multi-level clarification procedure in place, the phenomena of a therapeutic and diagnostic misconception are to be expected. These misconceptions restrict participants' ability to undertake a realistic risk-benefit assessment of their participation at imaging examinations.

16.6 Health Economic Assessment of Individualized Medicine

IM is still in its infancy. However, based on our own research and complementary papers which were published so far, including literature concerning related themes, as well as considering numerous discussions with colleagues, who are experts in this area, we can draw some important conclusions.

16.6.1 Individualized Medicine: From Potential to Macro-Innovation

Corresponding to experience so far, the comprehensive implementation of IM into clinical routine will take a long time. This also becomes clear, if we take the perspective of health economics. Decision making must essentially be based on the comparison between costs and benefits. This argument is appropriate for technological problems as well as for individual decisions. IM can be regarded as a novel health technology which has the potential to displace the existing health technology which can totally or in parts be described by the current medical paradigm. To become a so-called macro-innovation, the innovation has to prove that it is superior at the macro level. Under the corresponding setting this can only be the consequence of many isolated and often individual decisions. Therefore this can take a long time. In comparison to a private good, like a mobile phone, which has to prove its superiority only to the prospective buyers, decision making in health care is much more complex. There are many stakeholders with their own interests. Private agents do not necessarily need comprehensive evidence for decision making. Fashion and herd behavior also play an important role. Stakeholders in the health care system do have their own interests and they consider costs and benefits against the background of different time horizons. They also differ in their openness about new ideas. Thus there are different barriers at the level of stakeholders, which can hamper the penetration of IM.

16.6.2   Assessing Individualized Medicine—The Example
of Immunoadsorption

According to our current knowledge, IM is no panacea. IM is not a homogenous product, but a set of rather different products which must prove in a distinct way, that they are superior in comparison to the solution of the problem used so far. This can be shown by the results of cost-effectiveness analyses. Currently there is evidence, that IM is not superior in general. In addition, cost-effectiveness ratios can vary over time. If costs can be reduced, the results from evaluations can be different. Thus we have to consider carefully, under which circumstances or settings the IM solution is superior. Furthermore, effectiveness can change because of improved therapy outcomes. However, there is still a lack of evidence, which is the consequence of a rather low number of patients included in IM studies, as the target groups are much smaller. There is especially a need for long-term evidence. Thus, there is a need for further research.

Even in the case of promising IM arrangements, there needs to be incentives at the level of health care providers to implement IM into the clinical routine. This is a major conclusion from the analysis of the case of Immunoadsorption therapy with subsequent IgG substitution (IA/IgG). Health care providers must think economically. DRG reimbursements are calculated on the base of average costs. In some cases hospitals have the chance to agree upon additional charges (Zusatzentgelt) with the health insurance companies for special treatments. If innovative IM diagnostic plus therapy tandems are reimbursed by these charges it has to be guaranteed that the whole package is covered. Otherwise disincentives can be installed with the consequence, that not all services are provided. This is the case concerning the diagnostic tests for deciding about the predicted response for IA/IgG. Adequate incentives for health care providers for implementing promising concepts into clinical routine are generated by an eligible reimbursement system.

The perspectives of implementing IM into clinical routine can potentially be improved if some IM providers specialize in certain services. The calculation of actual costs of IA/IgG and the corresponding gene expression analyses showed that in this case a high share of total costs is caused by durables. Thus cost degression can occur through declining costs per sample. This can be the case for specialized providers, e.g. laboratories, which are able to provide these services or products much cheaper than health facilities, which rarely need them and suffer from high fixed costs.

16.6.3   How Individualized is Medicine Today? The Case
of Heart Failure in the G-DRG System

The comprehensive implementation of IM could be a major challenge for DRG based hospital reimbursement systems. Complex syndromes like heart failure are already treated in a very individualized way. As a consequence of the demographic aging their importance will probably increase in the near future. Unlike that, the current DRG system is based on the idea of standardization. According to that, demographic characteristics, diagnoses, and clinical interventions of a patient can define medically and economically homogeneous groups. Hospital payment is based on the corresponding average costs. Individualized treatment thus implicates a higher risk of deviation from average, as corresponding high spreads provide a risk of loss for service providers. This contradiction can be a major barrier for implementing IM in a comprehensive way. A solution could be to differentiate the DRG system according to the needs defined by IM. However, the German DRG system is already quite complex. The number of DRGs and additional charges has risen during the last years. There are currently signs that the level of detail and differentiation of the German DRG system shall not be extended much more. Thus there is an urgent need to think about appropriate solutions.

Index

© Springer International Publishing Switzerland 2015   321
T. Fischer et al. (eds.), *Individualized Medicine,* Advances in Predictive,
Preventive and Personalised Medicine 7, DOI 10.1007/978-3-319-11719-5

Printed in the United States
By Bookmasters